NYĀYAKUSUMĀÑJALI
Hindu Rational Enquiry into the Existence of God

NYĀYAKUSUMĀÑJALI

Hindu Rational Enquiry into the Existence of God

Interpretative Exposition of Udayanācārya's Auto-commentary with Translation of *Kārikās*

BHASWATI SINHA

Aryan Books International
New Delhi

Nyāyakusumāñjali

ISBN: 978-81-7305-151-7

First Published: 1999

Reprinted: **2025**

Published by:
Aryan Books International
Pooja Apartments, 4B, Ansari Road
Daryaganj, New Delhi - 110 002 (INDIA)
Tel: 011- 23255799/41099562
Email: aryanbooks@gmail.com
www.aryanbooks.com

Printed in India

In loving memory of
my teacher, late Professor T.M.P. Mahadevan
and
my father, late Dr. B.C. Bhattacharya

CONTENTS

PREFACE

The *Nyāyakusumāñjali,* authored by Udayanācārya, the 10th-11th century logician and redoubtable philosopher of the Nyāya-Vaiśeṣika system of philosophy, is a critical and constructive, polemical and philosophical treatise, dealing with the proofs for the existence of God. It is a great work both in dialectics as well as in style, and hence, it occupies the foremost place among the Indian philosophical classics.

Most religions establish the existence of God on the authority of scriptures and the religious experience of the prophets. In Indian philosophy Vedānta in all its forms too establishes the existence of God from the deliverances of the scriptures. God is hailed as a trans-empirical and super-sensuous Being that cannot be perceived, much the less inferred. Hence, its existence is said to be revealed. Further, human reasoning is considered as finite and not competent to fathom the mystery of God.

Some systems of philosophy in India, viz. the Cārvāka, the Bauddha and the Jaina, set aside the case for accepting God as the Supreme Principle. The Sāṅkhya and the Mīmāṁsā systems reject the concept of God as Ultimate Reality.

Against this atheistic, anti-rationalistic and pro-scriptural authoritarian background Udayanācārya, the first Indian theologian, attempts to establish the existence of God with the help of austere logic and refutes the criticisms made by opponents. In the *Nyāyakusumāñjali* Udayana with matchless skill has shown in great detail as to how it is possible to prove the existence of God and as to how the position of non-Īśvaravādins, viz. the Cārvāka, the Bauddha and others, is untenable. Udayana's originality is revealed in his independent line of thinking and his characteristic subtleties of interpretation of Nyāya-Vaiśeṣika and other philosophical systems. He treats in details all the matter relating to God, His nature, existence and other aspects. Udayana's method of treatment surpasses the methods adopted by his predecessors and is exemplary for the philosophers of other systems of thought. With sharp critical acumen he analyses the arguments against the existence of God by adopting the methods of syllogistic reasoning, refutes them and finally puts forth his arguments in support of God's existence and nature. Though he examines critically all the atheistic systems, his chief opponent is the Bauddha. The Bauddha denies God taking recourse to rationalism, and so Udayana takes up the position of an ultra-rationalist just to prove that even an uncompromising rationalist cannot but has to accept the existence of God. He refutes the arguments of the Bauddha on their own grounds.

The intention of Udayana is not only to prove that God comes legitimately within the scope of syllogistic reasoning, but also to acknowledge that God is realized in direct experience. At the end of the fifth chapter of the treatise he mentions how the direct perception of God is possible. As the atheists resort to reasoning to prove that God cannot be established by reasoning, so Udayana endeavours to certify that even from a purely rationalistic stand-point the preference is in favour of God's existence as against His non-existence.

There are three stages to God-realization. First, one has to hear about God from scriptures and from those who have attained God-realization. Secondly, one should reason about the validity of such testimony. Thirdly, after ascertaining the validity one is required to worship and to contemplate upon God. Udayana claims that his sole concern in the *Nyāyakusumāñjali* is with the second stage.

While ascertaining Udayana's contribution to theism, it may be noted that he is in no way the founder of any theory on theism. Many doctrinal points on theism in general and Nyāya theism in particular have been in existence long before Udayana. The *Nyāya-sūtra* (4.1.19) of Gautama refers to God's existence. Vātsyāyana, Udyotakara and Vācaspati Miśra discuss this topic defending thereby the Nyāya position against the onslaught of the opponents. But their contributions are only incidental to the main issues of the Nyāya system. It is Udayana who writes an independent treatise on Nyāya theism. His esteemed work, *Nyāyakusumāñjali*, reveals how doughty a champion he is in the cause of theism.

It is an acknowledged fact that the *Nyāyakusumāñjali* is not an easy text to comprehend. Udayana compresses his arguments in cryptic *kārikās,* most of which are not understandable without his auto-commentary, and even with the help of his auto-commentary some of the *kārikās* remain incomprehensible. To work out an exposition of the *Nyāyakusumāñjali* was suggested to me by late Professor T.M.P. Mahadevan, Director, Centre of Advanced Study of Philosophy, University of Madras. I took up some portions of the text for my Ph.D. dissertation and worked under the supervision of Professor Mahadevan. I greatfully acknowledge late Professor Mahadevan for inspiring me to work on this difficult topic.

After joining the Department of Religious Studies, Punjabi University, Patiala for teaching Hinduism and Comparative Religion, I decided to expound the whole text in detail, so that the students of religion and philosophy may understand the valuable contribution of Udayana to rational theology. It took many years of hard work to complete the present work.

While preparing translation of *kārikās* and interpretative exposition of Udayanas auto-commentary all the available commentaries and translations are taken help of. The author of the present undertaking is specially indebted to Sri Uttamur T. Viraraghavacarya of Madras and Sri N.S. Ramanujan Tatachariar of Tirupati at whose feet the author had the good fortune of studying the text, *Nyāyakusumāñjali.* Sri Viraraghavacharya's commentary, *Kusumāñjali-vistara,* is the only elaborate and exhaustive commentary on the *Nyāyakusumāñjali,* without which many portions of the text would have remained incomprehensible. Hence, the present interpretative exposition of the *Nyāyakusumāñjali* is mainly based on the *Kusumāñjali-vistara.*

I take this opportunity to express my gratitude to Dr. N. Veezhinathan, Professor of Sanskrit, University of Madras and Dr. V. Swaminathan of Kendrīya Sanskrit Vidyāpīṭha, Tirupati who helped me variously while I was researching at the Centre of Advanced Study of Philosophy, University of Madras. I cannot forget the help bestowed upon me in many ways by my dear friend Dr. R. Bhagyaleela, while I was studying at Tirupati. I am profusely thankful to my husband and my senior colleague, Dr. Atul N. Sinha, for providing me all possible co-operation in the completion and publication of the work in its present form.

I am thankful to Mr. Umesh Upadhyaya of Eureka Computer, Nehru Place, New Delhi for preparing the typeset and to Mr. Vikas Arya of Aryan Books International, New Delhi for the publication of the book.

Patiala Bhaswati Sinha (*née* Bhattacharya)

STABAKA - I

GOD — THE CONTROLLER OF *ADṚṢṬA*

Udayana begins his treatise, *Nyāyakusumāñjali*, without any dogmatic formulation of his position as fundamentalists do; he arrives at his position by following the critical and dialectical method of refutation of the opponents' points of view. Udayana's theism is argued or reasoned theism. In his confrontation with other schools, he takes up first for examination the arguments of the extreme position held by the Cārvāka who stands for the denial of all extra-empirical concepts, viz. God, soul, merit and demerit *(adṛṣṭa)*, etc. The Cārvāka is a radical empiricist; to him reality is matter and motion, and nothing else. All the things of the world are modifications of matter. He relies on perception as the sole instrument of knowledge and rejects other instruments of knowledge, like inference, verbal testimony, etc. Udayana's chief aim, on the other hand, is to establish the existence and reality of trans-empirical values. He is convinced of the operation of the just law of *karma*. One reaps what one sows. There must be connection between an act and the result it produces. Acts are short-lived; they produce, according to the Naiyāyikas, a potency called *adṛṣṭa* which inheres in the soul, the agent. By itself it is inert, super-sensuous and extra-empirical. It cannot be perceived by human beings. Udayana posits the existence of God to account for the functioning of *adṛṣṭa* in yielding its reward to the agent. God is omnipotent; He alone can cause the *adṛṣṭa* to move; He is conceived of as the guiding principle of *adṛṣṭa*; He is the dispenser of the fruits of *adṛṣṭa*. This is the moral argument Udayana advances for the existence of God.

सत्पक्षप्रसरस्सतां परिमलप्रोद्बोधबद्धोत्सवो।
विम्लानो न विमर्दनेऽमृतरसप्रस्यन्दमाध्वीकभूः।
ईशस्यैष निवेशितः पदयुगे भृङ्गायमाणं भ्रम-
च्चेतो मे रमयत्वविघ्नमनघो न्यायप्रसूनाञ्जलिः।।१।।

May this handful of flowers of faultless logic, endowed with petals fully blossomed, delighting the wise with its fragrance, remaining uncrumbled while crushed, the source of honey wherefrom flows the essence of nectar, offered at the feet of the Lord, rejoice uninterrupted my bee-like hovering mind.

1.i. The *Nyāyakusumāñjali*, the composition of which Udayana considers as the reflection *(manana)* on God, starts with an auspicious opening verse (*maṅgalācaraṇa*) worded in the above *kārikā*.

Here *amṛtarasaprasyandamādhvīkabhūḥ* signifies the purpose *(prayojana)* of writing the treatise; *īśasya* stands for the subject *(viṣaya)*; *niveśitaḥ* suggests the relation *(sambandha)*; *bhramacceto me* points towards the aspirant *(adhikārī)*. The word *nyāyakusumāñjali* stands for handful of inferential arguments establishing marks *(liṅga)* which are free from all defects; faultless inferential arguments are here called flowers *(kusumas/prasūnas/puṣpas)*.

1.ii. The *kārikā* admits of twofold interpretation, from the point of view of *puṣpāñjali* (handful of flowers) as well as that of Nyāya.

In respect of *puṣpāñjali*, *satpakṣaprasaratva* means properly blossomed flowers *(samīcīnadalavikāśavattvam)*. *Satāṁ parimalaprodbodhabaddhotsavatvaṁ* stands for the delight derived from inhaling the fragrance of flowers by the wise possessing faultless (free from all defects) olfactory sense-organ. *Vimardane 'pyavimlānatvaṁ* may imply that the flowers remain uncrumbled even while crushed, owing to their being divine. *Amṛtarasaprasyandamādhvīkabhūtva* signifies the source of honey wherefrom springs the immortality like nectar. *Īśasyaiṣa niveśitaḥ padayuge* means offering of flowers of logic in an act of worship.

Bhramaccetaḥ signifies hovering of mind here and there owing to the lack of discriminative knowledge between honey and absence of honey.

From the point of view of Nyāya the *kārikā* may be interpreted in the following manner. The word *añjali* here stands for the treatise, *Nyāyakusumāñjali*. Logical arguments are gathered and collected in the treatise just as flowers are gathered and collected within the fold of hands (with a view to offering at the feet of the Lord, that means, to lend support to the *pramāṇas* and *tarkas* that establish the existence of God). *Satpakṣaprasaraḥ* in this interpretation stands for faultless *pakṣa* (subject) possessing *pakṣatā* (the condition constituting a subject). A valid *hetu* (probans) is characterized by five marks, viz. *pakṣasattva* (existence of *hetu* in the subject), *sapakṣasattva* (existence of *hetu* in the similar instance), *vipakṣāsattva* (non-existence of *hetu* in the contrary instance), *abādhitatva* (absence of incongruous *hetu*) and *asatpratipakṣitatva* (absence of counter-balanced *hetu*). *Satpakṣaprasaraḥ* signifies the existence of *hetu* in the *pakṣa (pakṣasattva)*. By *parimala* is to be understood *anvaya-vyatireka-vyāpti* (concomitance of probans, i.e. *hetu,* with probandum, i.e. *sādhya* in presence as well as in absence) resulted from *sapakṣasattva* and *vipakṣāsattva*. By *satāṁ parimalaprodbodhabaddhotsavo* is implied the delight derived from the proper knowledge of *vyāpti*. *Vimlāno na vimardane* signifies that the *hetu* is characterized by *abādhitatva* and *asatpratipakṣitatva*. *Amṛtarasaprasyandamādhvīkabhūḥ* stands for that which gives rise to longing for *mokṣa* which is compared with honey (immortality). Faultless inferential arguments in support of the existence of God dispel contrary thoughts about God's existence and thus generates intense longing for *mokṣa*.

The treatise, *Nyāyakusumāñjali*, consists of five clusters *(stabakas)*; the logical reasons *(hetus)* stated in the *kārikās* in each cluster are compared with flowers *(prasūnas)*; petals *(dalas)* are the other reasons mentioned in the discussion part. *Īśasyaiṣa niveśitaḥ padayuge* signifies establishing the existence of God. Here a question may arise that in Nyāya God is admitted to be incorporeal, how can God possess a pair of feet *(padayuga)*? It may be replied that even though it is not a logical necessity that God should possess a body in order to create, yet He may assume a temporary body in order to set up a tradition. Or, *padayuga* may refer to feet meditatively visualized as belonging to images that represent God. Or, by *padayuga* may mean words referring to God *(pada)* and *pramāṇa* and *tarka (yuga).*

Three adjectives in the *kārikā*, viz. *satpakṣa...*, *satāṁ parimala...* and *vimlāno na...*, refer to the absence of such defects in the arguments, viz. the absence of *pakṣadharmatāvirodha* (fallacy of the absence of the character of the subject), the absence of *vyāptivirodha* (fallacy of the absence of pervasion) and the absence of *anumitivirodha* (fallacy of the absence of inferential knowledge) respectively. Absence of these fallacies implies the absence of *arthadoṣa* (defective subject).

The word *anagha* in the *kārikā* signifies the absence of *śabdadoṣa* (fallacious form of expression). The word *niveśita* suggests that the words should convey their proper meanings. Hence, it refers to the absence of misuse of words *(apaśabdarāhitya). Bhramacceta* may mean contrary knowledge or doubtful knowledge preceding the logical enquiry into the existence of God. The expression *ramayatvavighnam* may suggest that even if one acquires erroneous knowledge about the existence of God owing to the lack of discrimination between erroneous knowledge and true knowledge, the inferential arguments advanced in support of God's existence should contribute to ascertaining the true knowledge by removing all impediments coming in the way of attaining true knowledge. Or, it may signify that the handful of flowers of inferential arguments would offer unobstructed delight to readers, swayed by contrary or doubtful knowledge. By *avighnam* here the successful completion of the treatise may also be intended.

स्वर्गापवर्गयोर्मार्गमामनन्ति मनीषिणः।
यदुपास्तिमसावत्र परमात्मा निरूप्यते।।२।।

The Supreme Being whose worship is regarded by the wise as the pathway to heaven and liberation is being considered here.

2.1 It has been contended that the logical arguments set forth in the *Nyāyakusumāñjali* in support of God's existence lead to the attainment of liberation, that means, the logical enquiry into the existence of God is admitted as the way to liberation. The question arises here that it is the knowledge of *ātman (jīvātman)* which is accepted on the scriptural evidence as the way to liberation. Bondage continues for the *jīvātman* so long as it is ignorant about its real nature; when the *ātman* is known in its real nature liberation is attained. Hence, the knowledge of God *(paramātman)* does not figure as the means to liberation. This contention, however, is not pertinent. The worship of God is regarded by the scriptural authorities as the cause of *dharma, artha, kāma* as well as *mokṣa.* In order to worship God one needs to attain to the knowledge of God; hence, the Supreme Being is to be ascertained. Without the grace of God, resulting from the worship of God, the self-realization is also not possible. As the knowledge of *ātman* comes as a grace from the worship of God and the knowledge of God precedes the worship of God, the existence of God needs to be considered.

2.ii. One more apprehension is raised against the logical consideration of the existence of God. Inferential arguments are employed in a case where the existence of the thing to be inferred is considered of doubtful character. Inferential arguments never find application in an instance where the thing to be inferred is never cognized or where the thing to be inferred has already been ascertained. So far as God is concerned, there does not exist any doubt about His existence. For, all worship the Supreme Being whatever *puruṣārtha (dharma, artha, kāma, mokṣa)* may be the object of their desire. Thus, the followers of the Upaniṣads accept the Being as pure consciousness; the Sāṅkhyas regard Him as the first knower. The followers of Patañjali worship the Being as one who is unaffected by such afflictions *(kleśas)* as ignorance *(avidyā)*, egoism *(asmitā)*, attachment *(rāga)*, aversion *(dveṣa)* and clinging to life *(abhiniveśa)*, who is unaffected by prescribed and interdicted actions, who is untouched by three modifications *(vipākas)*, such as birth *(jāti)*, life-span *(āyu)* and enjoyment *(bhoga)*, who is unaffected by the residues of merit and demerit that reside in the soul till their fructification, and who assumes a body in order to create and reveal the tradition of the Vedas and is gracious in dispensing with fruits. Further, the Pāśupatas, the followers of great vows, worship the Supreme Being as the Creator of the world who is untainted even by such actions which are opposed to secular and Vedic laws. The Śaivas call the Being as Śiva; the Vaiṣṇavas pay obeisance to Him as Puruṣottama (Highest Person); the followers of the Purāṇas call him as Pitāmaha (Brahmā); the ritualists *(yājñikas)* regard the Being as the sacrifice personified *(yajñarūpī)* who is to be worshipped by the sacrifice *(yajñārādhya-puruṣa);* the Bauddhas accept the omniscient Being *(sarvajña);* the Jainas take the Being, the Immaculate Self, who is free from the karmic veil *(nirāvaraṇa);* the Mīmāṁsakas accept the Being in the form of prescribed *mantra,* etc., who is to be worshipped by sacrifices *(homa),* chanting *(japa),* etc. The Cārvākas also go by the convention of the world which worships the Supreme Being in the form of the ruler of the world, or in the form of images, etc. The Naiyāyikas worship the Supreme Being as one who is endowed with what is rationally attributed to Him. Leaving aside the seekers of Reality, even the artisans worship the Supreme Being as the architect of the universe (Viśvakarmā). Thus, the greatness of the Lord Bhava (Śiva) is acknowledged as long as there exists the world with its caste *(jāti),* family *(gotra),* sub-family *(pravara),* Vedic school *(caraṇa),* race *(kula),* duty *(dharma),* etc.; there does not arise any doubt about the existence of God. This being the case, it may be enquired what is it that is to be considered?

न्यायचर्चेयमीशस्य मननव्यपदेशभाक्।
उपासनैव क्रियते श्रवणानन्तरागता।।३।।

This logical enquiry into the existence of God may be called reflection (manana); *this indeed is the worship that follows the hearing (of the* Śruti*).*

3.i. Even though there does not exist any doubt about the existence of God, yet the act of worship in the form of logical reflection on the existence of God that takes place after hearing about God from teachers is not redundant.

The Śruti prescribes that 'He is to be heard of, to be reflected upon.' Hence, the Lord who has often been heard, mentioned in the Śruti (scripture), Smṛti (law-texts), Itihāsa (epics) and Purāṇas (ancient lores) and has also been heard from teachers, is to be reflected upon henceforth. Smṛti too refers to the need of reflection as follows:

> One attains to the highest *yoga* by cultivating one's intellect in threefold manner — through scripture *(āgama)*, through inference *(anumāna)*, and through the delight of continued meditation. (Mention of inference here refers to reflection which stands for constant thinking with the help of logical reasons.)

Thus, the treatise *Nyāyakusumāñjali* is an act of reflection on the existence of God grounded on logical arguments and reasons.

3.ii. Udayana now points out fivefold objection *(vipratipatti)* advanced by the opponents against the existence of God. It is admitted that even though there is no doubt about the existence of God, yet His existence may still be investigated upon in the form of reflection *(manana)* which has also been approved by the Śāstras. But as there does not remain any doubt about the existence of God, further enquiry into His existence does not stand to reason, and hence Udayana's introduction of the fivefold objection appears to be impertinent.* This (introduction of the fivefold objection) leads to the fallacies of shifting the premise *(pratijñā-tyāga)*, viz. there does not remain any doubt about the existence of God; of admitting the investigation into what has been discarded *(punastyaktanirūpaṇasvīkāra)*; and of introducing doubt which does not exist *(avidyamānasandehopanyāsa)*. But the above-mentioned fallacies do not affect the point under consideration. The premise here is 'to enquire into the existence of God', and this has not been given up; hence, the first two fallacies, viz. shifting the premise and investigating into what has been discarded, do not arise. It also cannot be said that the

* Vīrarāghavācārya in his *Kusumāñjalivistara* raises this question and points out certain fallacies, and he refutes these fallacies.

absence of doubt invalidates the logical investigation into the existence of God, otherwise in the absence of logical enquiry how can one do *manana* (reflection) on God?

Those who set forth objections against the existence of God possess some kind of knowledge about God; hence, they do not have doubt about Him. The author also does not possess any doubt about God's existence, because he has scriptural knowledge of God. Reflection may take place in this manner also. Hence, the scriptural injunction *śrotavyo mantavyaḥ* joins that reflection should be done on that which has been heard from the scripture. It, however, may not be raised that this establishes what has already been established, involving the fallacy of *siddhasādhana* (establishing what has already been established). The Nyāya puts forth the syllogistic reasoning for the sake of other's knowledge; the fallacy of *siddhasādhana* does not arise in that case. So far as one's own knowledge is concerned, one can establish one and the same thing through various means of knowledge, provided one has the desire to do that. Moreover, it is the natural human tendency to adduce various proofs and arguments in favour of an entity which has already been determined. Hence, reflection takes place on what has been heard of. Again, there are those who find it difficult to engage in reflection on object of auditory perception owing to inability to cope up with the complicated subtleties of the science of reasoning. Such aspirants will be benefitted by the method of reflection as enunciated in the body of the *Nyāyakusumāñjali*.

The next question is that as the contention that there is no doubt about the existence of God has already been ascertained, how can then the author set forth objections about God's existence subsequently? In reply to this question it may be said that here *vipratipatti* (objection) does not mean propositions that give rise to doubt. *Vipratipatti* signifies two propositions that give rise to contradictory meanings. The word *vipratipatti* in the text stands justified, because the author advances both standpoints — *pūrvapakṣa* (opponent's objection) as well as *siddhānta* (proponent's thesis). The use of the word *vipratipatti* signifies the contrary propositions that contradict the propositions to be listened to and to be pondered over. Further, even though there is no doubt about the existence of God, yet the knowledge of God may not be attained in the proper manner; there is possibility of the rise of erroneous knowledge in such manner as silver is mistaken in the mother o'pearl. The contingency of such erroneous knowledge too needs to be rooted out. Hence, the ascertainment of the existence of God with the aid of logical reasoning is done as *manana* (reflection) to remove the erroneous knowledge of God too. Therefore, the stand that in the absence of doubt about the existence of God the ascertainment of God is redundant, stands rejected. The ascertainment of the existence of God is in the fitness of things, and this in turn lends validity to the mention of doubt generating contrary statements advanced by others about the existence of God.

3.iii. In connection with the ascertainment of the existence of God, Udayana sets forth five objections negating the existence of God. The five objections which have been refuted in the five *stabakas* of the *Nyāyakusumāñjali* respectively by Udayana are as follows:

A. There is absence of *adṛṣṭa* which is considered as the means of the attainment of the other world *(alaukikasya paralokasādhanasyābhāvāt).*

B. Even in the absence of God the performance of sacrifices, etc. which is the means of attaining the other world, is possible either on the consideration that the injunction of performing sacrifices, etc. is entailed by the eternal flawless Vedas or on the consideration that Kapila and others are the authors of the Vedas, implication being that God is not warranted either as the dispenser of the fruits of *adṛṣṭa* accruing from the performance of sacrifices or as the author of the Vedas *(anyathā'pi paralokasādhanānusṭhānasambhavāt).*

C. There are *pramāṇas* which prove the non-existence of God *(tadabhāvā'vedakapramāṇasadbhāvāt).*

D. Even if God exists, there is no *pramāṇa* to prove His existence *(sattve'pi tasyāpramāṇatvāt).*

E. There is no *sādhaka-pramāṇa* (supporting argument) to prove the existence of God *(tatsādhakapramāṇābhāvāt).*

A. God is postulated as the controller of *adṛṣṭa*, as the author of the Vedas that set forth the means for the attainment of the other world, and as the cause or creator of the world. The Cārvāka is of the opinion that as there does not exist *adṛṣṭa*, God cannot be proved as the controller of *adṛṣṭa*. With the Cārvāka, there does not exist any such thing as the other world for the attainment of which means is to be sought for. Again, as there is no *pramāṇa* for the Vedas which are supposed to enjoin the means for the attainment of the other world, *adṛṣṭa* cannot be established as the means for the other world on the basis of the Vedic injunction. The Cārvāka denies the cause-effect relation; hence, with him it is illogical to assume *adṛṣṭa* as the cause of the other world. Further, as the very causal relation is denied, there is no scope for establishing God as the cause of the world. Thus, the chief intention of the Cārvāka is to disprove the existence of God on all counts.

The first objection entails five more sub-arguments as follows:
a. An effect is not dependent on a cause, as the relation of causality is invalid.
b. Even if in some cases an effect depends on a cause, the causality is not a universal law.
c. Provided causality is accepted as a universal law, an effect will be produced from the same cause always.
d. Provided an effect is produced from different causes, there is no necessity of admitting a cause for the attainment of the other world.
e. Even if there is a cause for the attainment of the other world, performance of sacrifices, etc. can be that cause and there is no *pramāṇa* for proving *adṛṣṭa* inherent in the soul as the cause of the other world.

B. The second objection comes from the quarters of the Mīmāṁska and the Sāṅkhya. Under the second objection there are the following sub-arguments:

a. The validity of knowledge being admitted as intrinsic the Vedas possess intrinsic validity; hence, there is no need of admitting a trustworthy person (God) as the author of the Vedas in order to ascertain the validity of the Vedas.

b. Even if elsewhere knowledge is established as extrinsic, the Vedas being eternal their validity cannot be extrinsic.

c. The validity of the Vedas may be ascertained from their being accepted by the great men.

d. If the Vedas were considered as non-eternal, the world being eternal and along with it its traditions being continually existent the Vedas remain continually existent; hence, God cannot be established as an agent of introducing the Vedas.

e. If it is said that the Vedas are not eternal owing to their being got lost during the period of interval between creation and dissolution, then the validity of the Vedas may be attributed to the authorship of such great compassionate seers as Kapila and others.

Among these five sub-arguments the first four are advanced by the Mīmāṁska, the last one is set forth by the Sāṅkhya.

C. The third objection is attributed to the Mīmāṁska. Just as the inference, 'fire is cold because it is an object of knowledge *(prameya),*' is invalidated by perception, so also the inferential arguments advanced in support of God's existence are invalidated by perception, inference, analogy, verbal testimony, presumption and non-cognition. Under this objection there are six sub-arguments keeping in view the six means of knowledge.

D. Even if the inference of God as the author of the Vedas were not sublated, then also the validity of the Vedas cannot be established on the basis of their divine authorship. The validity of the Vedas is derived from their being accepted by the trustworthy persons. Since there is no *pramāṇa* for God's existence, the validity of the Vedas cannot be based on the divine testimony. This objection is put forth by the Mīmāṁska. Under this objection there are the following sub-arguments:

a. God's knowledge is not valid *(pramā),* because it does not consist in apprehending what is unapprehended.

b. Since the knowledge of God is all-comprehensive, it also entails the erroneous knowledge that we possess.

c. Were the knowledge of God valid, it cannot be caused by a *pramāṇa* owing to its being eternal, and the knowledge that is not caused by a *pramāṇa* is not *pramā.*

E. The aforesaid four objections have been discussed and refuted by Udayana in the first four *stabakas* respectively. Udayana's main concern in these *stabakas* is to refute the objections against the existence of God, but in the fifth, the last *stabaka,* he advances positive proofs *(pramāṇas)* for the existence of God in reply to the objection that there is no supporting argument *(sādhakapramāṇa)* to prove the existence of God. Inferential arguments are advanced to establish the existence of God; the fallacies that are presumed to vitiate the arguments are rebutted; and the position of the atheists that there is no supporting argument to prove God's existence is invalidated.

सापेक्षत्वादनादित्वाद्वैचित्र्याद्विश्ववृत्तितः।
प्रत्यात्मनियमाद् भुक्तेरस्ति हेतुरलौकिकः।।४।।

There is a super-natural cause (adṛṣṭa) *on account of dependence, eternity, diversity, universal observance, and the experience of the result of action being confined to each individual.*

4. The first objection is advanced by the Cārvāka against the existence of God. The Cārvāka denies the existence of God on the ground that there does not exist any extra-empirical cause, i.e. *adṛṣṭa*, leading to the other world. Hence, God cannot be established as the dispenser of the fruits of *adṛṣṭa.*

When analysed, the above argument given by the Cārvāka consists of five factors: (a) the relation of cause and effect is not admitted, so effect does not depend on a cause; (b) although effect in certain cases depends upon a cause, it is not invariably so; (c) assuming that it is invariably so, it is quite possible that all the effects ensue from a cause that is of uniform nature; (d) even assuming that the cause is manifold, it is not necessary to admit either the other world or the cause to be the means of the other world; and (e) even admitting the existence of the other world, it is quite possible that sacrifice, etc. are themselves the cause of the other world, and we need not assume *adṛṣṭa* as resulting from the scrifice, etc. and as present in the soul through the relation of inherence.

The above arguments are critically examined in the first *stabaka,* and finally *adṛṣṭa* is established. Udayana is convinced that when *adṛṣṭa* is established, the postulation of God as the controller of *adṛṣṭa* follows logically.

The thesis to be established is that there exists a supernatural cause in the form of merit and demerit *(adṛṣṭa).* So, first of all, to establish the class of causes in general Udayana advances the reason 'on account of dependence'. Dependence here means that the effect is occasional *(kādācitka).* This transmigratory existence, characterized by manifold happiness and misery, cannot be independent; it cannot come into existence without any cause. We have to point out the cause or the aggregate of causes that bring these opposites, happiness and misery, into existence. A cause cannot be denied, since everything in this world is subject to occasional existence. An occasionally existing thing cannot but be perishable, and it is absurd to say that a perishable thing is uncaused. Further, to say that things are uncaused would constrain one to explain why they are not eternal entities, like *ākāśa* (ether) and the like, or why they should not be treated as eternally non-existent, like the horn of a hare, or a flower sprung from the sky. In fact, happiness and misery are neither eternal entities nor absolute non-existents, they are experienced occasionally. Hence, we have to show some antecedent phenomena with which they are inseparably connected. They are not directly experienced by human beings, but come forth as a result of their manifold activities. Activities produce *adṛṣṭa*, and *adṛṣṭa* gives rise to happiness or misery.

हेतुभूतिनिषेधो न स्वानुपाख्यविधिर्न च।
स्वभाववर्णना नैवमवधेर्नियतत्वतः ।।५।।

(By saying that the world is accidental) neither the cause nor the production is denied; nor can it mean that the effect itself or an absolute nothing (unreal thing) is the cause; nor can it be said that (effect

is produced) by its own nature, since (an effect) invariably possesses limits.

5.i. It may be noted that the postulation of *adṛṣṭa* is based on the principle of causality. The Cārvāka denies the principle of causality itself. He may argue that this world is uncaused; it is *ākasmika — akasmāt bhavati,* i.e. the world is accidental. Udayana refutes the *ākasmika-vāda* which in the ultimate analysis reduces itself to *svabhāva-vāda* (theory of spontaneous origination).

The expression *akasmāt bhavati,* says Udayana, is riddled with inconsistencies, and it admits of five interpretations. In the first place, it may mean that the world comes into existence without a cause *(na kasmāt bhavati).* This interpretation negates a cause with reference to an effect. Secondly, it may be maintained that an event is not produced at all *(kasmāt na bhavati).* It is ordinarily understood that an object is produced. But this, according to this interpretation, is not the actual state of things. The very act of origination is denied. Thirdly, the expression *akasmāt bhavati* may mean that an effect is produced by itself, i.e. an effect is not caused by something different from it *(svasmāt bhavati).* Fourthly, *akasmāt* may mean absolute nothing *(anupākhya),* i.e. the world comes into existence from an absolute nothing. Fifthly, *akasmāt* may mean *svabhāva* (nature). That is, the expression *akasmāt bhavati* means that the world evolves out of itself. It is the nature of the world to come into existence on its own accord.

Udayana critically examines all these arguments and finally rejects them. With regard to the first alternative he argues, if it be held by the Cārvāka that an event happens without a cause, then the Naiyāyika would ask him to explain what prevents an effect from appearing at all times and what determines its appearance at a particular moment of time. The Naiyāyika explains this position by postulating a cause that determines the character of an effect. Unless a cause as the determinant is posited, it is indeed difficult to account for the specific character of an effect. A particular event happens at a particular moment of time, neither earlier nor later. The Cārvāka has to explain why the said event occurs at that specific moment of time. If no cause is admitted to account for the emergence of that event at that specific moment of time, it may just happen at any other time earlier or later.

According to the second meaning of the expression *akasmāt bhavati*, origination of the object itself is denied. In this case, the object would not originate at all as it was before creation, because there is no criterion, according to this alternative, whereby the origination of an object could be established.

The third alternative that an effect comes into existence from itself is not intelligible. It is because an effect before its creation is not existent according to the Nyāya school. Being non-existent it cannot become the cause of its origination. To say that an event would happen by itself is tantamount to saying that the event would not happen at all. It is an admitted fact that an event does not exist prior to its appearance. So, it is not conceivable, remarks Udayana, how it can exert causal efficiency upon the event. Further, the identification of the cause with its effect is an absurdity, which goes against the testimony of perception. For producing a jar one collects a lump of clay, the material that produces the jar, not the jar itself. If the effect were identical with the cause, one would have collected the jar and not a lump of clay.

Time sequence is also an important factor in the causal nexus. The relation of cause and effect means the invariable relation of antecedence and subsequence. And an object cannot be the substratum of antecedence and subsequence at the same time. Antecedence and subsequence, on the other hand, can exist only in different objects at different times. Since an object cannot be the substratum of antecedence and subsequence, and since the relation of antecedence and subsequence alone is the relation of cause and effect, an object cannot be the cause of itself.

The fourth alternative, i.e. the expression *akasmāt bhavati* means that an effect comes into existence from absolute nothing, is also not free from defects. It is because that the effect which is said to come into existence from an absolute nothing at a given time, could have come into existence even earlier as there is no factor which could prevent its rise from an absolute nothing earlier. The result of this argument is that the effect should have to be treated as eternal.

The Cārvāka may argue that the expression *akasmāt bhavati* does not mean any of the four alternatives mentioned above. It only means that an object comes into existence without depending on anything. The argument in favour of this view is that it is the nature of an effect to exist only in a particular place, just as, according to the Nyāya school, an effect, (say) pot, which is admitted to be arising from an aggregate of causes, like stick, potter's wheel, pot-sherds, etc. is admitted to be existing only in pot-sherds and not in stick, etc. In the same way the Cārvāka argues that it is the very nature of a particular object to come into existence only at a particular time. This is what is meant when it is said — *kaścinniyata-deśavanniyata-kālasvabhāvaḥ.*

The above view is unsound. The Naiyāyika indeed admits that an object, pot, exists only in a particular substratum, viz. pot-sherds; but this existence is not the nature of the thing, that is, it is not independent of anything. On the strength of experience it is admitted that it is the pot-sherds that are capable of creating the pot in themselves, and stick, potter's wheel, etc. are assumed to be possessing the power to create pot in the pot-sherds only. So, the existence of an effect in a particular place is not natural, but only caused by something else. The opponent's argument that just as existence in a particular place for an effect is natural so also the existence of an effect at a particular time is natural to it, is not tenable. Moreover, if it is insisted that it is the nature of an object to exist at a particular time, then it may be asked whether any antecedent limit is admitted. If no antecedent limit is admitted, the effect cannot be occasional. The opponent has to explain whether he does not admit any antecedent limit at all, or having admitted an antecedent limit he does not hold that a particular limit is invariably antecedent to a particular effect. If the first alternative is maintained, it should be conceded that the effect exists at all times. If the second is held, anything may become the antecedent limit of a particular effect. Hence, there arises the contingency of the effect remaining always.

5.ii. It may be argued that an object may exist for all times, yet it may be viewed as occasional. It is because an effect is considered to be occasional only in this sense, viz. it is related to a subsequent moment. Since an object that exists is related to a subsequent moment, there is nothing wrong in viewing it to be occasional.

Udayana criticizes this on the ground that even etheric space, in this sense, can be referred to as an occasional entity, because it is related to a subsequent moment. But no Naiyāyika would admit the etheric space to be occasional. So, we have to say that an object is occasional only when it is absent before its creation and is related to a subsequent point of time. If no antecedent limit is admitted, it is impossible to predicate the antecedent negation of a particular effect. When an antecedent limit is admitted, that alone is referred to as the cause. So, it is decided that an effect has a cause and it does not come into existence naturally or spontaneously.

5.iii. The opponent may argue: "Let the effect be dependent upon a cause. But that cause is merely the antecedent negation of a particular effect and not the existent objects, such as stick, potter's wheel and the like." This contention is unfounded. Prior to the time of origination of a pot, the stick and the like also exist along with antecedent negation. So, there is no specific reason to decide that antecedent negation alone is the cause and not the stick and the like. Moreover, the antecedent negation of pot cannot be explained satisfactorily in the absence of stick and the like. The antecedent negation of a pot is perceptually comprehended in the form 'here a pot is going to come into existence', when aggregate of causes like stick, etc. are perceived there. So, it is not proper to exclude stick, etc. from the purview of causal aggregate. Hence, antecedent negation alone is not the antecedent limit or the cause of pot.

5.iv. It may be argued that antecedent negation must be admitted to be the cause, as otherwise there will arise the contingency of the rise of another pot in the pot-sherds wherein a pot has already come into existence. This contingency would be avoided by pointing out that one condition for a pot to come into existence in pot-sherds is its antecedent negation. There does not arise another pot in the pot-sherds where a pot has already come into existence, because there is no antecedent negation of the pot. So, the opponent contends that antecedent negation must be admitted to be the only cause of an effect.

Udayana criticizes this on the ground that antecedent negation *(prāgabhāva)* is beginningless *(anādi),* if it were admitted to be the only cause of a particular effect, then the effect would ensue even before the aggregate of causes like stick, etc. is brought in. Since no effect like the pot ensues without stick and the like, it may be concluded that antecedent negation is not the only cause.

5.v. The opponent may argue that apart from the antecedent negation other existent objects, such as stick and the like, may be the antecedent limits of an effect. But he admits only this much that the effect has those things as its antecedent limits. He does not admit that the effect is dependent upon those antecedent limits. So, when it is said *svabhāvataḥ bhavati,* it means that the effect comes into existence without depending upon those antecedent limits.

The question arises as to what is meant by absence of dependence. If it means that the antecedent limits need not be invariably present before the production of an effect, then like fire an animal also may become the antecedent limit of smoke. If, on the other hand, it means that the antecedent limits are invariably present, but they do not contribute to bringing about an effect, then it does not differ from the *siddhāntin's* standpoint. For, the *siddhāntin* also does not admit that whatever is invariably present as the antecedent limit, contributes to the production of an effect. But he admits that invariable presence itself is known as dependence which is another name for a cause. So, if the second alternative is maintained, the opponent adopts the line of argument of the *siddhāntin.*

5.vi. The opponent rejoins that he does not admit any cause whatsoever. Even without admitting any cause there is absolutely no difficulty in accounting for the occasional character of a particular object. The occasional character is the very nature of the thing like *nitya-svabhāva*. It is explained thus. Occasional nature is defined as that which does not depend upon any cause. In that case *ākāśatva* which is present in *ākāśa* must be treated to be occasional, as it does not come into existence from any cause. *Ākāśa* is eternal, so it does not depend upon any cause. It is, therefore, reasonable to hold that *ākāśatva* which exists in *ākāśa* also does not depend upon any cause. If an entity exists without a cause, it is termed *ākasmika* or occasional. When viewed in this light, *ākāśatva* is also occasional. Yet it exists only in *ākāśa* and not in *ātman* or *kāla*. This is because *ākāśatva* is the essential nature of *ākāśa*, and not that of *ātman* or of *kāla*. In the same way the opponent argues that *kādācitkatva* or occasional existence is the nature of pot, etc. So, pot, etc. may come into existence without depending upon a cause and yet they could be occasional.

The above contention is answered thus. The illustrative example has been explained well. But it is not reasonable to apply in the same case of pot, etc., what has been arrived at by the analysis of illustrative example. If *ākāśatva* exists in *ātman*, it cannot be considered as the essential nature or *svabhāva* of *ākāśa*. *Svabhāva* is a positive and unique characteristic present in a particular object. The characteristic like *ākāśatva* cannot, therefore, be the essential nature of many objects. So, the contention of the opponent that *ākāśatva* is the essential nature of *ākāśa* only, is quite correct. But *kādācitkatva* or occasional existence cannot be the essential nature of a particular object, viz. pot.

5.vii. Earlier the *siddhāntin* has mentioned that if the antecedent limit of an effect is not admitted, then the effect would exist at all times, that is, the effect would be eternal. The opponent says that an effect cannot be viewed as eternal, as eternal existence of an effect contradicts its occasional existence. *Ākāśatva* which is invariably related to *ākāśa* cannot be invariably related to *ātman*, for such a relation would be contradictory. In the same way pot, etc. that are subject to occasional existence cannot be postulated as eternally existents, because such postulation would be contradictory.

The *siddhāntin* replies that *kādācitkatva* cannot be explained by *ākāśatva*, because the same logic does not hold good in both the cases. The point at issue is whether the opponent who is of the view that pot, etc. do not depend upon cause, does admit any limit or not. If no limit is admitted, it means that the effect, viz. pot, exists prior to its creation. Occasional existence implies absence of an object prior to its creation. So, this occasional existence cannot be the essential nature of the object, because it exists prior to its creation. If a limit is admitted for an object, then it may be asked whether that limit is variable or invariable. If the former, it means that, that limit exists at a time different from the one at which pot-sherds and other factors that are admitted to be the causes of pot, exist. And that variable antecedent limit is contrary to *kādācitkatva* or occasional existence. If an invariable limit is admitted, it amounts to admitting a cause for an effect.

5.viii. The opponent contends that even if it is admitted that an effect depends upon its antecedent limit, yet this is not the case with all effects. An effect which is positive in character and which is perceived to have only antecedent limit, need not be admitted to have a subsequent limit. Similarly, for an effect which is known to have only a subsequent limit, an antecedent limit need be admitted. Both antecedent and subsequent limit may be admitted in the case of those effects that are perceived to have both the limits. The fact that some effects have only one limit, either antecedent or subsequent, cannot be gainsaid. The prior non-existence *(prāgabhāva)* has only the subsequent limit and posterior non-existence *(pradhvaṁsābhāva)* has the antecedent limit only. Just as in the case of non-existence *(abhāva)* only one limit is admitted, so also in the case of positive entities *(bhāva-padārtha)* only one limit may be admitted. If an object is not perceived to have an antecedent limit, then no antecedent limit should be admitted for that object. If an antecedent limit is admitted in respect of an object even if that object is not perceived to have an antecedent limit, there arises the contingency of infinite regress.

For, that antecedent limit being an effect would require some other antecedent limit, and so on *ad infinitum*. Antecedent limit here means the aggregate of causes that give rise to an effect.

Udayana criticizes this viewpoint by stating that if an object does not have an antecedent limit, then just as a jar comes into existence today without any antecedent limit whatsoever, it could have come into existence yesterday also. For, there is nothing to restrict the production of the jar only to the present day and not to yesterday.

It cannot be said that it is the nature of the jar to come into existence only on a particular day. For, that particular day too may exist at any time, one day or one hundred million days before. According to the Nyāya time is eternal; days, months, etc. are determined by the *upādhi* that delimits time. Thus, the *upādhi* which delimits a day may exist at any time earlier, since there is nothing to restrict its existence to a particular period of time. The result of this argument is that the object will be eternal like etheric space, etc.

प्रवाहोऽनादिमानेष न विजात्येकशक्तिमान्।
तत्त्वे यत्नवता भाव्यमन्वयव्यतिरेकयोः ॥६॥

This course (of causes and effects) has no beginning, nor has it one power (śakti) *abiding in different classes of things; one must try to determine the true nature of agreement* (anvaya) *and difference* (vyatireka).

6.i. The infinite regress pointed out by the opponent is not a fallacy *(doṣa)* here, because a *regressus ad infinitum* ceases to be a fallacy when it is proved by the evidence of our senses as in the case of seed and sprout. The stream of cause and effect is beginningless. Just like the continued series of seed and sprout, the succession of causes and effects is perpetual.

The Cārvāka further argues that causality may subsist in individual objects, but it cannot subsist in class character.

The Nyāya is of the view that cause-effect relation subsists not only in definite individuals, but all individuals coming under a class are causally related. The opponent says that there may be causal relation between one particular fire and one particular smoke, but this causal relation does not hold good between smoke as a whole and fire as a whole. The Cārvāka's denial of causality in respect of the class is due to the fact that there arises lack of correspondence to the invariable concomitance between fire as a whole and the causes of fire as a whole. For example, the effect-fire arises from three causes, viz. straw *(tṛṇa),* the fire-producing wooden stick *(araṇi),* and fire-producing stone *(maṇi).* If causal relation is admitted between fire as a whole and the causes of fire as a whole, there arises lack of correspondence in this way — fire arises from the fire-producing wooden stick in the absence of fire-producing stone; in the same way the former arises from the straw in the absence of the other two factors. So, it is logical to maintain individual causal relation between fire and fire-producing wooden stick, between fire and straw, etc. instead of maintaining universal causal relation between fire as a whole and the causes of fire as a whole. By arriving at this conclusion the Cārvāka aims at proving that if at all there could be causal relation, it could be only between particulars and not between universals.

To this Udayana rejoins that the denial of causality in respect of class-character will break down the very nature of a thing belonging to a definite class. It would not be possible then to determine a particular cause from a particular effect. In this case something, by being produced from a mere anything, can have the nature of belonging to a particular class. If it is maintained that it is the nature of a thing to belong to a particular class even if it is produced from anything, then effects produced from any and every cause will belong to all classes or to the same class. Thus, a jar being produced by the causal aggregate of cloth will belong to the class of cloth, and things other than jar being produced by the causal aggregate of jar will belong to the class of jar.

With the Nyāya, causality is ascertained by the method of agreement *(anvaya)* and difference *(vyatireka),* and the universal character determines both a cause as well as an effect. Unless causal relation exists between anything that is seed and anything that is sprout, it is not possible to determine whether a particular sprout is the effect of a particular seed. Seedness determines the cause-character of a seed, and sproutness determines the effect-character of a sprout.

6.ii. The Cārvāka objects that if class-character determines both a cause and an effect, then how it is possible that the fire belonging to the same class is produced by different causes, viz. straw, fire-producing wooden stick and fire-producing stone. Here positive instance *(anvaya)* is available, i.e. when straw or fire-producing wooden stick or fire-producing stone is present, fire is also present. But the negative instance *(vyatireka)* is doubtful. Nobody can definitely say that the absence of straw would cause the absence of fire, because fire may be produced from fire-producing wood or fire-producing stone. Similar is the case with fire-producing wood and fire-producing stone. In the absence of fire-producing wood, straw or fire-producing stone may produce fire, and in the absence of fire-producing stone, fire may be produced from straw or fire-producing wood.

This objection may be met in two ways, from the viewpoint of the Mīmāṁsā and that of the Nyāya. Udayana first puts forth the Mīmāṁsā answer to the objection, criticizes it and lastly gives his own answer.

The Mīmāṁsā answer is based on its theory of *śakti (śaktivāda)*. He says that the same effect may be produced by more than one cause, as all the causes possess one uniform power. Straw, fire-producing wood and stone, all these possess the power favourable to fire-production *(vahnyanukūla-śaktimatva)* and give rise to the same fire. Thus, *anvaya-vyatireka* relation is sought to be established between fire and the power favourable to fire-production. When the power favourable to producing fire is present, fire is produced from straw, fire-producing wood and stone, and in the absence of it none of them can produce fire.

This answer to the Cārvāka objection does not find favour with Udayana. He rejects the very concept of power. Different kinds of causes, says Udayana, cannot have a uniform power.

Even if things belonging to different classes can have the same power of producing things belonging to the same class, nowhere can there be the inference of a particular cause from a particular effect. Nor is it possible to infer the non-existence of an effect of the same class from the non-existence of its cause. For, even if the cause is absent, the effect may be produced from another having the power of producing a thing of that class.

It cannot be said that when an effect is seen, the specific cause is inferred from the specific nature of the effect. For, the cause of the effect cannot be determined, as the *vyatireka* relation between power and effect is not available. A cause is not determined merely by co-presence, that is, the presence of the cause and the effect in the same subject *(pakṣa).* Co-absence *(vyatireka),* that is, when the effect is absent, the cause also is absent, is necessary in determining a cause. *Vyatireka* relation between an effect and power is not available, because power is never seen. It cannot be maintained that straw possessing the power favourable to fire-production is the cause of fire, because *vyatireka* relation between fire and the straw possessing power cannot be obtained. If such straw were the cause of fire, it would be absent when fire is absent. But this is not so, as in the absence of straw, fire may be produced from fire-producing wood.

The argument that an effect may be produced from a subtle unseen *jāti* also stands rejected. If an effect is produced from a subtle *jāti,* then smoke may be produced by something other than fire, because that subtle *jāti* may exist in that thing.

The Mīmāṁsā may argue that straw, fire-producing wood, etc. are efficient causes *(nimittakāraṇa)* of fire. Efficient causes do not determine the difference and non-difference of the class of the effect. What determines the difference and non-difference of the class of the effect is the difference and non-difference of the material cause *(samavāyikāraṇa).*

This reasoning, however, cannot be sustained, because in this case fire may be produced even in the absence of straw, fire-producing wood, etc. Hence, they cannot be the cause of fire; but this is not true, fire is produced out of straw, etc. By a cause is not meant the presence of the effect when it is there, but its presence only when it is there. Difference in the class of effect is not determined by the difference in the material cause only; it is determined by the difference in the entire causal aggregate. If material cause were the determinant of the class of effect, there cannot be a difference in class in spite of their having the common material cause.

It cannot be said that difference in the material cause determines the difference only in that class of effects that are substances *(dravyas)*. This position is taken up just to avoid the previous difficulty of the three categories having the common material cause. It is observed that the same atoms constituting milk produce curd also. Here both milk and curd are substances, yet they belong to different classes in spite of their having the non-different material cause, viz. atoms constituting milk. Therefore, it is not proper to say that difference in the material cause determines the difference only in that class of effects that are substances.

By the same line of argument the Bauddha theory of *apohavāda* is also refuted. The Bauddha does not admit *jāti*; hence, according to him causality does not subsist between fireness and smokeness. Causality subsists in two particulars and not in two classes.

This position is not accepted by the *siddhāntin*. He maintains that the class of effect is produced by the class of cause, and the difference in the class of effect is determined by the difference in the causal aggregate. Thus, fire produced from the friction of fire-producing wood is different from that which is produced from refracted sun-rays, as the causal aggregate, viz. fire-producing wood, etc. in one case is different from the causal aggregate, viz. lense, etc. in the other. That one kind of fire is different from the other is supported by the evidence of the senses also. The fire in a lamp illumines the entire apartment of a mansion, while fire from fire-producing wood, though having flames, produces little light, and still less is produced by the cow-dung fire. If it is not possible to comprehend the subtle difference in the various kinds of fire and their causal aggregates, then the fire in general may be inferred from smoke in general.

It may be argued that difference in various kinds of fire is determined by a factor in smoke which cannot be comprehended. Thus, causal relation is between fire and that unseen factor and not between fire in general and smoke in general. Were this the case, the *vyatireka* relation between fire and that factor cannot be ascertained. In *vyatireka* (negative instance) the absence of *sādhya* is followed by the absence of *hetu.* Here the *hetu,* viz. the unseen factor, being invisible, it cannot be ascertained whether the absence of *sādhya* is invariably followed by the absence of *hetu,* for when the causal aggregate of one kind of fire is absent fire may be produced from the other kind.

The Bauddha holds that when the *hetu* is produced from the *sādhya (tadutpatti),* there is cause-effect relation between them. If the *hetu* were invisible, it cannot be determined whether the *hetu* is produced from the *sādhya* or not. Hence, the Bauddha cannot infer cause-effect relation on the ground that the *hetu* is produced from the *sādhya.* Causality, according to the Bauddha, is not based on *anvaya-vyatireka* relation. The Bauddha also maintains that there is cause-effect relation between *hetu* and the *sādhya,* when the *hetu* is identical with the *sādhya (tādātmya).* For example, in the inferential argument 'this is *śiṁśapā,* because this is tree' — there is the relation of identity between *śiṁśapā* and tree, because both are tree. Hence, tree and *śiṁśapā* are causally related. The relation of identity here is based on the fact that the causal aggregate that produces a tree, produces *śiṁśapā* also. But if the *hetu* were invisible, then the causal relation between *śiṁśapā* and tree cannot be established. In that case, the Bauddha would not be able to infer causal relation on the ground of the identity between the *hetu* and the *sādhya.* Therefoe, inference of causality on the basis of *tradutpatti* and *tādātmya* cannot be established.

According to the Bauddha theory of *apoha,* fire is that which is not not-fire. In that case, it cannot be definitely said that fire is the cause of smoke, because smoke may be produced from anything which comes under the category of the negation of non-fire.

The Cārvāka points out that as the *vyatireka* relation between the class of effect and the class of cause cannot be obtained, causality cannot be established; consequently, inference which is based on the causal relation would have to be abandoned. The *siddhāntin* says that this position is actually disadvantageous to the Bauddha.

The Bauddha does not admit *anvaya-vyatireka* relation. In that case, he would not be able to establish causal inference, for, in the absence of *anvaya-vyatireka* relation it could not definitely be ascertained whether the *hetu* is produced from the *sādhya* or the causal aggregate that produces *sādhya* produces the *hetu* also. With the Naiyāyika when two things falling within two particular classes have *anvaya* and *vyatireka* relation, then as between them, their relation of cause and what is caused is determined by those classes. Causality can be established between sub-categories also on the same principle of *anvaya-vyatireka.*

It is observed that particular cause, (say) straw, is responsible for producing fire. It may be asked what is responsible for producing fire in general. Udayana replies that for the production of any kind of fire wind is the efficient cause *(nimittakāraṇa),* the conjunction of parts of fire is the non-inherent cause *(asamavāyi-kāraṇa),* and the parts of fire are the inherent cause *(samavāyi-kāraṇa).* Gold, according to the Vaiśeṣika, is one kind of fire; gold is produced by the same causal aggregate mentioned above accompanied by a substance having weight. The same with the help of produced touch existing in fire causes fire-burning. All the causal aggregate mentioned above conjoined with water produces celestial fire (lightning), conjoined with earthy substance produces terrestrial fire, and conjoined with both, viz. water and earthy substance, produces gastric fire.

Out of the five reasons adduced by Udayana for proving *adṛṣṭa*, two reasons, viz. dependence *(sāpekṣatvāt)* and eternity *(anāditvāt)* have been discussed. The third reason is diversity *(vaicitryāt).*

The opponent points out that there may be a common cause which belongs to the one and the same class. Thus, there is no need of differentiation in the cause. He cites an example of a lamp which without any differentiation produces more than one effect different from one another, viz. removing darkness, producing modification in the wick, and revealing different objects in the room. The purport of the argument is that in this there is no need to establish God as the controller of *adṛṣṭa*, since the supposed causes will evolve their own effects, and so, it is needless to assume *adṛṣṭa* as the special cause of the world.

एकस्य न क्रमः क्वापि वैचित्र्यञ्च समस्य न।
शक्तिभेदो न चाभिन्नः स्वभावो दुरतिक्रमः ॥७॥

There can be no succession of what is one, there can be no variety of what is uniform, and it is not (owing to) a difference in potency; since this is not different (from the substratum), nature can hardly be transgressed.

7. Udayana critically examines and finally discards the above view of the opponent. Diversity is an objective fact that cannot be explained by a unitary cause. A single cause without being assisted by the accessories cannot produce diverse effects. Effects are produced in sequence. Were the cause an unitary entity, it would be devoid of accessories that produce sequence; hence, effects would not be produced in sequence.

It may be argued that it is the nature of the cause to produce effects in sequence, just as one lamp is the cause of several effects, viz. dispelling darkness, producing modification in the wick, etc.

Now the question arises whether the cause at the time of producing the first effect possesses the nature of producing the second effect too or not. If the first alternative is held, then the second effect too may be produced at the time of the production of the first effect, since the cause possesses the nature of producing the second at the time of the production of the first one. This, however, is in conflict with ordinary experience.

In the second alternative, if the cause does not possess the nature of producing the second effect at the time of the production of the first one, the cause itself will cease to exist. For, a thing devoid of its nature is a non-entity, as the concept of fire without its nature, namely heat, is an absurdity.

The Bauddha explains sequence in the production of effects with the theory of momentariness *(kṣaṇabhaṅgavāda)*. Effects are produced successively and not simultaneously, because a thing lacks the casual efficiency with regard to all objects at the same time as it ceases to exist at the next moment. The Bauddha does not admit sequence in the execution of causal efficiency in connection with one permanent cause.

But the doctrine of momentariness too fails to explain succession in the production of effects unless it admits the Nyāya *sahakārivāda* (theory of supplementation). For the Bauddha an object exists for one moment, hence the object is equated with the moment. If it is the nature of the moment to produce the next moment without depending on the prior one, then the next moment too may be produced at the time of the production of the prior one. Thus, there will be no order of succession. If it is admitted that effects are produced in definitive order, then it must also be admitted that production of the next moment depends on the production of the prior one. Were it so, various causes must be admitted for the production of the various effects.

The cause belonging to one category cannot produce effects belonging to various categories. If the cause of fire without depending on anything more or less than the cause of fire produces non-fire too, then the product will be neither fire nor non-fire. For, fire and non-fire being contradictory, their causal aggregates are also contradictory.

Thus, at the time of the production of fire, the causal aggregate of non-fire will oppose the causal aggregate of fire, and *vice versa*. Consequently, neither fire nor non-fire will be produced. If, on the other hand, it is maintained that fire and non-fire are not contradictory, and hence their causal aggregates do not oppose the production of either, then both fire as well as non-fire will be produced. This contradicts experience, because what determines non-fire lacks the capacity of producing fire. One and the same object can never possess the capacity of both production and non-production of fire.

It may be said that the same cause produces various effects possessing different latent potency called *śakti*. It may be asked whether *śakti* is different from the cause or identical with it. If the first alternative is held, then both *śakti* and the cause would possess the same causal efficiency, and effect would follow from the concomitance of *śakti*. Thus, *śakti* would possess the characteristics of the cause, and the postulation of a unitary cause has to be given up. If *śakti* were identical with the cause, then the earlier defect, viz. the difficulty in explaining the diversity of effects by the self-identical causal principle, arises.

It may be argued that it is not necessary to postulate various causes for the production of various effects; various effects may be produced from the same cause on the ground that it is the very nature *(svabhāva)* of the cause to produce mutually exclusive effects, e.g. fire and non-fire. This position is untenable. Were this the case, at the time of the production of fire nothing will prevent the production of non-fire too, since a cause cannot transgress its own nature *(svabhāvaduratikramaḥ)*. As the cause the very nature of which consists in producing mutually exclusive effects, is present at the time of the production of fire, non-fire may equally be produced at that very moment. The denial of production of non-fire at the time of the production of fire will turn the cause to be devoid of its own nature. Hence, nature-theory will not solve the issue under consideration.

Udayana holds that different effects are produced by different causes. He explains sequence in the production of effects with the help of *sahakārivāda*. One permanent cause produces effects successively in association with successive subsidiary factors. The presence of every condition is needed to produce an effect; causality consists in the assemblage of all conditions.

As the postulation of one cause cannot account for the diversity of this phenomenal world, due consideration should be given to the postulation of *adṛṣṭa* which explains the diversity. The opponent says that he may at best admit those auxiliary causes that are seen, but he cannot postulate an unseen *adṛṣṭa*. As sacrifice, etc. may be regarded as the cause of various effects, there is no need of postulating *adṛṣṭa*.

To meet this objection Udayana adduces the fourth reason, viz. universal observance *(viśvavṛtti)*, that establishes *adṛṣṭa*.

विफला विश्ववृत्तिर्नो न दुःखैकफलाऽपि वा।
दृष्टलाभफला नापि विप्रलम्भोऽपि नेदृशः ।।८।।

Universal observance cannot be fruitless, nor can it have suffering as the sole fruit, nor can it have some visible gain as fruit, deception too is not of this kind.

8.i. The denial of *adṛṣṭa* in fact goes against the universal observance of ritual acts. If it is said on the basis of the example of seed giving rise to sprout, sprout to tree and tree to fruit, etc. that what determines the posterior and the still posterior effects is merely the succession of the prior and the still prior modification of elemets, then people desiring heaven would not have performed sacrifices. Such sacrificial activities cannot be fruitless, nor can they have suffering as the sole fruit, as no one in that case will engage onself in sacrifices for the sake of heaven. Even a single sagacious person will not be involved in an activity which is fruitless or which has suffering as the sole fruit, then what to say of the whole world.

To say that people perform sacrifices for some visible gain, viz. profit, fame, etc., would be met by the argument that profit, fame, etc. are aspired for by political councillors and councillors in amusements. A spiritual aspirant, living a deserted life in a forest and performing sacrifices, does not care for such things.

Nor can it be argued that only pleasure-mongers get engaged in sacrificial activities, because in that case nihilists too may be prompted to act in the same manner. But nihilists do not perform sacrifices.

It may be argued that obligatory duties, although their

performance does not yield any fruit, are performed on the ground that they are enjoined in the scripture. In the same way, sacrifices also are performed not because they give rise to extra-empirical merit, but because they are enjoined in the scripture. Udayana replies that such a doctrine is advocated by the teacher Prabhākara and not by the teacher of the Nyāya (seeker after truth). Therefore, it is not proper to mention such point of view in this context.

8.ii. The Cārvāka further argues that sacrifices are performed by the elders to dupe the youngsters. But the fact is that the elders too perform them, and it is not reasonable to say that dupes dupe themselves. No sane person would offer sacrifices knowing that he is being duped by the scripture.

Should the Cārvāka argue that some fraudulent person first devised the custom of offering sacrifices and compelled the rest of the mankind to follow this example, Udayana replies that there cannot be such a deception. For, nobody would practise austerities at the cost of renouncing all his possessions, relatives, enjoyments, mortifying himself throughout his life, just to dupe others. No wise man would follow him too. Surely, the pleasure derived from duping is not greater than the sufferings to such an extent. Udayana asks the opponent to point out the ground on the basis of which he determines such a person to be fraudulent. The opponent may say that he determines a person observing ritual acts to be fraudulent on the ground that such observances are seen even among heretics. It is not so, observances of the heretics cannot be equated with the Vedic observances. The observances of the heretics are motivated by the attainment of some visible result, viz. less labour, enjoyment of pleasure, etc. But the Vedic observances which are of fatiguing nature are not motivated by the attainment of such result. Moreover, duping is possible only when some observances are already in vogue. When such observances continue eternally, duplicity may take place by resorting to other method, through the expenditure of immense wealth or through mere preaching. But duping is never done through actions, that is, through the performance of the ritual acts. A duper does not preach duplicity by performing the rituals himself. If the heretics too perform the Vedic ritual acts, there would be no criterion to ascertain dupery in their performances, and so their observances would be as authentic as the Vedic ones. Consequently, the performer of such ritual acts would not be called heretic also. So far as the Vedic observances are concerned, the performer after performing himself such laborious ritual acts, preaches to others asking them to follow him. Hence, it is not duping. The Vedas are authentic as they are derived from the authentic source. The Bauddha scriptures contradict the Vedic doctrines; hence, they are established as heretical.

8.iii. The Cārvāka further states that he could trace the variety of the world to such variegated causes as gifts, study, etc.

But gifts, study, etc. are momentary, and what is sought for is what will happen at a later period. Gifts, etc. are destroyed soon, fruits are attained at a later period; hence, the former cannot be the cause of the latter. There must be a transcendental factor in between them as a connecting link that leads to the attainment of fruits.

चिरध्वस्तं फलायालं न कर्मातिशयं विना।
सम्भोगो निर्विशेषाणां न भूतैः संस्कृतैरपि।।६।।

An action destroyed long ago cannot produce the result except through some trans-empirical excellence (ensuing from actions). The souls having no distinction cannot have enjoyment even through elements (objects, etc.) ceremonially purified (by the unseen factor).

9.i. It may be said that elements in the form of excellences are produced by the potentiality of their causes, whereby they become the means of definite enjoyments. It may be asked whether the excellence is the super-sensuous form of elements or a variety of contributories of elements. It cannot be the super-sensuous form of elements which are within the scope of the senses, because sensuous elements cannot have super-sensuous form. If it is the second alternative, *adṛṣṭa* is established, because *adṛṣṭa* is a super-sensuous variety of contributories towards the production of effects.

9.ii. The Pūrva Mīmāṁsaka is of the view that *adṛṣṭa* resides in the thing to be enjoyed, and not in the experient. The Cārvāka may, following the Mīmāṁsaka, maintain that if at all *adṛṣṭa* exists, it must be residing in the thing to be enjoyed and not in the experient.

If *adṛṣṭa* does not reside in the soul, the difficulty is that there would be no distinction between the souls. As a result it would be impossible to account for the varying degrees of enjoyment enjoyed by the different souls. This is expressed in what Udayana says that 'the souls having no distinction cannot have enjoyment even through elements ceremonially purified by the unseen factors'. In fact, the person who performs actions, enjoys the fruits also. Therefore, it is assumed that some transcendental factor which ensues from actions resides in the soul, through which the soul experiences the fruits of actions. This is the fifth reason Udayana offers for the existence of *adṛṣṭa.*

9.iii. The Mīmāṁsaka puts forth the view that enjoyment in each person is produced by *adṛṣṭa* residing in the objects to be enjoyed, just as the same fire produces different effects in respect of two people owing to the presence and absence of *śakti*, an imperceptible potency, in the fire. Fire burns when *śakti* abides in it, but it does not burn when there are counteracting factors, such as incantation *(mantra),* fire-extinguishing gem, and the like, that destroy the *śakti.* Otherwise, if fire burns of itself without depending on *śakti,* there would have been the effect of burning wherever there is contact between a hand and the fire, even though the former wore the fire-extinguishing gem. The incantation or the gem obstructs the burning capacity; hence, it is termed *pratibandhaka.*

It may be argued that causality implies the non-existence of counteracting agent *(pratibandhakābhāva);* hence, *śakti* has nothing to do here. Against this the Mīmāṁsaka raises a series of objections. First, causality cannot imply the non-existence of counteracting agent, because causation always implies presence and existence, whereas non-existence is non-existent. Secondly, the non-existence of counteracting agent does not invariably accompany the production of an effect. It is seen that when a neutralizing agent *(pratibandhakottambhaka)* is present, the effect is produced even without the non-existence of counteracting agent. Thirdly, non-existence being of various types, viz. prior *(prāgabhāva)*, posterior *(pradhvaṁsābhāva)*, etc., that are mutually incompatible, non-invariability in respect of a cause is unavoidable.

It may be said that the various kinds of non-existence are somewhat similar, since all of them possess some kind of non-existence. It does not matter whether it is prior or posterior. Even then, the Mīmāṁsaka objects, a cause would not be invariable, since in spite of the presence of the counteracting agent, the effect may be produced as non-existence of some other counteracting agent belonging to the same class may be present there. Therefore, the Mīmāṁsaka concludes that there is some super-sensuous *śakti* in the effect, e.g. fire, which is conducive to burning. What destroys it is a counteracting agent; when the fire is free from such defect, burning is produced. The series of *śakti* is different from each other owing to origination and destruction. But as causes they do not involve non-invariability, since all of them belong to the same class.

भावो यथा तथाऽभावः कारणं कार्यवन्मतः ।
प्रतिबन्धो विसामग्री तद्धेतुः प्रतिबन्धकः ।।१०।।

As existence, so also non-existence is accepted as a cause as well as an effect; counteraction is a defect in the causal aggregate (visāmagrī), *(and) its cause is the counteracting agent.*

10.i. In reply to these objections, Udayana observes that all these arguments are unconvincing. A positive entity could be both a cause and an effect. It is a cause of another object and an effect of some other entity. Non-existence which is known to be only an effect could well be admitted to be a cause also. Surely, nobody would deny that non-existence is an effect, since non-existence is the destruction of a thing previously existing. The destruction is always preceded by some cause, viz. hitting with hammer in the case of a jar's destruction. Similarly, non-existence can be a cause also. The *anvaya-vyatireka* relation that determines the causality of a positive thing, determines the causality of a non-existent thing too. Thus, wherever there is non-existence of the fire-extinguisher, there is burning; wherever there is absence of such non-existence, there is no burning. Non-existence is not a mere nothing, it is not something like the horn of a hare. It has a definite existence from the point of view of existence *(bhāva)* as well as of non-existence *(abhāva)*.

To argue that causality is pervaded by positiveness *(bhāvatva)*, absence of positiveness would occasion the absence of causality, is unintelligible. Causality in respect of a positive thing cannot be determined unless it is pervaded by non-existence.

The Mīmāṁsaka may say that the proximity of non-existence is an inevitable factor and not a cause *(abhāvasya avarjanīyatayā sannidhiḥ na tu hetutvena)*, just as *daṇḍatva* (stickhood) in respect of a *ghaṭa* (jar). But the same argument, Udayana remarks, applies to positiveness also. The proximity of positiveness may be an inevitable factor, not a cause. In fact, no law can be formulated to the effect that a positive thing is the cause only of a positive thing and only a positive thing can be a cause.

10.ii. The opponent's second objection that non-existence of a counteracting agent is not an invariable cause is groundless. The point he stresses is that an effect is produced in the presence of the counteracting agent; hence, the non-existence of the counteracting agent is not an invariable cause. The non-invariability lies in the effect's being produced in the presence of that in the presence of which it was not produced before.

Udayana meets this objection that the cause here is the non-existence of the counteracting agent attended by the non-existence of the neutralizing agent. The effect is produced in the presence of the neutralizing agent in spite of the presence of the counteracting agent, because there is non-existence of the counteracting agent attended by the non-existence of the neutralizing agent. Thus, there is no non-invariability in respect of the cause. The non-existence of the neutralizing agent attended by the non-existence of the counteracting agent is present even when simply the neutralizing agent is present, or both the neutralizing agent and the counteracting agent are present, or both are absent, just as the qualified knowledge *(viśiṣṭa-jñāna),* viz. absence of man without a stick, can be attained when there is the presence of mere stick, or the presence of both man and the stick, or the absence of both.

The opponent may ask why should not an aggregate of neutralizing agent, or an aggregate of counteracting agent and neutralizing agent, or an aggregate of the non-existence of both, be the cause of the effect separately? This suggestion, however, is not sound, because three separate causal aggregates cannot produce the same effect, e.g. fire, which is of unitary character.

The opponent may say that if straw, fire-producing wood, etc. can produce the same fire, the three causal aggregates too may produce the same effect. This argument is futile; straw, etc. do not produce the same fire; different kind of fire is produced by them. The difference in the effect is made out by the *anvaya-vyatireka* relation. But in the latter case, i.e. the production of the same fire from the three different causal aggregates, there is no valid reason for assuming a difference in the effect.

10.iii. The Mīmāṁsaka's third objection that the non-existence of a counteracting agent is not an invariable cause on account of there being prior and posterior non-existence is baseless. The cause, according to Udayana, is relational non-existence *(saṁsargābhāva).* The effect is not produced in that relation where the counteracting agent exists, but it is produced in that relation where the counteracting agent does not exist. There is no question of non-invariability in respect of the cause, since the cause is only one, viz. relational non-existence; prior non-existence, etc. are the varieties of the relational non-existence only and not different kinds of non-existence.

10.iv. The last objection of the opponent that the effect may be produced even in the presence of a counteracting agent, since there may exist non-existence of some other ocunteracting agent belonging to the same class, does not convey any sense. This contradicts experience, there is no reason to presume that there is non-existence of a counteracting agent where the counteracting agent is actually present. When it is said that a cloth is produced in the presence of yarns, by implication it is also understood that a cloth is not produced in the absence of yarns. Just as in the presence of yarns, the cause of cloth, the non-existence of other yarns does not prevent the production of cloth, so also in the presence of one *pratibandhaka* the non-existence of other *pratibandhakas* cannot produce the effect.

10.v. According to Udayana, a defect in the causal aggregate is the chief meaning of the term counteraction *(pratibandha)*. In the case of the production of fire, fire-extinguishing gem, etc. are the defects in the causal aggregate of fire. But they are not the counteracting agents *(pratibandhakas)*. Those who employ the fire-extinguishing gem and the like, are the counteracting agents.

A section of the Naiyāyika says that counteraction is the non-production of the effect, and fire-extinguishing gem, etc. that produce counteraction are the counteracting agents. Udayana does not admit this. Non-production of the effect may be viewed as either prior non-existence or prior non-existence coming into existence at a later time. It is not the first, because prior non-existence being beginningless cannot be produced. It is not the latter also, because time being one cannot be differentiated as earlier and later. It may be said that time can be differentiated by adjuncts, and prior non-existence may come into existence at a time differentiated as later by an adjunct. But differentiating adjunct of time, viz. movement of the sun, depends on its own cause irrespective of fire-extinguishing gem, etc. Prior non-existence being beginningless exists at that time; fire-extinguishing gem, etc. have no control over it.

It may be argued that differentiating adjunct of mere time may not depend on gem, etc., but the adjunct of time which demarcates the prior non-existence depends on them. It is not so, because the adjunct of time that demarcates the prior non-existence exists even before the employment of gem, and the like. Hence, the former does not depend on the latter.

The question may arise if prior non-existence cannot be produced, then how it is possible to establish causal relation between non-existence of effect and that of cause on the ground that non-existence of cause gives rise to the non-existence of effect.

Udayana in reply says that priority and posteriority in respect of the non-existence of cause and that of effect cannot be determined. Since there is the rule of priority and posteriority in respect of the causal aggregate and the effect, there is assumed in a secondary sense a rule about the priority and posteriority in respect of their non-existence also. But this is not true; in fact, there is only simultaneity.

Therefore, non-production of effect cannot be the counteraction.

10.vi. If the Mīmāṁsaka does not admit Udayana's view that gem, incantation, etc. are the *pratibandha* and the person who employs them is *pratibandhaka*, then the theory of *śakti* also would not solve the problem. The point at issue is that the counteracting agent destroys either the *śakti* favourable to burning or its attributes, or produces another attribute unfavourable to burning, or does not produce anything. Among the four alternatives the last one is not acceptable, since what is not capable of doing anything cannot reasonably be a counteracting agent. In the case of the production of another attribute, on the other hand, the effect can be produced only when there is its non-existence. Hence, non-existence as a cause has to be admitted, and once it is admitted, prior non-existence, etc. also come into the purview. Again, if the counteracting agent destroys the *śakti* or the attributes of *śakti*, when the effect is generated in the presence of neutralizing agent, the cause turns out to be non-invariable. First, before the employment of the counteracting agent the effect is produced by the causal aggregate, and later it comes into existence on the strength of the neutralizing agent. Hence, there is non-invariability in respect of the cause. This difficulty cannot be evaded by arguing that the causal aggregate and the neutralizing agent possess the same *śakti*. Udayana is of the view that things belonging to different classes cannot have similar *śakti*.

It is not proper to maintain that the neutralizing agent obstructs the *śakti* of the counteracting agent only, but does not generate *śakti* for production. For, in that case effect would not be produced. If the effect is admitted to be produced from a particular time, there arises the difficulty of non-invariability of the cause.

10.vii. The opponent argues that inherent *śakti (sahaja-śakti)* may not be the cause of effect, but infused *śakti (ādheya-śakti)* is surely the cause of effect. The argument supporting the view is that sacrificial oblation is made with certain grains that undergo ceremonial purification through sprinkling, etc. An infused *śakti* must be admitted in these grains, that leads to the attainment of particular effects at a definite period. Otherwise, there would be no difference between purified and unpurified grains, and sacrificial oblation could be made by unpurified grains as well.

It may be argued that the chanting of hymns, purification, etc. act as auxiliary causes, without producing any *śakti* in the grains. But hymns, etc. cannot, the opponent says, act as auxiliary causes, because the effect is produced even when they are destroyed long ago. Nor can the grains, etc. produce the effect through the posterior non-existence *(pradhvaṁsābhāva)* prevailing after the purificatory modification. Were it so, heaven, etc. could have been brought about by the posterior non-existence of sacrifices and other actions, and the postulation of *adṛṣṭa* could be done away with.

It may be argued that as the cases of posterior non-existence of sacrifices and other actions are infinite in number, in the case of their producing the effect there would result infinity for the number of currents of fruits. But even if *adṛsta* is postulated, says the opponent, there is no criterion to determine the number of fruits. If it is maintained that it is the very nature of *adṛṣṭa* to produce series of effects enjoyed in a definite period of time, the same may be maintained about the posterior non-existence also. Only difference between *adṛṣṭa* and posterior non-existence is that in the former case the last of the series of effects produced from *adṛṣṭa* destroys the *adṛṣṭa* itself, whereas in the latter case none of the series of effects destroys it.

10.viii. It may be said that sprinkling, etc. are *upalakṣaṇa* (casual specification) and not *viśeṣaṇa* (qualificatory specification). *Upalakṣaṇa* is simply *vyavartaka-dharma* and *viśeṣaṇa* is *kāryānvayi-vyavartaka-dharma*. In the case of *upalakṣaṇa*, *vyavartaka-dharma* is not present in the substrate at the time of predication; its presence in the substrate at some time is necessary, but its presence in the substrate at the time of predication is not necessary. In the case of *viśeṣaṇa* the *vyavartaka-dharma* is present in the substrate at the time of predication. Grains, etc. casually specified *(upalakṣaṇatayā)* by sprinkling, etc. are used when sacrificial oblation is made even though sprinkling, etc. do not exist in the grains. Just as the *Śābarabhāṣya-ṭīkā* is specified as *guru-ṭīkā* after Guru Prabhākara even though Prabhākara is no more present, so also the grains specified by sprinkling, etc. are used for the sacrificial oblation even though sprinkling, etc. are absent in the grains. And thus, the sacrificial grains are differentiated from other grains. For this reason an excellence need not be postulated in the grains.

The above position is untenable. *Upalakṣaṇa* cannot be the cause when *upalakṣita* things or their *vyāpāra* are absent at the time of the production of effect. Again, the usage of the name *guru-ṭīkā* depends on the knowledge of *guru-ṭīkā* and not on *guru*. Thus, *guru* is not established by the usage of the name *guru-ṭīkā*. And the knowledge of *guru-ṭīkā* which is of the form of inference, memory, etc., depends on its own cause and is not produced by *upalakṣita* which is destroyed long ago without leaving a link behind.

If *upalakṣaṇa* were the cause, then sacrifices, etc. also can be *upalakṣaṇa* and either time or the sacrificer being *upalakṣita* by the sacrifices, etc. can bring about heaven, etc. *Adṛṣṭa* need not to be postulated for that purpose.

It may be argued that *adṛṣṭa* is postulated not because sacrifices, etc. cannot otherwise be explained, but because there is inferential proof for the existence of *adṛṣṭa*. The inference runs thus: 'The body, sense-organ, mind, etc. of Devadatta, assembled together by his own *adṛṣṭa*, are fit for enjoyments, because they are the means for the enjoyments thereof, like garland, etc.' The opponent may ask what does the statement, 'assembled together by his own *adṛṣṭa*', mean? It may mean that *adṛṣṭa* is the contributory cause for producing enjoyment, or it may mean that *adṛṣṭa* produces the body, etc. It is not the first, because desire, volition and knowledge of human beings can produce enjoyments, and thereby inference is possible from the viewpoint of desire, volition, etc. The second alternative is untenable; though *adṛṣṭa* can produce body, it cannot produce mind which is an eternal entity.

If a little modification is made to the argument by saying that *adṛṣṭa* produces those bodies, etc. that are effects, thus exluding the mind which is not an effect, still the purpose would not be served. Knowledge, desire and volition of past birth can produce the body, etc.; hence, *adṛṣṭa* need not be postulated. It cannot be objected that knowledge, etc. of past birth cannot produce the body, etc., because they do not exist now. For, regarding sprinkling, etc., it has been said that sacrificial oblation can be made by *upalakṣita* grains, even in the absence of grains that are purified by sprinkling, etc. If the knowledge, etc. of past birth being absent are incompetent to produce the body, etc., then sacrificial oblation also cannot be made by *upalakṣita* grains in the absence of grains that are modified by sprinkling, etc.

The Mīmāṁsaka, thus, concludes that sprinkling, etc. generate an excellence of a positive nature in the grains that leads to the attainment of certain fruits at a later time. Similar is the case with agriculture, medicine, etc. Seeds, drug, etc. can be reduced to the final atoms which are devoid of sub-classes. Unless some positive excellence is posited in the atoms, it would be difficult to explain why one kind of atom generates sprout and the other heals disease.

संस्कारः पुंस एवेष्टः प्रोक्षणाभ्युक्षणादिभिः।
स्वगुणाः परमाणूनां विशेषाः पाकजादयः।।११।।

We accept a purificatory modification produced in the person through spraying on, sprinkling on, etc.; the qualities of atoms as produced by contact with heat, etc. are the distinctive features (of the final atoms).

11.i. Udayana observes that there is no excellence or *śakti* of a positive nature in the final atoms of seeds, drug, etc. The attributes of the final atoms are produced by the contact with heat, and these are the distinctive features of the final atoms. The ultimate atoms qualified by particular contact with heat produce respective effects. Just as paddy-seeds differ from oat-seeds in class or cow-milk from buffalo-milk, similarly their ultimate atoms too which are the ultimate causes differ from each other by what are produced by contact with heat. It is not possible for the cow-milk to be fragrant, sweet and cooling, if its ultimate cause-atoms possess the opposite characteristics. The ultimate atoms qualified by the contact with heat of a particular nature determines the differentiation in diads as well as in sprout.

It may be asked how in the beginning of the cycle sprout was generated, since at that time there was no seed and no human being to till the soil, to water the field, etc. It may be replied that sprouts were generated at the beginning of creation by the ultimate atoms of the same peculiarity as at present. Only difference is that now sprouting takes place with the help of auxiliary causes, viz. cultivation, etc., whereas at the beginning of the cycle ultimate atoms of the same kind were brought together by *adṛṣṭa* inherent in the soul. It is not proper to ask why sprouting should not take place now only with the help of *adṛṣṭa* independently of cultivation, etc. For, if the activities of cultivation, etc. are dispensed with, there will result the dispensing with of the enjoyment that is produced by cultivation, etc. Moreover, where there are visible causes, the function of *adṛṣṭa* is to bring together those causes. *Adṛṣṭa* becomes the means for enjoyment through settled order of actions that is seen.

Thus, particular effects are generated by the ultimate atoms qualified by particular results of contact with heat and particular relative position. The effects are qualified by particular associations with fire, air and water, which again are qualified by actions. And action is qualified by push, impact, weight, velocity, fluidity and contact with soul having *adṛṣṭa* in different appropriate ways. Hence, there is no necessity for assuming any *śakti* in the ultimate atoms.

11.ii. Udayana opposes the view that the transcendental factor generated by sprinkling, etc. resides in the grain. The transcendental factor inheres in the soul which is the substratum. Just as the offerings of oblations put into the fire along with chanting of hymns, create a purificatory modification in the person and neither in the fire nor in the deities, similarly sprinkling, etc. create a purificatory modification only in the person and not in that grain.

It is unfair to argue that if grains were devoid of purificatory modification, unpurified grains also can serve the purpose in sacrifices. It is a well-known fact that as a result of *kārīrī* sacrifice, a ceremonial purification is generated in the person who performs the sacrifice. And through the contact with that person the water-clouds move and rain down water. In the same way, Udayana points out, through the contact with the person who is the substratum of purificatory modification, the grains yield their fruits. Just as in the former case the result is not achieved if there is defect in the doer, in the deed or in the means employed for the deed, so is it in the latter case also.

It may be rejoined that, that which is performed in respect of someone produces something in that person. For example, sacrifice on the birth of a son and that for the departed souls produce something in the son and the departed souls. Similarly, chanting of hymns, etc. is performed in respect of grains, etc.; hence, something must be produced in the grains, etc.

This rejoinder cannot be substantiated. For, offerings of oblations that are made in respect of fire do not produce anything in the fire, that is conducive to the generation of the fruit later on. Again, it may also be noted that *vyāpāras* of the sense-organs, *liṅga* and of *śabda,* operated in respect of the objects of valid knowledge, produce something only in the agent of the valid knowledge and not in the objects of valid knowledge.

निमित्तभेदसंसर्गादुद्भवानुद्भवादयः।
देवताः सन्निधानेन प्रत्यभिज्ञानतोऽपि वा।।१२।।

Production and non-production (of touch in air and obstruction of fluidity in water) arise from contact with different causes; the deities are worshipped through their presence (in the images) or through the worshipper's recognition (that images have been worshipped earlier and have been duly consecrated).

12. It has been said that particular effects are generated by ultimate atoms qualified by particular results of contact with heat. But no particularity is produced by contact with heat in air and water, so it must be explained how touch is produced in air and natural liquidity of water is obstructed in ice, etc. Again, if no excellence is admitted in the images of deities made of earth, it cannot be explained how merit is produced through worship of those images even though images are consecrated by installation, etc., and demerit is produced if installed images are not worshipped, and neither is produced when there is no installation. The opponent says that there must be an excellence produced in the image of a deity by the ceremony of consecration that renders the image a fit object of worship.

In reply to the first objection, Udayana says that ultimate atoms, with the assistance of *adṛṣṭa* which brings them together, produce particular substances. On account of the particularities of ultimate atoms there arises particularities in production.

Thus, particularities of the ultimate atoms of air are responsible for producing manifested *(udbhūta)* and unmanifested *(anudbhūta)* touch in air. Similarly, water, though fluid by nature, gets its fluidity obstructed by particular causes. Images acquire eligibility to be worshipped through the presence of the deities in the images by the consecration ceremony, etc. Deities are conceived of as residing in different images, and their presence is established in the images by ceremonial acts. By presence of the deities in images is meant the conceit of 'I' and 'mine', viz. 'I am this', 'this is mine', in images.

The Mīmāṁsaka does not admit sentience in deities. With him, the recognition of the fact that images have been worshipped earlier and that images have been duly consecrated constitutes the images' fitness to be the objects of worship. And the ceremony's importance lies in its contributing to produce this recognition.

जयेतरनिमित्तस्य वृत्तिलाभाय केवलम्।
परीक्ष्यसमवेतस्य परीक्षाविधयो मताः।।१३।।

In order to bring out the auxiliary (cause) of the cause of victory or defeat, which cause inheres in the person subjected to the ordeal, are the rules of the ordeal instituted.

13.i. With a view to proving the fact that *adṛṣṭa* resides in the object of enjoyment the opponent puts forth one more argument. In olden times various ordeals were prescribed in order to find out whether a person is really guilty of committing certain crime or not. One such ordeal is weighing ordeal. According to this ordeal, after the utterance of *mantra,* etc. the suspected person is asked to climb a balance, and if the balance sinks down the person is convicted of guilt, but if it goes up the person is declared to be innocent. The opponent says that a *śakti* must be admitted to be produced in the balance by the ceremony of ordeal owing to which there results the rising or sinking of the defendant in the balance.

This objection is replied to thus. Ordeals are prescribed, according to Udayana, with a veiw to acquiring the auxiliary cause of victory or defeat. The principal cause of the victory or defeat is the *adṛṣṭa* that resides in the convict. Ordeals are meant for the acquisition of the auxiliaries of the *adṛṣṭa.* The consciousness in the convict's mind in the form 'I who according to the rule of the ordeal now mount the balance am innocent or guilty' is the auxiliary cause of *adṛṣṭa.*

There may be another explanation of the ordeals. The rules of ordeals are instituted in order to produce merit and demerit. Thus, merit is produced when the convict is innocent the result being the victory, and demerit is produced when the convict is guilty the result being the defeat.

13.ii. The Mīmāṁsaka finally questions the authority on the basis of which the Naiyāyika denies the theory of *śakti.* Udayana, in reply, observes that there is no authority as such. Rather, he admits *śakti*, but in a different sense. *Śakti* is causality, it is the fact of belonging to a class determined at a time earlier, or non-generation of an effect due to a defect in the contributory. The assumption of *śakti* over and above causality is unwarranted, since there is no authority for that. One cannot even entertain any doubt whether there is *śakti* or not. To doubt requires the previous cognition of the character of the thing doubted. But in the case of *śakti* one does not have previous cognition of its character. It is not also a subject-matter of dispute, because contributories, as they are conducive to the production of effects, are also referred to by the term *śakti.* When an effect is produced by different contributories, there is something even there over and above the contributories, which is conducive to the generation of the effect and which is removed by the *pratibandhaka.* If it is accepted as *śakti*, it virtually substantiates the thesis of Udayana.

13.iii. It may be asked how it is ascertained that it is the sentient being wherein resides the effect of purificatory modification and not the external objects. It may be replied that since the enjoying souls are eternal and all-pervading, they are connected with all external objects. If the effect of purificatory modification resides in the external objects, not in the souls, it will be difficult to explain definite enjoyments for each soul. External objects being common to all souls, all the souls will experience the same enjoyment. This goes against experience.

To establish definite enjoyment for each soul, it is assumed that the effect of purificatory modification that determines the experience of each individual soul resides in the soul and not in the external objects.

It cannot be argued that only a particular body, a particular mind and particular sense-organs belong only to a particular soul, because there is no criterion to regularize so. And a certain attribute of the external objects is not special for a sentient soul, since attributes of external objects are common to all souls which are eternal and all-pervading.

13.iv. It may be said that as the number two that is produced in objects is cognized by a definite individual being, so also *adṛṣṭa* residing in the external objects produces enjoyment for definite individual.

This position is untenable. For, number two, etc. are also produced in the objects by the enumerative cognition of the sentient being, and enumerative cognition is an attribute of the soul. Thus, the cognition of the number two, etc. by definite individuals is determined by the attribute of the sentient soul and not by its mere residence in the object. If the cognition of the number two, etc. were not determined by the attribute of the sentient soul, the causal relation between the enumerative cognition and the knowledge of the number two, etc. would break down. If the fact of the cognition of the number two, etc. by definite individual were not determined by the attribute of the soul, viz. the enumerative cognition, the knowledge of the number two, etc. would not arise at all. For, it is an established fact that an attribute of the soul is produced by another attribute of the soul. Thus, volition is produced by desire, desire is produced by knowledge, and so on.

13.v. The view that *adṛṣṭa* is *bhūta-dharma,* i.e. *adṛṣṭa* resides in external objects, is rejected on the ground that if *adṛṣṭa* resides in external objects, all external objects being connected with all eternal, all-pervading souls definite enjoyment for each soul will not arise.

The Sāṅkhya holds that *adṛṣṭa* resides in the *antaḥkaraṇa;* this view is refuted by Udayana. The Sāṅkhya doctrine is as follows. The world consists of twenty-five principles — *puruṣa, prakṛti, mahat, ahaṅkāra,* eleven sense-organs, five *tanmātras,* viz. *śabda* (sound), *sparśa* (touch), *rūpa* (sight), *rasa* (taste), *gandha* (smell), five *bhūtas,* viz. earth, water, air, fire and ether. *Puruṣa* is of the nature of sentience, immutable, neither a cause nor an effect. *Prakṛti* is the primeval cause, uncaused, non-sentient and subject to modification. From *prakṛti* are evolved *mahat* and other principles gradually.

Mahat and other evolutes are admitted for the following reasons. If the *puruṣa* who is of the nature of consciousness were related with objects and if that relation were natural, then the *puruṣa* will always cognize objects. As a result the *puruṣa* will not attain liberation. Again, if the relation between the *prakṛti* and the *puruṣa* were natural, then the eternal *prakṛti* being ever related with the *puruṣa,* liberation for the *puruṣa* will not be possible. Further, if the *puruṣa* were directly related to objects, there will not arise such distinction as 'this is seen', 'this is not seen', since all objects being connected with consciousness there will remain nothing which is unseen. To obviate all these difficulties sense-organs are admitted. The *puruṣa* comes in contact with objects through the channel of the sense-organs; hence, the above-mentioned defects do not arise.

In spite of the sense-organs being in contact with objects, sometimes the knowledge of some object arises immediately, sometimes it does not. To account for the delay in the production of knowledge the mind is admitted. The knowledge arises only when the mind comes into contact with the sense-organ which is in contact with objects.

In the state of dream one has the conceit *(abhimāna)* of being a boar or a tiger, but one does not comrpehend oneself as a man. If the mind were the cause of this conceit, one would have comprehended oneself as a man even when one conceives in the form 'I am tiger'. One does not comprehend oneself as a man in the state of dream, because mind is not the cause of the *abhimāna* of 'I' *(aham).* It is *ahaṅkāra* which is the cause of self-sense. Due to *ahaṅkāra* sometimes one comprehends oneself as man and sometimes does not.

In the state of deep sleep there is the cessation of the function of *ahaṅkāra,* etc. But in that state there continues the series of breathing activity, and *antaḥkaraṇa* is admitted as the cause of this activity. The function of *antaḥkaraṇa* continues in the state of waking, dream as well as of deep sleep. Impression produced by *anubhava* also resides in the *antaḥkaraṇa*. Among the mind, *ahaṅkarā* and *antaḥkaraṇa*, *antaḥkaraṇa* occupies principal position. Object residing in the *antaḥkaraṇa* becomes the object of knowledge of the *puruṣa*. The *puruṣa* who is of the nature of consciousness is free from agency. When the *puruṣa* feels agency thinking 'I am doing this', this feeling is due to non-discrimination between the *puruṣa* and the *mahat*. Agency that exists in the *mahat* is mistakenly felt by the *puruṣa* owing to non-discrimination. Similarly, *mahat*, the agent, due to non-discrimination thinks itself as the substratum of consciousness which, in fact, is the nature of the *puruṣa*. Since *mahat* is the substratum of agency, it is the substratum of *adṛṣta* also. The Sāṅkhya does not admit that the *puruṣa* is the substratum of *adṛṣṭa*. The *puruṣa* is untainted like the leaf of a lotus and is free from agency.

According to the Sāṅkhya doctrine objects residing in the *antaḥkaraṇa* come in contact with the *puruṣa*, and the *puruṣa* feels agency owing to non-discrimination between the *puruṣa* and the *mahat*. The question may arise as to how the *antaḥkaraṇa* becomes the substratum of objects, and how there arises non-discrimination between the *puruṣa* and the *mahat*, and how again objects come into contact with the *puruṣa*.

It may be replied that *nirvikalpaka* apprehension is the *vyāpāra* of the sense-organ, and *savikalpaka* apprehension is the *vyāpāra* of the mind. The *vyāpāra* of the *ahaṅkāra* is the feeling of 'I'. Determination of volition in the form 'I should do this' is the *vyāpāra* of the *mahat*. This determination *(adhyavasāya)*, viz. 'I should do this', consists of three parts — (a) 'I' represents the cognition of the self, (b) 'this' represents the cognition of objects, (c) 'should do' represents the cognition of volition *(kṛti)*. The cognition of the self that arises in *mahat* occurs due to *cetana-uparāga*. *Cetana-uparāga* stands for the identity between the sentient *puruṣa* and the *mahat*. When the sentient *puruṣa* is reflected in the *mahat*, due to non-cognition of difference between the *puruṣa* and the *mahat*, the *puruṣa* identifies itself with the *mahat*. Due to this identification there arises the cognition of the self in the form 'I' in the *mahat*. This identification is unreal.

Just as the relation between the reflected face and the mirror on which the face reflected is unreal, so also the relation of identity between the *puruṣa* and the *mahat* is unreal. The *antaḥkaraṇa* coming out through the channel of the sense-organs transforms into objects with which it comes into contact. The relation of the *mahat* to objects through the modification of the *antaḥkaraṇa* is *viṣaya-uparāga* represented by 'this'. The relation of the *mahat* with objects is real, just as dimness of mirror when one blows one's breath on it is real. *Mahat* is related with volitional activity through its relation with the *puruṣa* and objects, *Mahat*'s relation with objects is knowledge *(jñāna)*. The relation of the reflection of the *puruṣa* on the *mahat* with objects is cognition *(upalabdhi)*. According to the Sāṅkhya, *buddhi, jñāna, upalabdhi* these three are diffeent and not the same.

As the *puruṣa* is immutable and eternal, volitional activity resides in the *mahat* which is called *buddhi*, and *adṛṣṭa* produced by the volitional activity also is an attribute of *budhi*. Eight dispositions, viz. virtue, vice, pleasure, pain, desire, aversion, knowledge and volition, all are attributes of the *mahat* and not of the *puruṣa*.

कर्त्तृधर्मा नियन्तारश्चेतिता च स एव नः।
अन्यथाऽनपवर्गः स्यादसंसारोऽथवा ध्रुवः।।१४।।

Attributes of the agent determine (the experience), and to us the agent himself is the sentient being; otherwise there will be no liberation, or the world will certainly not originate.

14. The Sāṅkhya view is criticized by Udayana. It must be admitted that one who is the agent of action, is the substratum of virtue, vice, pleasure, pain, etc. also. Sentience is also the attribute of the agent, and not that of *buddhi*. The agent is a sentient being; hence, it is not proper on the part of the Sāṅkhya to postulate different substrata for agency and sentience, the experience, viz. 'I, the sentient, am doing this', results from the fact that sentience and volition are experienced as co-existing in the same substratum. The validity of this experience cannot be questioned, because there is nothing to invalidate the experience.

To invalidate the contention that the agent *buddhi* is the substratum of sentience, an inferential argument may be offered. The *buddhi* is devoid of consciousness, because it is subject to modification, like a pot. But this argument may be countered by the inference that the *buddhi* is devoid of agency, because it is subject to modification, like a pot. Obviously, this position would not be accepted by the Sāṅkhya who maintains that agency is the natural attribute of the *buddhi*.

The opponent may advance one more argument to prove absence of sentience in the agent. The agent *mahat* is insentient, because it is the product of insentient *prakṛti*. This inference is fallacious, because *hetu*, viz. being the product, is unestablished in the subject. The agent is not an effect, it is an eternal substance. The agent being eternal is not subject to production and destruction; hence, it cannot be an effect.

The *Nyāya-sūtra (vītarāgajanmādarśanāt,* 3.1.25) adduces proof for the eternality of the soul who is the agent. This *sūtra* means that when a person is born, he is born with longings. This longing could be due only to the recalling to mind of things previously experienced. The soul remembering the things experienced by it in its previous body, longs for them. This longing forms the connecting link between its two lives. There are similar links between its previous life and the life preceding that, and so on *ad infinitum*. This shows that the connections of the soul with bodies and with longings are without beginnings. From this beginningless series of attachments and consequent bodies it follows that the soul is eternal.

As in the experience 'I, the sentient, am doing this' the agent is experienced as the substratum of sentience, it is established that agency and sentience reside in the same substratum. The Sāṅkhya calls this substratum the *buddhi*, and the Naiyāyika the soul.

If sentience is not attributed to the agent *(buddhi)* and a sentient principle other than the agent *(puruṣa)* is admitted, then bondage and liberation become an impossibility. The Sāṅkhya may argue that the bondage is due to non-cognition of difference between the *puruṣa* and the *buddhi*. It may be asked whether the *buddhi* is eternal or non-eternal. If it is the former, *buddhi* being ever related with the *puruṣa*, the *puruṣa* can never attain liberation. In the latter case *buddhi*, being non-eternal, is perishable, and hence is subject to origination. Along with the destruction of *buddhi*, the attributes of *buddhi* too being lost, prior to the origination of *buddhi* there being no determinant to produce the *buddhi*, the *buddhi* will not be produced. In the non-origination of *buddhi*, the respective body, sense-organs, etc. will also remain unproduced, and thus there arises the non-origination of the world. To argue that *buddhi* always remaining in a subtle form determines origination is untenable. For, in that case the liberated too will tend towards worldliness. It may be rejoined that the *buddhi* qualified by *vāsanā* (residues) is the cause of bondage; the *buddhi* in its subtle form being devoid of *vāsanā* cannot be a hindrance to liberation. Were it so, the *buddhi* existing always, the *buddhi* qualified by *vāsanā* may be termed as *buddhi*, and the *buddhi* in its dormant form may be called *prakṛti*; thus, there does not arise any necessity of admitting a separate *prakṛti*. With the non-acceptance of a separate *prakṛti*, *ahaṅkāra* and *mana* also become redundant, as the special functions of *ahaṅkāra* and *mana* may be attributed to the *buddhi* itself. Thus, *buddhi* being non-different from *prakṛti*, the probans, viz. 'the product of insentient *prakṛti*', advanced to prove insentience in *buddhi* remains unestablished. Since there is no invalidating *pramāṇa* against admitting sentience in the *buddhi*, the *buddhi* qualified by *vāsanā* determines the origin of the world, and the non-functioning of the *buddhi* by the cessation of *vāsanā* leads to liberation.

Objection may be raised that the soul being the substratum of volition *(kṛti)*, etc. which are subject to destruction cannot be eternal. This objection may be rejected on the ground that there having been absolute difference between attributes and the substratum, the substratum soul can remain eternal even when its attributes, viz. volition, etc. are destroyed. Thus is established the sentient principle, viz. soul as eternal, agent, enjoyer and the substratum of *adṛṣṭa*.

नान्यदृष्टं स्मरत्यन्यो नैकं भूतमपक्रमात्।
वासनासङ्.क्रमो नास्ति न च गत्यन्तरं स्थिरे।।१५।।

One cannot remember what another has seen; the elemental body cannot remain the same, as it decays; residue of impression cannot be transferred; in the acceptance of eternity there is no other way.

15. The Cārvāka rejects the conception of an eternal and all-pervading soul which enjoys the fruits of its actions produced by *adṛsta* inherent in it. According to him a certain kind of element modified in the form of the body is the substratum of sentience. That elemental body is the substratum of sentience is proved by concomitance and non-concomitance. When there is an invariable concomitance between two things, they are causally connected, one being the material cause of the other, the effect. Observation goes that there is an invariable concomitant relation between a body and the consciousness; hence, body is the material cause of the consciousness. The soul is body endowed with consciousness. Cognition, impression, desire all inhere in the body, because they are perceived as identical with it. 'I cognize', 'I recollect', 'I desire' all these mean that body cognizes, recollects and desires.

In this context it may be asked whether the cognition exists in the whole of the body, or in some part of it. If the former, as the body changes everyday due to accumulation and diminution of parts, the experience of one body today cannot be remembered by the other tomorrow. In the latter case, if one hand or one leg were the instrument of cognition, when it drops off, no other part would recollect what was experienced by it.

Nor is it that just as the residue of the musk *(mṛgamadagandha)* can be transferred to different clothes through association, so also the residue of impression of one can be transferred to the other through association. Were it so, a baby in the mother's womb could remember the mother's experience, since due to association with the mother the residue of the mother's experience would have been transferred to the baby.

The Cārvāka may argue that the transference of experience from one to the other is possible where the former is the material cause and the latter its effect. A baby in the mother's womb cannot recollect the mother's experience, because there is no such causal relation between them. This position can be maintained from the viewpoint of the Bauddha doctrine of momentariness which holds that there is no such thing as whole; there are only momentarily existing atoms, and the prior atoms are the material causes of the later atoms, the effects. The Cārvāka does not subscribe to the doctrine of momentariness; hence, he has to admit hands, legs, etc. as the material causes of the body. Then on his hypothesis, if a person's hand is cut off, the maimed body will not remember a prior experience of that hand, since that will no longer be the material cause of the body.

न वैजात्यं विना तत्स्यान्न तस्मिन्ननुमा भवेत्।
विना तेन न तत्सिद्धिर्नाध्यक्षं निश्चयं विना।।१६।।

Without difference there cannot be that (momentariness), if it (difference) is there, inference is not possible; without that (inference), that (your hypothesis) cannot be established, nor can there be perception (nirvikalpaka) *without determinate perception* (savikalpaka).

16.i. The Cārvāka may argue with the Bauddha that all things dissolve every successive moment, and each heap of atoms, the material cause, generates a succeeding heap, its effect. To the Bauddha existence means practical efficiency or adaptability to practical needs of life *(arthakriyākāritva)*. The collection of atoms of elements possessing causal efficiency regulates recollection of experience. As the prior and still prior atoms are the material causes of the later and still later atoms, there arises the gradual transference of the residues of cognition and action from the prior atoms to the later ones. The Cārvāka adopts the Bauddha doctrine of momentariness and declares that there is no sentient being over and above the elements.

Udayana refutes the doctrine of momentariness in the following manner. According to the doctrine of momentariness every object differs from moment to moment. As the object exists for one moment, the object is equated with the moment. The earlier and still earlier moments give rise to the later and still later moments. The object existing in a particular moment is destroyed in the very next moment. The cause possessing an efficiency called *kurvadrūpatva* gives rise to the effect immediately after it. For example, the causal efficinecy in the seed that exists for one moment, gives rise to the sprout at the next moment.

Difference in the seeds that exist in succeeding moments cannot be perceived. Just as the difference between two flames cannot be ascertained owing to their similarity, though both are different in different moments, so also due to similarity of different moments efficiency that is present in succeeding momentary seeds and owing to which seeds differ from moment to moment cannot be perceived. Causal efficiency in the previous moment is inferred from the perception of the effect in the next moment. For example, observing that a seed in the granary does not sprout whereas a seed in the field sprouts, a causal efficiency is inferred in that seed which exists immediately prior to the sprout. If such causal efficiency is not admitted, then the seed in the field would be as good as the seed in the granary, and the former would not give rise to the sprout like the latter. Therefore, it must be admitted that the seed in the field possesses some causal efficiency which is not possessed by the seed in the granary, and thus is established the difference between seeds existing in different moments.

Momentariness is inferred by the following inference. Positive things are momentary, because they exist; that which exists is momentary, like the flame of a lamp. Existence here stands for causal efficiency. Causal efficiency consists in producing an effect. Causal efficiency that produces the effect is momentary, as it ceases to exist in the next moment after producing the effect.

The doctrine of momentariness does not stand the test of critical examination; hence, it it refuted. Momentariness cannot be established without assuming causal efficiency that distinguishes one individual from another, e.g. causal efficiency of the seed in the field distinguishes it from the seed in the granary which lacks causal efficiency. But cauasal efficiency itself is unproved. Production and non-production of the effect, e.g. sprout can be explained with respect to the same seed with the presence and absence of contributories. Thus, the seed sprouts when it is assisted by the contributories, viz. water, soil, etc., and in the absence of such assistance it does not sprout.

It is not reasonable to argue that even in the absence of causal efficiency momentariness in respect of seeds may be established by the postulation of many seeds of the same kind. For, in the absence of causal efficiency it will be impossible to determine whether the seeds belong to the same kind or to different kinds. As production and non-production can be explained in respect of the same seed with the help of the presence and absence of contributories, there is no need for assuming causal efficiency in different individuals.

16.ii. The presumption of causal efficiency would lead to the abolition of all inference. The Bauddha admits three kinds of inference — *kārya-anumāna, anupalabdhi-anumāna,* and *svabhāva-anumāṇa.*

In *kārya-anumāna* the cause is inferred from the effect. If the cause possesses an efficiency, then the effect too which is produced by such causal efficiency may possess some efficiency. Thus, what would be produced by the causal efficiency is the efficiency in the effect and not the effect that is seen.

If it be said that efficiency in the effect is not perceived, hence not produced, the reply would be that efficiency in the cause also is not perceived, and so it does not exist in the cause.

As the presumption of causal efficiency in the seed occasions the presumption of efficiency in the sprout, the causal efficiency in the seed will produce only efficiency in the sprout, and not the sprout. Efficiency in the sprout being imperceptible cannot be ascertained; hence, the cause-seed cannot be inferred from the presumed efficiency in the effect. Therefore, the cause cannot be inferred from the effect. Consequently, *kārya-anumāna* stands rejected.

Anupalabdhi-anumāna also cannot be established. *Abhāva* is inferred from *anupalabdhi.* The seed existing immediately prior to the sprout possesses two *jātis*, *bījatva* and *kurvadrūpatva. Jāti* according to the *Nyāya* is a positive entity. With the Bauddha *jāti* is that which distinguishes one thing from the other. Two *jātis* never co-exist, as both are mutually exclusive. If they co-exist, one must be the *vyāpya* (pervaded) and the other *vyāpaka* (pervader). If this is not admitted, then both *jātis* should either belong to the same class or to different classes. Two *jātis* belonging to the same class cannot co-exist, because if they belong to the same class, they would not be accepted as two jātis. *Ghaṭatva* and *kalaśatva* both belong to the same class; hence, they are not accepted as two *jātis.* The second alternative is also untenable. If two *jātis* belonging to two different classes co-exist, it amounts to saying that there can be co-existence of two classes that have mutual exclusion. Thus, two contradictory *jātis,* viz. *gotva* and *aśvatva,* can co-exist in the same locus. But *gotva* and *aśvatva* do not co-exist in the same locus; that they do not co-exist in the same locus may be due to the fact that the causal aggregate of *gotva* is different from the causal aggregate of *aśvatva,* or that the effect *gotva* is different from the effect *aśvatva.* The first alternative stands rejected, because difference in causal aggregates cannot be ascertained. It cannot be said that causal aggregates are different, because *gotva* and *aśvatva* are never seen to be mutually related. For, it is not possible to know by perception that *gotva* and *aśvatva* are never mutually related. In fact, owing to their animality both possess similar features, and so there is co-existence of the causal aggregates of their features. Possibility is there for the co-existence of all the causal aggregates also.

If it is said that absence of co-existence is due to mutual difference between the effects *gotva* and *aśvatva*, there arises the contingency of the absence of co-existence of *kampatva* (quivering) and *śiṁśapātva (śiṁśapāness); kampatva* and *śiṁśapātva*, however, co-exist in a quiverish *śiṁśapā* tree. It may be said that *kampatva* and *śiṁśapātva* are not perceived to co-exist in a *śiṁśapā* which does not quiver; hence, they do not co-exist. Then, were the perception the proof, *gotva* and *aśvatva* may co-exist, since on the basis of perception it cannot be ascertained that *gotva* and *aśvatva* never co-exist.

Prior to the production of the sprout *kurvadrūpatva* is present in the non-seed that does not sprout, and *bījatva* is present in the seed in the granary that is devoid of *kurvadrūpatva*. As *kurvadrūpatva* exists in the substratum where *bījatva* is absent, and *bījatva* exists in the substratum where *kurvadrūpatva* is absent, both are contradictory *jātis*. If two mutually exclusive *jātis* co-exist in the same substratum, then the absence of one cannot be inferred from the presence of its contradictory *jāti*. On the perception of *gotva* the absence of *aśvatva* which contradicts *gotva*, is inferred. If *gotva* and *aśvatva* that have mutual exclusion co-exist in the same substratum, their mutual contradiction will not be established; hence, the perception of *gotva* will not give rise to the inferential knowledge of absence of *aśvatva*. Thus, non-cognition which consists in *anukūla-anupalabdhi* and *pratikūla-upalabdhi* will not produce the knowledge of *abhāva*. Therefore, *anupalabdhi-anumāna* is unestablished.

Svabhāva-anumāna is *tādātmya-anumāna* which also stands refuted. In *tādātmya* inference on the basis of identity between the *hetu* and the *sādhya*, the *sādhya* is inferred from the *hetu*. There is the relation of identity between *vṛkṣatva* (treeness) and *śiṁśapātva*; hence, *vṛkṣatva* is inferred from *śiṁśapātva*. If two mutually exclusive *jātis* can co-exist in the same substratum, then *śiṁśapātva* and absence of *vṛkṣatva*, viz. *avṛkṣatva*, also may co-exist; hence, *vṛkṣatva* cannot be inferred from *śiṁśapātva*. Thus is rejected *tādātmya* inference.

It may be argued that there is *bādhaka* for the co-existence of *śiṁśapātva* and *avṛkṣatva* in the same substratum. *Tādātmya* inferences are as follows: 'This is tree, because this is *śiṁśapā*'; 'this is designated as tree, because this is *śiṁśapā*'. Both these inferences prove identity between *śiṁśapā* and tree. The causal aggregate of *śiṁśapā* does not fall outside the causal aggregate of tree; hence, there arises absence of the causal aggregate of *śiṁśapā* in the absence of the causal aggregate of tree. If *śiṁśapā* were not tree, in the absence of the causal aggregate of tree the causal aggregate of *śiṁśapā* would be absent. Thus, *śiṁśapā* would fall outside of its own causal aggregate.

Again, if *śiṁśapā* were not designated as tree, there arises the same difficulty. The designation of a tree follows from what has branches, etc. in general, and the designation of *śiṁśapā* follows from what has particular branches, etc. If *śiṁśapā* were not designated as tree, there would arise the absence of what has branches, etc. in general, and in the absence of what has branches, etc. in general, there would arise the absence of what has particular branches, etc. Hence, the designation of *śiṁśapā* which has particular branches, etc. would fall outside of its own causal aggregate. Thus, the *bādhaka* for the co-existence of *śiṁśapātva* and a*vṛkṣatva* consists in the fact that if *śiṁśapātva* does not exist with *vṛkṣatva* in the same substratum, but exists with a*vṛkṣatva*, *śiṁśapā* will fall outside of its own causal aggregate.

This position cannot be substantiated. For, in that case there may arise the contingency of such inferences as 'this is *śiṁśapā*, because this quivers'; 'this is designated as *śiṁśapā*, because this is designated as that which quivers'. *Śiṁśapā* has been designated as that which has particular branches, etc. The designation of quiverish *śiṁśapā* consists in its being a variety of what has particular branches, etc. Hence, the causal aggregate of quivering is included within the causal aggregate of *śiṁśapā*. Therefore, if quivering exists falling outside the causal aggregate of *śiṁśapā*, it will fall outside of its own causal aggregate.

In fact, *śiṁśapā* cannot be inferred from the act of quivering found in *śiṁśapā*, since quivering may be produced by some other extraneous factors connected with that *śiṁśapā* which quivers and not by the causal aggregate of *śiṁśapā*.

It may be said that the causal aggregate of *śiṁśapā* is included within the causal aggregate of tree, but the causal aggregate of quivering is not included within the causal aggregate of *śiṁśapā*. The act of quivering is determined by such casual factors as impact, and the like. As the Bauddha does not admit that production and non-production of effect are due to the presence and absence of contributories, he has to admit a moment which possessing causal efficiency produces quivering during the state of impact. This moment, as it produces quivering, is included within the causal aggregate of quivering; hence, it should fall within the causal aggregate of *śiṁśapā* also, as quiverish *śiṁśapā* is a variety of *śiṁśapā*.

If again impact, etc are regarded, following the Nyāya, as contributories with the help of which the causal aggregate of *śiṁśapā* produces quivering, then quivering being produced with the help of contributories without presuming any efficiency in the causal aggregate of quivering, momentariness will be unestablished.

It may be argued that impact, etc. belong to the nature of *śiṁśapā*, but they produce quivering with the assistance of adventitious contributories, viz. air, etc. Like the causal aggregate of *śiṁśapā*, the causal aggregate of other things also may produce quivering with the help of adventitious contributories, viz. air, etc. Hence, the causal aggregate of quivering falls outside the causal aggregate of *śiṁśapā*. Were it so, it may be said that *śiṁśapā* is also produced by the causal aggregate of tree when some contributories are introduced into the causal aggregate of tree. Like the causal aggregate of tree, the causal aggregate of other things also may produce *śiṁśapā* with the help of such contributories. Hence, the causal aggregate of *śiṁśapā* falls outside the causal aggregate of tree.

It is true that *śiṁśapā* may be produced with the help of contributories. But there is a difference in the two cases. In the case of the production of *śiṁśapā*, the adventitious contributories assist the causal aggregate of tree only. But in the case of the production of quivering, even a pot is seen to sway, so the adventitious contributories assist not only the causal aggregate of *śiṁśapā*, but the causal aggregates of all tangible things *(mūrta-dravya)* that quiver.

This is not the point of concern here. The point at issue is whether the causal aggregate of quivering is a variety of the causal aggregate of *śiṁśapā* or not. If the first, the causal aggregate of quivering will not be different from the causal aggregate of *śiṁśapā*. If it is the second alternative, impact, etc. will be established as contributories. In that case, production and non-production of quivering can be explained by the same causal aggregate with the help of presence and absence of contributories. Thus, the causal aggregate that produces non-quiverish *śiṁśapā* will produce quivering in *śiṁśapā* with the help of contributories. Hence, the causal aggregate of quiverish *śiṁśapā* will not be different from the causal aggregate of non-quiverish *śiṁśapā*, and in the absence of difference in the causal aggregates, momentariness remains unestablished.

As sometimes quivering is produced in the same *śiṁśapā*, quivering is not a different *jāti;* hence, there does not arise the contingency of two different *jātis,* viz. *kampatva* and *śiṁśapātva*, co-existing in the same locus. Therefore, two mutually contradictory *jātis* do not co-exist in the same locus. If two *jātis*, like potness and earthness, co-exist in the same locus, then one is the *vyāpya* and the other the *vyāpaka*. Those that co-exist without being *vyāpya* and *vyāpaka* are not *jātis*, but *upādhis,* like *kāryatva, bhūtatva, mūrtatva, kampatva*, etc. *Upādhis* are of the form of substance, quality, action, etc. For example, *kāryatva* is substance, *bhūtatva* is quality, and *mūrtatva* and *kampatva* are actions. *Upādhis* do not co-exist on account of the difference of the individual, and *jātis* do not co-exist on account of differences in the substrata.

Therefore, as in the case of quiverish *śiṁśapā*, all productions and non-productions can be explained with the help of contributories. As there is no proof for the causal efficiency over and above the contributories, momentariness stands rejected. If an efficiency is admitted in the cause, there being the contingency of the production of an efficiency in the effect also, all *kārya-anumāna* will be abolished. Again, there being the contingency of two contradictory *jātis,* co-existing in the same locus, *anupalabdhi* and *svabhava anumānas* also remain unestablished. Therefore, inference as such will be abolished.

Without inference, the doctrine of momentary flux cannot

be established. On the basis of inference in the form 'this is momentary, because this exists', the Bauddha establishes the doctrine of momentariness. Hence, inference being invalidated the doctrine of momentariness will also be nullified. Nor can the Bauddha maintain that momentary flux is proved by perception. He admits only *nirvikalpaka* perception, whereas the Nyāya admits both *nirvikalpaka* and *savikalpaka* perceptions. In perceptual knowledge when a sense-organ comes in contact with the object, e.g. a jar, there arises two distinct ideas in the mind, the idea of a jar and the idea of the nature of jar. This is called *nirvikalpaka* perception. Later the mind combines both the ideas into one, 'a jar possessing the nature of jar' — this is *savikalpaka* perception. We are not conscious of the *nirvikalpaka* perception, it is logically inferred from the subsequent *savikalpaka* perception as the cause of the latter. Since *nirvikalpaka* perception cannot be established without *savikalpaka* perception, *nirvikalpaka* has no authority without *savikalpaka.* Perception certifies the validity of that object which is cognized by *nirvikalpaka* and is determined by *savikalpaka; nirvikalpaka* without *savikalpaka* is devoid of authority. Therefore, *nirvikalpaka* perception is not an authority for momentary flux, otherwise the horn of a hare can be established by *nirvikalpaka* perception.

It may be said that the present moment in which a jar exists can be apprehended by *savikalpaka* perception. The persent moment is different from the prior and the posterior existing moments. Being present is determined by being different from the prior and the posterior moments. This proves that an object is different from moment to moment, and thus is established momentariness.

The question arises as to what this present moment stands for. If by present moment is meant the *dharmī,* viz. a jar, then it is only the jar that is apprehended by *savikalpaka* perception, and not its difference from the prior and the posterior moments. Hence, difference cannot be established by *savikalpaka* perception.

If again by the present moment is meant the attribute *(dharma)* of jar, its difference from the prior and posterior attribute may be established. But difference in *dharma* does not make the *dharmī* different, since *dharma* and *dharmī* are not identical according to the Bauddha, *dharma* is admitted to be different from the *dharmī*.

It may be argued that the prior and the posterior moments are absent; being present and being absent are contradictory; hence, they cannot exist in the same substratum. Therefore, the substrata of present and absent entities must be different.

This argument is untenable. Contradiction arises only when by being present and being absent are meant existent and non-existent. In the knowledge produced by *pratyabhijñā*, viz. 'this is that jar', the jar is apprehended as existent in both the prior and the posterior cognitions. As the jar is apprehended as existent in both the prior and the posterior cognitions, and not as non-existent, there does not arise any contradiction.

If it is said that by being present and being absent are meant being related to present and past times, then also there does not arise any contradiction. Relation with present time does not contradict the relation with past time. In the *pratyabhijñā*-knowledge there arises the knowledge of object that is related to past as well as present time.

It may be said that knowledge is simultaneously related to past and present times. But a jar cannot be so related, its relation with past and present times arises in succession. In reply Udayana says that relation arises in succession only when the causal aggregate of relation arises in succession. There is nothing wrong, if the jar is related to past and present times in succession.

The Bauddha may rejoin that *pratyabhijñā* is not valid knowledge; hence, on the authority of *pratyabhijñā* it cannot be ascertained whether the same object is related to past and present times.

Udayana replies that the question whether *pratyabhijñā* is valid or not would be considered. But it is a fact that perception proves the existence of an enduring object; momentariness does not become the object of even erroneous perception. Hence is unestablished the doctrine of momentary flux.

स्थैर्यदृष्ट्योर्न सन्देहो न प्रामाण्ये विरोधतः।
एकतानिश्चयो येन क्षणे तेन स्थिरे मतः।।१७।।

There is no doubt in respect of either continuance or its cognition, nor in respect of its validity, as there is contradiction; the same (proof) that determines endurableness in what is momentary, determines endurableness in what is continuant also.

17. The Nyāya refutes the doctrine of momentariness. The Bauddha, on the other hand, seeks to establish it. As this is the position, the Cārvāka may say that the position of the flux hypothesis is still undetermined and not refuted. There may be doubt about its vlaidity, but outright rejection is not possible, since at any rate it is not proved to be an impossibility. Thus, absence of means to prove momentariness does not establish endurableness *(sthairya)*. What it proves is that momentariness is doubtful, and not that there is endurance.

Against the flux doctrine the Nyāya endeavours to prove endurance on the ground of recognition *(pratyabhijñā)*, e.g. 'this is the same jar I saw yesterday'. The opponent may reject it on the ground that there is doubt as to its being the same jar. It may be asked doubt is about what, is it in respect of the subject of recognition, i.e. the continuance of the jar, or in respect of the recognition itself, or in respect of its validity? There can be no doubt about the continuance of the jar, since it is perceived by recognition. No doubt can be there about recognition also, since its correctness can be ascertained by the consciousness of having possessed of this knowledge. About the last alternative, is the doubt in respect of the validity of all cognitions or that of recognition only? Doubt regarding the validity of all cognitions involves self-contradiction, because if one doubts the validity of one's own consciousness of doubt, the very existence of the doubt will be unproved. Doubt about the validity of recognition also cannot be sustained. In order to prove recognition unauthentic an example may be cited. When hair is removed and new hair grows, nobody recognizes that the same hair which was removed has come into existence. Hence, recognition is unauthentic, and by this endurableness cannot be established. But this example is without any weight. The same proof that determines the endurableness of momentary entities determines the endurableness of continuant entities also. The criterion of the absence of inimical properties *(viruddha-dharma-viraha)* determines, according to the Baudha, that the jar that exists only one moment is the same during that moment and not a different one. The same criterion determines, says Udayana, that the jar that continues on from moment to moment is also the same jar and not a different one. Just as one cognition may have connection with many different objects, e.g. a chair with things on it, so also an object may be related, without inconsistency, with many different moments. An object's relation with those moments is successive, since it is dependent on the succession of its own causes, viz. the moments.

हेतुशक्तिमनादृत्य नीलाद्यपि न वस्तु सत्।
तद्युक्तं तत्र तच्छक्तमिति साधारणं न किम्।।१८।।

Unless the potency of cause is admitted, even blue colour, etc. cannot have actual existence; it (cause) becomes endowed with the ability of producing (an effect) when associated with something else (concurrent conditions), why should it not be common then?

18. The arguments advanced so far purport to prove the existence of *adṛṣṭa* as the cause of the world beyond. The Cārvāka throws up one more objection to refute causality. He asks whether causality is the nature of a positive thing or its extraneous adjunct. If it is the nature of a positive thing, it will be common to everything just as blueness is common to all blue colours. As a result, there arises the production of everything. In the second alternative if the extraneous adjunct is allowed to have natural causation, it would be subject to the aforesaid objection. If again, its causation depends on extraneous adjunct, it ends up in infinite regress as an infinite series of such adjuncts would be required. Hence, since causality is neither natural nor adjunctive attribute, the opponent concludes that it is not an entity at all.

If it is argued that it is the nature of causality not to be common to everything, then the cause should start producing all the effects simultaneously commencing from the time of its own generation, since an endurable object has the same nature. A child at the very moment of its birth should produce its son also. As in reality this does not happen, the obvious conclusion is that there is no such relation as the relation of cause and effect.

In reply to the objection, Udayana observes that unless causality is admitted, even 'blue colour' which the opponent cites as an example, will not have actual existence. Both the Bauddha and the Cārvāka do not admit eternal things; so, 'blue colour' cannot be eternal. In the absence of a cause it cannot be non-eternal also. To be real is to be determined by cause-effect relation.. If causal relation is unreal, blue colour, etc. would not be real. Therefore, the very illustration of the opponent proves to be unreal. Further, a cause cannot produce all the effects simultaneously from the time of its own generation. A cause is endowed with the ability of producing an effect only when it is associated with the concurrent conditions. Therefore, causality does not mean that a cause should be in action from the very first moment of its existence. Again, commonness means universality of cognition. Blueness is common to all blue colours, because it is universally cognized so. Such kind of universality is present equally in a cause also when associated with its concurrents, e.g. fire with fuel, seed with water, air, etc. It is universally cognized that there is causal relation between fire and smoke when the former is associated with wet fuel, between seed and sprout when the former is associated with water, etc., and so on.

पूर्वभावो हि हेतुत्वं मीयते येन केन चित्।
व्यापकस्यापि नित्यस्य धर्मिधीरन्यथा न हि।।१६।।

(Invariable) anticedence is causality, it can be proved by any argument, even for the eternal all-pervading being; otherwise, the substratum of the attribute cannot be known.

19. The Cārvāka makes one more attempt to refute the Nyāya position. *Adṛṣṭa* resides, according to the Nyāya, in the soul producing respective enjoyments, and the soul is admitted as the material cause of *adṛṣṭa*. The Cārvāka argues that the soul cannot be the material cause of *adṛṣṭa*, since the soul being eternal and all-pervading no negative instance *(vyatireka)*, either in point of space or of time, is available. Causality is determined by *anvaya* and *vyatireka* relation, since cause is defined as that the presence and the absence of which necessitate the presence and the absence of its effect. In the case under consideration, the *anvaya* relation, such as 'when the soul is pesent, *adṛṣṭa* is also present', is available. But the *vyatireka* relation, such as 'when the soul is absent, *adṛṣṭa* is also absent', cannot be proved.

It may be argued that an eternal and all-pervading soul qualified by the extraneous adjuncts, such as body, etc. may have temporal and spatial *vyatireka* relation. Then, the question arises whether the extraneous adjunct is generated by the soul or is eternal. Eternal and all-pervading soul is devoid of *vyatireka* relation; hence, such soul cannot be proved as the generator of the adjunct. If the extraneous adjunct were eternal, the soul qualified by that adjunct would be devoid of *vyatireka* relation.

It may be said that there is *vyatireka* relation in respect of *viśiṣṭa-ātmā*, that is, the soul qualified by extraneous adjunct which is not pervaded by eternality, and not in respect of *śuddha-ātmā*, that is, eternal, all-pervading soul devoid of extraneous adjunct. But the Naiyāyika does not admit a *viśiṣṭa-ātmā* which is different from *śuddha-ātmā*. And even if he admits, causality will be attributed to *viśiṣṭa-ātmā* which has *vyatireka* relation and not to *śuddha-ātmā*.

Udayana, in reply, says that causality is not always determined by *anvaya-vyatireka* relation. The definition of a cause is invariable antecedence to the effect *(kāryātniyataḥ pūrvabhāvaḥ)*. A cause in this sense is determined sometimes by *anvaya-vyatireka* relation and sometimes by the means of knowledge that makes known the substratum of the attributes *(dharmigrāhakāt pramāṇāt)*. If the *vyatireka* relation of a thing which is to be causally determined is not known, causality should be determined in accordance with the attributes associated with which *anvaya-vyatireka* relation is observed.

If causality is determined only by *anvaya* and *vyatireka* relation, then nowhere can a cause be inferred from an effect, since an effect is not seen to have *anvaya* relation with a cause. For example, stick, one of the causes of jar, is inferred from the jar, but the jar is not seen to have *anvaya* relation with stick. And if an effect is seen to have *anvaya* relation with the cause, the cause need not be inferred from the *liṅga*-effect, since their causal relation can be established by the perception of the *anvaya* and *vyatireka*.

It is not proper to argue that since causality is determined generally by *anvaya* and *vyatireka* relations, the causality of eternal and all-pervading soul also must be determined in the same manner, for there is *bādhaka-tarka* (*reductio ad absurdum* argument) against this position. The *bādhaka-tarka* is as follows. If the effect does not depend on the material cause, it will not be produced in a definite substratum. An effect is seen to have a material cause, and the *anvaya-vyatireka* relation between the material cause and the effect is also seen. The effect is produced in the presence of material cause; in the absence of the latter the former is not produced. Though the material cause of the cognition, etc. is not seen, yet they must have some material cause. Otherwise, in the absence of material cause they will be devoid of non-inherent *(asamavāyi)* and efficient *(nimitta)* causes, since non-inherent cause, etc. can function as causes only when acting in close proximity to the material cause. And in the absence of all causes either there will be non-production of the cognition, etc., or there will be production always and everywhere. If the production takes place in a definite place just by the efficiency of the efficient cause, then there is established a definite place in general inhering in which the effect is produced. And an inherent or material cause is also established. The five elements, viz. earth, water, fire, air and ether, cannot be the material causes of cognition, etc., since cognition, etc. are not produced inhering in the earth, etc. Nor are the cognition etc. qualities of dimension, time and mind, because only general qualities *(sāmānyaguṇa)* inhere in them, whereas cognition, etc. are specific qualities. As the earth, etc., the eight substances, cannot be the material causes or the substrata of cognition, etc., the ninth substance, i.e. the soul, is established as the material cause of cognition, etc. Thus, *anvaya-vyatireka* relation is not the only means of proving causality; causality can be proved by evidence that establishes the substratum of the attributes.

The *Kusumāñjali-kārikā* subscribes to the view that the soul can be proved as the material cause of *adṛṣṭa* through *anyoṅyābhāva-vyatireka*. Thus, 'that which is not soul has no pleasure inherent in it, just as that which is not yarn has not cloth inherent in it'. Here, the absence of pleasure in the absence of the soul is proved by the *anyonyābhāva-vyatireka*. Similarly, it can be said that 'that which is not soul has no *adṛṣṭa* inherent in it'; here, the absence of *adṛṣṭa* in the absence of the soul is proved by the *anyonyābhāva-vyatireka*. Therefore, even if one of the determinants of causality were the *vyatireka* relation, the soul can be proved as the material cause of *adṛṣṭa* through *anyonyābhāva-vyatireka*.

इत्येषा सहकारिशक्तिरसमा माया दुरुन्नीतितो
मूलत्वात्प्रकृतिः प्रबोधभयतोऽविद्येति यस्योदिता।
देवोऽसौ विरतप्रपञ्चरचनाकल्लोलकोलाहलः
साक्षात्साक्षितया मनस्यभिरतिं वध्नातु शान्तो मम। ।२० ।।

May that God whose unique concurrent power has been described as māyā *owing to the difficulty of its being unravelled, as* prakṛti *due to its being the root cause, and as* avidyā *owing to its being destroyed by right knowledge, the serene, arresting the turmoil of the waves of mundane creation, Himself being the direct witness bring joy in my mind.*

STABAKA - II

GOD — THE AUTHOR OF THE VEDAS

The first *stabaka* establishes God as the controller of *adṛṣṭa* which is the cause of our happiness and miseries. *Adṛṣṭa* is accomplished by the observance of the sacrificial rites. The nature of the scrificial rites can be known only through the Vedas. The Vedas consist of sentences; hence, they must have an author. The second *stabaka* proves that the author of the Vedas is God who is omniscient. The knowledge, viz. that such and such action produces the *adṛṣṭa* giving rise to such and such fruits, precedes the performance of the sacrificial rites. Such knowledge is not attained by any other *pramāṇa* excepting the Vedas, and the author of the Vedas is God.

> प्रमायाः परतन्त्रत्वात्सर्गप्रलयसम्भवात्।
> तदन्यस्मिन्ननाश्वासान्न विधान्तरसम्भवः।।१।।
>
> *As (validity) of valid knowledge is extrinsic, as there is possibility of creation and dissolution, and as there can be no reliance on anybody other than Him, there is no other course left open.*

1.i. The chief objector to God's authorship of the Vedas is the Mīmāṁsaka. *According* to him the origin of the Vedas cannot be traced back to any personal being. The Vedas are *apauruṣeya* (not of human origin); hence, they are free from all flaws and errors. The knowledge of the Vedas is attained through the tradition of the teacher and disciple, and by this tradition is established the *adṛṣṭa* and the fruits produced by the *adṛṣṭa*. There is no need for postulating God either as the controller of *adṛṣṭa* or as the author of the Vedas.

Even if the Vedas are admitted as *pauruṣeya,* that is, composed by a person or persons—Kapila and others, who have attained omniscience by the practice of *yoga* and the sacrificial rites, may be considered as the author or authors of the Vedas. And they may teach the Vedas to the disciples. In that case, God need not be postulated as the author of the Vedas. This view is held by the Sāṅkhya.

This position, when analysed, is untenable. The validity of knowledge depends on excellence *(guṇa).* The knowledge produced by the Vedas also depends for its validity on excellence. Excellence consists in the author's having the valid knowledge of the meaning of the Vedas, as one can compose something after understanding the meaning of what one composes. As the validity of the Vedas depends on the author's having the valid knowledge of the Vedas, the Vedas must have an author. And that author is God. The seer Kapila and others cannot be the author, since there is no proof to testify to the fact that they possess the valid knowledge of the Vedas which is required to compose the Vedas. Thus, one cannot rely on Kapila and others.

Moreover, owing to periodical creation and dissolution words are produced at the time of creation and destroyed at the time of dissolution. Thus, words are not eternal, and the Vedas which consist of words are also not eternal. Non-eternal Vedas are subject to production and destruction; hence, there must be an author to produce the Vedas.

1.ii. The following inference is adduced to prove God's authorship of the Vedas. 'The valid knowledge is produced by something more than the general causal aggregate which gives rise to knowledge, because being an effect it is a particular knowledge.' There is a *vyāpti* to the effect that, that which is a particular knowledge and is an effect is produced by some extraordinary cause apart from the general causal aggregate of knowledge. For example, an invalid knowledge is produced by an extraordinary cause, viz. defect. Similarly, valid knowledge is produced by excellence which is an extraordinary cause other than the general causal aggregate of knowledge. Excellence in the case of the verbal knowledge produced by the Vedas stands for the author's possessing the valid knowledge, and God is established as the substratum of this knowledge.

It may be said that invalid knowledge arises from defect, and valid knowledge arises from absence of defect. Thus, validity of knowledge does not depend on excellence. This is not so, because it may be counter argued that valid knowledge arises from excellence and invalid knowledge from absence of excellence. Thus, invalidity of knowledge is not caused by defect.

To say that defect is the cause of invalid knowledge, because there is *vyāpti* between presence of invalid knowledge in the presence of defect and the absence of invalid knowledge in the absence of defect, is not tenable. For, it may be equally argued that excellence is the cause of valid knowledge as there is *anvaya-vyatireka-vyāpti* between presence of valid knowledge in the presence of excellence and the absence of valid knowledge in the absence of excellence.

The opponent may contend that the validity of that verbal knowldge which is of human origin is produced by excellence, and invalidity by defect. The validity of the Vedas which are established as not of human origin is established by the absence of defects. It is not so, because the absence of excellence in the author would invalidate the Vedas. The objection that absence of excellence is never seen to be the cause of invalidity would be replied to by saying that absence of defect is also never seen to be the cause of validity. Therefore, absence of defect alone cannot stand for the validity of the Vedas.

It may be argued that the cause of invalidity is defect in the form of error, etc. Defect is accompanied by absence of excellence, but absence of excellence is not the cause of invalidity. Were it so, it may be counter argued that the cause of validity is excellence, excellence is accompanied by absence of defect, but absence of defect is not the cause of validity. An example may be cited to corroborate this idea.

Attachment is the cause of *pravṛtti* (volition), and aversion is the cause of *nivṛtti* (withdrawal of volition). Attachment is accompanied by absence of aversion, but absence of aversion is not the cause of *pravṛtti.* Similarly, aversion is accompanied by absence of attachment, but absence of attachment is not the cause of *nivṛtti.* In the same manner though excellence is accompanied by absence of defect and defect by absence of excellence, the cause of validity is excellence and not absence of defect, and the cause of invalidity is defect and not absence of excellence.

It may be argued that since the Vedas are established as *apauruṣeya* (not of human origin), the validity of the Vedas should be determined by the absence of defects of the speaker. The inferential argument to this effect would be thus: 'The Vedas are valid, because they are free from the defects of the speaker'. Even in the absence of excellence of the speaker, the absence of defects of the speaker being there in the Vedas, absence of defects of the speaker must be accepted as the cause of the validity of the Vedas, and not the excellence of the speaker.

The inference set forth above is vitiated by a counter-probans; hence, validity of the Vedas cannot be established by the absence of defects of the speaker. The counter-probans is: 'The Vedas are not valid, because they are free from absence of excellence of the speaker'. The opponent may argue that it is not their intention to establish validity of the Vedas through inference, because the validity of the Vedas is intrinsic. Therefore, advancement of counter-probans does not affect the opponent's position. However, even though validity of the Vedas is intrinsic, yet sometimes invalid knowledge too being accepted as valid doubt arises as to the validity of that knowledge. As invalidity is always caused by presence of defects, what is established by absence of defects is only the removal of doubt about the validity of knowledge, and nothing more.

Absence of defects cannot remove doubt about invalidity of knowledge, because it is also observed that absence of excellence causes invalidity of knowledge. Thus, doubt remains about invalidity even in the absence of excellence.

To argue that absence of excellence causes invalidity only when there exists presence of defect, will be counter argued that absence of defects also causes validity only when there exists presence of excellence.

Therefore, as the validity of knowledge depends on

excellence, the validity of the verbal knowledge of the Vedas too must depend on the excellence. Excellence in respect of the Vedas stands for the author's possessing the valid knowledge of the Vedas, and God is established as the author of the Vedas.

1.iii. According to the Mīmāṁsaka validity of knowledge is not produced by any extraneous cause other than absence of defect This is not accepted by the Naiyāyika; with him validity is produced by excellence which is other than absence of defect (this has been discussed above).

The Mīmāṁsaka and the Naiyāyika differ in their opinion about the validity of knowledge. The Mīmāṁsaka holds that the validity of knowledge is intrinsic. All the three schools of the Mīmāṁsā, viz. the schools of Bhāṭṭa, Prābhākara, and Murārimiśra, advocate the intrinsic validity of knowledge. The causal aggregate that apprehends the knowledge apprehends the validity also inherent in that knowledge. According to the Bhāṭṭa school, knowledge is inferred by the *liṅga-jñātatā* (knownness) and along with knowledge its validity is also inferred. With the Prābhākara school knowledge being self-illumined apprehends itself. All knowledge, according to the Prābhākara, has within its sphere triple objects, viz. knowledge, knower and the object known. The knowledge takes the following form — 'I know the jar'. This knowledge comprises the knower (i.e. soul), the object known (viz. the jar) and the knowledge. As the knowledge is apprehended by the self-illumined knowledge itself, its validity is also apprehended by that knowledge.

According to the school of Murārimiśra, knowledge arises in the form 'this is a jar'. After that arises the reflective cognition 'I know the jar'. Reflective cognition *(anuvyavasāya)* apprehends the knowledge, and along with knowledge it apprehends the validity of knowledge also. The factor common in the view of all the three Mīmāṁsakas is that the validity of knowledge is apprehended by the same causal aggregate that apprehends the knowledge itself. But they differ in respect of the causal aggregate. With the Bhāṭṭa validity is apprehended by inference; with the Prābhākara it is apprehended by the self-illumined knowledge;

and with Murāri it is by reflective cognition. The view that the validity of knowledge is apprehended by the same causal aggregate that apprehends the knowledge is expressed in other words when it is said that the validity of knowledge is intrinsically known.

The Naiyāyika examines the theory of the intrinsic validity of knowledge and ultimately rejects it. He upholds the theory of the extrinsic validity of knowledge. Udayana offers the following syllogistic reasoning to establish the extrinsic validity of knowledge. 'Validity is extrinsically known, since at the stage of non-recognition it is doubted, like invalidity'. Non-recognition *(anabhyāsadaśā)* means absence of recognition; recognition is repetition *(āvṛtti)*. Recognition consists in the generation of knowledge, the like of which has already been produced. For example, when a person perceives water in a lake from a distance in which lake he has taken bath the other day, the knowledge of water he attains then is a knowledge that arises at the stage of recognition. For, this knowledge is similar to what he attained the other day. Doubt does not arise about the validity of this knowledge. When one perceives water in a lake for the first time from a distance, knowledge of water at that time arises at the stage of non-recognition. The validity of this knowledge is doubted. If validity is to be known intrinsically, such doubt must not arise. For, when the knowledge of water is apprehended either by its self-illumined character, or *by jñātatā-liṅga,* or by reflective cognition, its validity too must be apprehended along with it. In this case there should not arise the doubt whether this is a valid knowledge or not. But there arises doubt; therefore, it may be admitted that when knowledge is apprehended its validity is not apprehended. Validity is inferred from the successful activity; the inference is thus — 'this is a valid knowledge, because this leads to successful activity'. Hence, validity is not apprehended by the causal aggregate that apprehends the knowledge. It is, on the other hand, inferred by a different *hetu,* viz. successful activity. This is what is meant by extrinsic validity of knowledge.

The Mīmāṁsaka may contend that validity is apprehended along with knowledge in respect of both valid and invalid knowledge. Validity that is apprehended in respect of invalid knowledge is later on given up owing to some defect in *karaṇa*, or sublation, or unsuccessful activity. Valid knowledge is not affected by defective *karaṇa*, etc.; hence, the validity that is apprehended before in respect of valid knowledge is retained as such.

But this position also cannot do away with doubt. In the absence of any criterion to ascertain the difference between valid and invalid knowledge, doubt arises whether a particular knowledge is valid or not. It may be asked how doubt regarding the validity alone proves the extrinsic validity of knowledge. The position is this — when a valid knowledge is apprehended, the question arises whether the validity is apprehended along with the knowledge or not. If it is apprehended, there is no room for doubt. But doubt arises, and this means that validity is not apprehended when the knowledge is apprehended. This further proves that validity is not known intrinsically. Had it been known intrinsically, it would have been apprehended along with the knowledge.

1.iv. So far reasons have been adduced to establsih that validity of knowledge is known extrinsically. The reason that is set forth by the Mīmāṁsaka to prove intrinsic validity of knowledge is being refuted now.

Sometimes *pravṛtti* arises when there is doubt about the validity of an object, and validity is ascertained by the success of *pravṛtti*. Thus, the knowledge of validity is not the cause of *pravṛtti* in general. It is the cause only of intense *pravṛtti*. Intense *pravṛtti* arises only when the validity of an object is determined, and the validity of an object is ascertained only when the knowledge of validity in the form 'this knowledge is valid' is ascertained. Therefore, it is not proper to maintain that the knowledge of validity arises through inference, etc. *Pravṛtti* arises instantaneously towards an object when the knowledge of that object is attained, and along with knowledge its validity is also apprehended. One does not wait for inference, etc. to attain the knowledge of validity. This is the Mīmāṁsaka position.

This is refuted on the ground that the validity of an object can be determined even without the knowledge of validity. An effect cannot originate without a cause. Instantaneous *pravṛtti* too not being reasonable without the instantaneous accompaniment of its cause calls for that, but not for the knowledge of intrinsic validity. The cause of *pravṛtti* is desire, the cause of desire is the knowledge of *iṣṭasādhanatā*, that is, the knowledge that something is the means to a desired end. The cause of the knowledge of *iṣṭasādhanatā* consists in the knowledge that something belongs to the same category that was previously experienced as a means to a desired end. Nowhere in this causal series of *pravṛtti* has the knowledge of validity any place. Even if it has a place, it need not necessarily be intrinsic. Indeed visual perception of water does not apprehend water's capacity of quenching thirst. After perceiving water, inference arises in the form 'this water belongs to the same kind of water that quenched the thirst before; therefore, it must possess the capacity to quench thirst'. Thus, one has *pravṛtti* towards water after inferring water's capacity to quench the thirst. Similarly, when there arises the knowledge of water, the knowledge of validity is attained through inference. After apprehending validity through inference one has *pravṛtti* towards the object. Thus, it is proved that validity is not apprehended at the time of the generation of knowledge; it is apprehended later on through inference. Therefore, the validity of knowledge is known extrinsically.

1.v. The opponent raises an objection that if validity is inferred by the knowledge of successful activity, the *liṅga*, then there arises the fallacy of infinite regress. For, the validity of that inference by which validity is inferred must be ascertained by another inference and the validity of that one by another, and so on *ad infinitum*. To avoid this it is reasonable to hold that validity is known intrinsically.

This objection is unfounded. It is not warranted that the validity of each and every knowledge must necessarily be known. There may be doubt about the validity of a knowledge that arises for the first time. In order to remove that doubt the validity of that knowledge may be determined by inference. But it is not necessary that the validity of that inference too must be ascertained. Thus, there is no infinite regress, and even if it is there, intrinsic validity is also not free from infinite regress. Thus, in the Prābhākara doctrine when the knowledge 'I know the jar' apprehends validity of the jar, it must apprehend the validity of that validity also, since doubt may arise about the validity of that validity. Again the validity of the later one too must be known, and so on, so forth. According to the Bhāṭṭa validity is inferred through *jñātatā-liṅga,* then the validity of that inference too must be inferred by another inference and so on landing in infinite regress. With the followers of Murārimiśra validity is apprehended by reflective cognition. The validity of the reflective cognition also must be apprehended by another reflective cognition and the validity of that by another, leading thereby to infinite regress. To overcome this defect the Mīmāṁsaka may be constrained to hold that the validity of a knowledge is ascertained through inference only when there arises doubt about the validity of that particular knowledge. The validity of each knowledge need not be ascertained. The Naiyāyika of course subscribes to this position. It is, therefore, established that the validity of knowledge is produced by excellence which is different from the causal aggregate of knowledge. Validity is known by inference, etc. The Naiyāyika upholds that validity of knowledge is produced extrinsically and is known extrinsically.

1.vi. The opponent argues that the validity of knowledge may be extrinsic, but the validity of the Vedas is not extrinsic. The validity of the Vedas is established through acknowledgement by representative people. There is no need to postulate God to ascertain the validity of the Vedas. The Vedas are eternal; hence, they do not stand in need of a composer. So, God need not be postulated as the author of the Vedas. The seer and others acknowledge the validity of the eternal Vedas; the validity of the Vedas may be ascertained from such acknowledgement. God is not established either as the author of the Vedas or as the cause of the validity of the Vedas.

The view that the Vedas are eternal is not accepted by the Naiyāyika. The Vedas consist of letters; letters are not eternal, as they are subject to production and destruction. Since letters are not eternal, the Vedas that consist of letters cannot also be eternal. This leads to a prolonged discussion of the non-eternality of word or sound. On this issue there is a vital difference between the Naiyāyika and the Mīmāṁsaka. The former maintains that sound is a qulaity of ether *(ākāśa)* and non-eternal. With the latter sound is a substance and is eternal. There are two kinds of sound — letter-sounds and sound in general. According to the Naiyāyika both kinds of sound are non-eternal, since they are produced and destroyed. Their destruction is perceived thus — 'the sound *ga* heard before does not exist now', 'the tumult has ended'.

The perception of non-existence of sound *ga* that was heard before cannot be erroneous. Were it so, the cause must be pointed out owing to which the sound that was heard before is no more heard now. Non-apprehension of sound cannot be due to the fact that it has gone somewhere else. For, intangible sound *ga* cannot possibly go somewhere else. Nor can it be said that the sound is not heard, because it is not in contact with the sense of hearing. For, this cannot be proved, non-apprehension cannot be due to the hearer's lack of attention also, because it is not apprehended even with due attention. Nor the sense-organ is defective, because other sounds are heard. It is also not that due to the absence of auxiliary cause sound *ga* is not apprehended. The sense of hearing alone is the cause for the apprehension of sound; it does not stand in need of other auxiliary cause. Since the sound *ga* that was heard before is no more heard now in spite of the sense of hearing being in perfect condition, it may be concluded that the sound is destroyed and that is why it is not apprehended now. Thus is established non-eternality of sound.

I.vii. So far destruction of sound is proved by perception. To convince those who do not admit the perception of the destruction of sound, an inferential argument is offered to prove perceptibility of the destruction of sound. The argument is as follows. 'The destruction of sound is perceived by the sense-organ, because it is the non-existence of the counter-correlate of what is within the scope of sense-organ, like the non-existence of a jar.' The sense-organ that perceives an object perceives the non-existence of that object also. For example, the visual sense-organ that perceives a jar perceives the non-existence of that jar also. Similarly, sound is apprehended by the sense of hearing; hence, the non-existence of sound in the form of the destruction of sound also must be apprehended by the sense of hearing. Thus, this inference proves that the destruction of sound is perceived by the sense-organ.

The opponent observes that the inference mentioned above is conditioned by an *upādhi*. Contact with the sense-organ is the *upādhi*. There is a *vyāpti* that wherever there is sense-perception there is sense-contact, sense-contact being concomitant with sense-perception which is the *sādhya*. But destruction of sound which is the *pakṣa* does not have contact with the sense. As sense-contact is absent in the *pakṣa*, it is not concomitant with the *hetu*. The sense-contact which is concomitant with the *sādhya* without being concomitant with the *hetu* conditions the inference. Hence, the inference is *sopādhika*. Again, the inference is vitiated by the counter-probans *(satpratipakṣa)* also. The counter-argument is thus — 'the destruction of sound is not perceived by the sense-organ, because it does not have contact with the sense-organ'.

In reply to the first objection it may be said that the destruction of sound has contact with the sense of hearing, the contact being the relation of *viśeṣaṇa-viśeṣya-bhāva*. As sense contact is present in the *pakṣa*, viz. the destruction of sound, it is not non-concomitant with the *hetu*. Hence, it cannot be an *upādhi*, since an *upādhi* is that which is concomitant with the *sādhya* without being concomitant with the *hetu*.

The counter-argument is fallacious. The *hetu* in the inference is absence of sense-contact. This does not exist in the *pakṣa*, viz. destruction of sound, since the destruction of sound has contact with the sense of hearing in the form of the *viśeṣya-viśeṣaṇa-bhāva*. Since the *hetu* does not exist in the *pakṣa*, the counter-argument suffers from the fallacy, called *svarūpāsiddhi*. Therefore, it is established that destruction of sound is perceived by the sense-organ, and perception proves non-eternality of sound.

1.viii. One more counter-argument may be advanced to prove that the destruction of sound is not perceived by the sense-organ. 'The destruction of sound is not perceived by the sense-organ, because it exists in a super-sensuous substratum'. That which exists in a super-sensuous substratum is not perceived. Ether is super-sensuous; hence, the destruction of sound that exists in ether is not perceived by the sense-organ.

This counter-argument is untenable. Sound that exists in super-sensuous ether is perceived by the sense of hearing. Hence, existence in super-sensuous substratum cannot be offered as a *liṅga* to prove absence of perception of the destruction of sound. Moreover, perception of non-existence depends on the perceptibility of the counter-correlate. Perceptibility of the substratum is not required for perceiving non-existence. Since sound is perceptible, that is, fit to be perceived, its non-existence can also be perceived, even though the substratum of the non-existence, viz. ether, is not perceived. If the substratum is to be perceived before the perception of non-existence, then the knowledge, viz. there is no jar on the floor, would not arise at all. Absence of the conjunction of jar with the floor is the object of this knowledge. Like conjunction, absence of conjunction exists both in the floor and in the jar. Thus, absence of conjunction has two substrata, viz. the floor and the jar. If the floor is to be perceived before the perception of non-existence, then the jar too must be perceived. In fact, the perception of the absence of conjunction of the jar with the floor arises even without the perception of the jar. This proves that the perception of non-existence arises even without the perception of the substratum. Hence, non-existence of sound, that is, the destruction of sound, can be perceived even if it exists in super-sensuous ether.

1.ix. If it were a rule that non-existence can be perceived only when the substratum in which non-existence exists is perceived, then it may be pointed out how such knowledge as 'the sound heard before does not eixst now' arises at all. Perception cannot produce the knowledge of non-existence of sound, since the substratum ether is not perceived. This knowledge cannot take place through inference also. The inference may be as follows:

'the sound is destroyed now, because it is not apprehended'. The *pakṣa* in this inference is sound. When sound exists, the *hetu,* i.e. non-apprehension, does not exist in sound; this gives rise to the fallacy of *svarūpāsiddhi.* Again, when there is non-apprehension, the *pakṣa* sound does not exist; and this leads to *āśrayāsiddhi* fallacy.

The inference may be modified into the following form. 'The ether is devoid of sound now, because it is not apprehended as having sound'. The *pakṣa* ether being super-sensuous, it may be apprehended as devoid of sound even when there is sound. Thus arises the fallacy of *vyabhicāra,* since the *hetu,* viz. 'not being apprehended as having sound', exists in the *pakṣa* ether when the *sādhya,* viz. being devoid of sound, is absent.

One more inference may be advanced. 'I am at this moment in possession of an ear devoid of sound, since there is absence of apprehension of sound, like a deaf person'. This inference also is associated with defects. A deaf person who is the example here does not possess ear; hence, the *sādhya,* viz. possession of ear, is absent in the example. Again, the word 'deaf' itself means devoid of ear; hence, to be deaf and to possess ear are mutually contradictory.

It may be said that deafness does not mean absence of ear. Ear is the ether delimited by the ear-drum. Since ether is eternal, a deaf man too possesses such ear. As a deaf man possesses ear, the inference cannot be charged with the defect that the *sādhya* is not present in the example.

This argument is untenable. Even if a deaf man possesses ear, it cannot be proved that the ear is devoid of sound. Thus, the *sādhya,* viz. possession of ear which is devoid of sound, is still absent in the example.

It cannot be said that a deaf ear does not possess sound on the ground that had there been sound produced in the ear it would have been apprehended, for sound is not necessarily apprehended even when it is produced. For example, sound that is produced first is not apprehended, nevertheless it exists. The last one of a series of sounds that is produced from the first sound is apprehended when it reaches the ear. Thus, the inference mentioned above is defective, as the *sādhya* is absent in the example.

1.x. The following inference may be taken into consideration. 'Drum, harp, flute, etc. are devoid of sound now, because they are apprehended without the sound being apprehended even though they are fit to be the subject-matter of the same knowledge along with sound'. There is *vyāpti* that, that which is fit to be the subject-matter along with some other thing in the same knowledge, is apprehended even while the latter is not apprehended. For example, the floor which is fit to be the subject along with the jar in the same knowledge is apprehended as possessing absence of conjunction with the jar when the jar is not apprehended. Similarly, both sound and drum can be the subject-matter of the same knowledge, viz. 'there is sound in the drum, etc.' So drum can be apprehended even when sound is not apprehended. Thus, non-existence of sound may be inferred in this manner.

The above inference stands refuted. The *hetu*, viz. drum, etc. and sound being the subject-matter of the same knowledge, is unestablished. Sound being apprehended by the organ of hearing and drum, etc. by the organ of perception, both cannot be the subject-matter of the same knowledge.

It may be argued that just as in the perceptual knowledge of fragrant sandalwood, fragrance is apprehended through *jñānalakṣaṇa-pratyāsatti,* so also in the auditory knowledge of sound in the drum, etc., drum, etc. may be apprehended through *jñānalakṣaṇa-pratyāsatti.* Thus, both drum, etc. and sound may be the content of the same knowledge. Even then the inference is not valid. It may be asked whether the *sādhya* 'devoid of sound' stands for non-existence of sound or absolute non-existence of sound. The first is untenable; non-existence of sound exists in ether; hence, it cannot exist in the *pakṣa* drum, etc. It is not the second also, because absolute non-existence of sound being eternal exists all the time; hence, its existence in particular time in the form 'the sound heard before does not exist now' cannot be ascertained.

1.xi. It may be said that drum, etc. are the delimiting adjunct *(avacchedaka)* of ether, and sound is produced in the ether delimited by drum, etc. When it is said that sound is produced in the drum, it means that sound is produced in the ether delimited by the drum. When sound is denied in the drum, it is actually denied in the ether delimited by the drum. In the inference mentioned above the *pakṣa* where non-existence of sound is established, is not the drum, etc., but the ether delimited by drum, etc. Thus, the objection that non-existence of sound exists in ether and not in the drum, etc. may be avoided.

The inference in this modified form stands thus: 'The ether delimited by the drum, etc. is devoid of sound, because it is apprehended without the sound being apprehended even though it is fit to be the content of the same knowledge along the sound'. The *hetu* in this inference, viz. 'the ether being apprehended', is unestablished, because the ether being super-sensuous cannot be apprehended. Thus arises the fallacy of *svarūpāsiddhi,* as the *hetu* does not exist in the *pakṣa,* viz. the ether delimited by the drum, etc.

Therefore, as the knowledge, viz. 'the sound heard before does not exist now', cannot be produced by inference; it may be admitted that it is produced by perception.

If the perception of non-existence requires the perception of the substratum also, then non-existence of sound cannot be perceived, as its substrate the super-sensuous ether is not perceptible. Hence, it is stated that perception of non-existence depends only on the perceptibility of the counter-correlate. Substrate need not be perceived at all. Since sound, the counter-correlate of the non-existence of sound, is fit to be perceived, non-existence of sound is also apprehended by perception. Thus, the knowledge of non-existence of sound in the form 'the sound *ga* heard before does not exist now' is established by perception, and perception proves the non-eternality of sound.

1.xii. According to the *śāstra (sadbhyāmabhāva nirupyate),* perceptual knowledge of non-existence depends on the knowledge of the counter-correlate and the substrate that are existents. In that case, it may be objected that the perception of the non-existence of sound is not possible, since its substrate ether cannot be perceived. Moreover, to say that the knowledge of non-existence does not depend on the knowledge of the substratum goes against this *śāstra* also.

It may be replied to this objection that this *śāstra* was framed to apply to those cases where non-existence is determined by both the counter-correlate and the substrate. For example, in the instance 'there is no jar on the floor' absence of the conjunction of jar is apprehended in the floor. This absence of conjunction exists in both jar, the counter-correlate and the floor, the substrate. Hence, the knowledge of the absence of conjunction of jar does not arise without the knowledge of the jar and the floor.

The non-existence of sound exists only in the ether; hence, for its knowledge only the knowledge of the counter-correlate is needed and not the knowledge of the substrate. Further, the *śāstra* mentioned above is related to inference. In the inferential knowledge of non-existence, non-existence is established in the *pakṣa*. If the *pakṣa* is not known before, there arises the fallacy of *āṣrayāsiddhi*. To overcome this fallacy it is necessary that the *pakṣa* which is the substratum must be known. In the perceptual knowledge of non-existence, knowledge of the substratum is not needed. And even if it is needed, its perception is not necessary; mere knowledge of the substratum will serve the purpose.

If the importance of the perception of the substrate is what is stressed by the *śāstra*, then it would contradict the *Nyāya-sūtra*, viz. *vahnerdāhyaṁ vināśya anuvināśavat* (4.1.27). This says that just as fire after destroying the wood gets itself destroyed, so also the statement 'everything is non-eternal' after conveying the non-eternality of everything conveys its own non-eternality. Here destruction of fire is stated as an example. A thing can serve as an example only when both proponent and opponent have perceptual knowledge of it. The substrate of the destruction of fire is the series of parts of fire. Now if the perception of the substrate is to determine the perception of the destruction of fire, then the perception of the destruction of fire would not take place at all, since the series of the parts of fire cannot be perceived. As the destruction of fire is set forth as an example, it means that the destruction of fire can be perceived without perceiving its substrate. Similarly, non-existence of sound too can be perceived without perceiving the ether, its substrate. Thus is established that the knowledge of the non-existence of sound in the form 'the sound *ga* heard before does not exist now' arises through perception. This proves the non-eternality of sound. Therefore, perception is the proof for the non-eternality of sound.

1.xiii. Non-eternality of sound can be proved by inference also. 'Sound is non-eternal, because it is produced like a jar.' It may be argued that when the sound *ga* is heard, there arises a *pratyabhijñā* in the form 'this is that sound *ga* which was heard before'. Thus, an identity may be established between the sound heard now and the sound heard before; and on the basis of this identity it may be maintained that sound is eternal.

This line of argument is not correct. Just as in the *pratyabhijñā* 'this is that flame of light' what is established between earlier and present flame of light is similarity and not identity, so also in the *pratyabhijña* 'this is that sound *ga*' it is similarity that is established between the sound heard now and the sound heard before and not identity. This implies that the sound heard before is destroyed and a new sound is produced now.

It may be said that in flame there is assemblage of contradictory attributes in the form that some flame is longer and some is shorter. And the difference between earlier and later flame may be established on the basis of this contradictory attributes. Hence, in the *pratyabhijñā* of flame of light it is not identity, but similarity that is established between earlier and later flame. The sound *ga* is not found to have any contradictory attributes on the basis of which difference can be ascertained between earlier and later sound. And in the absence of the ascertainment of difference, it may be maintained that identity is established between two sounds.

This argument is without any basis. Sound also possesses contradictory attributes, as some sound is loud, some is louder, etc. Thus, the difference betweent the sound *ga* heard now and that heard before can be established on the ground of these contradictory attributes. As the difference can be ascertained between two sounds, it is clear that the *pratyabhijñā* 'this is that sound *ga*' does not convey identity, but similarity. Therefore, as eternality of sound cannot be established by *pratyabhijñā*, it is concluded that sound is non-eternal.

It may be observed that loudness, etc. are not the attributes of sound, they are attributes of some *upādhi* of sound and are imposed on sound. As loud, louder, etc. are not the attributes of sound, difference between the sound heard now and the sound heard before cannot be ascertained on the basis of these contradictory attributes.

This observation does not stand to reason. Were it so, coldness would not be the attribute of water and heat the attribute of fire, but they would be the attributes of some *upādhi* of water and fire.

In fact, if an attribute is perceived as abiding in a thing, that attribute constitutes the nature of the thing. Coldness perceived as abiding in water constitutes the nature of water. Similarly, loudness, etc. that are perceived as abiding in the sound *ga*, constitute the nature of the sound *ga*.

1.xiv. Nor is it proper to say that loudness, etc., abiding in some other thing, are superimposed on the sound. An attribute abiding in a thing is superimposed on another thing when that attribute is remembered. For example, while yellow colour abiding in bile is remembered, it is superimposed on conch. Remembrance of a thing takes place only when that thing is apprehended before. Loudness, etc. are never apprehended before as abiding in anything other than the sound *ga*, etc. In the absence of previous apprehension they cannot be remembered too, and in the absence of remembrance they cannot be superimposed on the sound *ga*.

It is not reasonable to argue that loud, louder, etc. are the attributes of *dhvani* (tone) and are superimposed on the sound. The question arises whether the *dhvani* which is the substrate of loudness, etc. is of the form of air or of ether. If it is the first, it may be pointed out whether loudness, etc., abiding in *dhvani,* are apprehended by the organ of touch or by the organ of hearing. Loudness, etc. cannot be apprehended by the organ of touch. For, the organ of touch apprehends only touch and the substance that is the substratum of touch. Loudness, etc. are neither touch nor the substrata of touch; hence, they are not apprehended by the organ of touch.

The organ of hearing is not of the form of air; hence, it cannot apprehend loudness, etc., the attributes of *dhvani* which is of the form of air. Just as visual sense-organ perceives neither air nor attributes of air because of its not being airy, so also the organ of hearing does not apprehend loudness, etc. which are the attributes of airy *dhvani.* As loudness, etc. are not apprehended either by the organ of touch or by the organ of hearing, they cannot be superimposed on the sound.

One may argue that just as letter-sound inheres in the ether, so also *dhvani* inheres in the ether. The organ of hearing being ethereal can apprehend ethereal *dhvani* and its attributes loudness, etc. Loudness, etc. that are apprehended by the organ of hearing are superimposed on the sound *ga,* etc.

This is not accepted. When the sound *ga* is apprehended as loud or louder, both letter-sound as well as loudness, etc., are apprehended, but *dhvani* is not apprehended. As *dhvani* remains unapprehended, loudness, etc. cannot be apprehended as the attributes of *dhvani.* Attributes are apprehended only through the particular object wherein inhere the attributes. If the particular object is not apprehended, its attributes are also not apprehended. For example, jarness *(ghaṭatva)* is not perceived in the absence of the perception of jar. Since *dhvani* is not apprehended, loudness, etc. cannot be apprehended as the attributes of *dhvani.* Loudness, etc. that are apprehended are the attributes of the sound *ga* which is also apprehended. Thus, as the difference between the sound *ga* heard before and the sound *ga* heard now can be ascertained on the ground of the contradictory attributes, viz. loud, louder, etc. that inhere in the sound, it is established that the sound that was produced before is destroyed. Thus, the inference, viz. the sound is non-eternal because it is produced, cannot be invalidated by the *pratyabhijñā*, viz. 'this is that sound *ga*'.

Nor is this inference affected by any counter-argument. First of all there does not exist a *hetu* that proves eternality of sound, and even if it exists, the sound cannot have two contradictory *hetus* proving eternality and non-eternality respectively. Hence, one of the two must be weaker. The *hetu* that may be offered to prove eternality of sound is weaker than the *hetu* that proves non-eternality of sound. Thus goes the counter-argument — 'sound is eternal, because it is a substance not having another substance as its material cause'. If the *hetu* were only 'being a substance', then it would be *vyabhicāri,* since it would apply to a jar which is a substance, but jar is not eternal. Hence, the *hetu* is 'being a substance not having another substance as its material cause'. A jar has another substance, viz. pot-sherds, as its material cause; hence, it does not come within the scope of the *hetu.* Sound is a substance which does not have another substance as its material cause, like ether; therefore, sound is eternal. The *hetu* in this argument is weak, because it does not exist in the *pakṣa.* Sound is a quality; hence, the *hetu* 'being a substance' does not exist in sound, as substance cannot exist in quality. As presence of *hetu* in the *pakṣa* which is one of the requirements of a valid inference is absent in this argument, the *hetu* is fallacious. Therefore, this cannot be offered as a counter-argument to invalidate the inference that proves non-eternality of sound.

1.xv: To establish sound as substance the following inference may be adduced. Sound is a substance, because it is apprehended by a direct relation; that which is apprehended by the direct relation of sense-organ is a substance, e.g. a jar. Sound that is inherent in the ether is apprehended by the direct relation of *samavāya*, contact being the inherence of sound in the organ of hearing. As this inference proves sound as a substance, the defect stated above that the *hetu* 'being a substance' does not exist in sound does not hold good.

This inference is also defective, as the *hetu*, viz. being apprehended by a direct relation, is unestablished in the *pakṣa* sound. Direct relation is either conjunction between two substances or inherence of attributes in the substance. So, unless sound is established either as substance or as the attribute of the organ of hearing, it cannot be apprehended by any of the direct relations.

It may be argued that sound is established as substance by the argument of elimination *(pariśeṣānumāna)*. Thus, sound is not *sāmānya, samavāya* and *viśeṣa,* nor is it *abhāva* as it is a *sattā-jāti.* It is not *karma* also, because it does not exist in *mūrta-dravya* (corporeal substance). Since sound is not included in *karma*, etc. and is apprehended by direct relation, it must be a substance. This position cannot be substantiated, as there is *bādhaka* argument against it. Thus, sound is not a substance, because there is restriction *(vyavasthā)* of the external sense-organ, like colour, etc. Just as colour, etc. that are established by the external visual sense-organ, etc. are not substances, so also sound which is established by the external organ of hearing is not substance. Therefore, as substancehood of sound is unestablished, the *hetu* 'being a substance' cannot establish eternality of sound.

1.xvi. The opponent avers that sound is not a quality. The Nyāya maintains that sound is the quality of the organ of hearing. This is not admitted by the opponent Mīmāṁsaka. To substantiate his position he argues in the following line. That which is cognized by a particular sense-organ is not the qualty of that sense-organ. For example, smell cognized by the organ of smell is not the quality of the organ of smell, as the organ of smell does not cognize the smell that inheres in it, but it cognizes the smell that inheres in other objects, viz. flower, etc. Similarly, sound apprehended by the organ of hearing is not the quality of the organ of hearing. Therefore, it is not proper to maintain that sound is the quality of the organ of hearing.

The argument that an attribute cognized by a particular sense-organ does not inhere in that organ is untenable. Smell which is cognized by the organ of smell may inhere in the organ of smell, there is no invalidating argument against it. But the inherence of smell in the organ of smell does not imply that like smell inherent in flower, smell inherent in the organ of smell too will be apprehended by the organ of smell. The organ of smell apprehends only that smell which is fit to be apprehended. Smell inherent in the organ of smell is not fit to be apprehended; hence, it is not apprehended by the organ of smell. If this is not admitted, then pleasure, etc. that are perceived by soul would not be the attributes of soul. Therefore, the *vyāpti* is that attributes whether one's own or those of others are apprehended only when they are fit to be apprehended, otherwise not. Sound can very well be the attribute of the organ of hearing, as there is nothing to invalidate it. Moreover, there is argument to lend support to the fact of its being an attribute. 'The organ of hearing apprehends a special quality, because it is a sense-organ, like the visual sense-organ.' That which is a sense-organ apprehends a special quality, e.g. visual sense-organ apprehends a special quality, viz. colour. Similarly, organ of hearing also must be apprehending a special quality, and this special quality must be sound as other special quality, viz. colour, etc., is not apprehended by the organ of hearing. If the organ of hearing does not apprehend sound, the creation of the organ itself would be useless. Nor can it be said that the use of organ of hearing lies in its apprehending a substance. For, sense-organ does not apprehend mere substance bereft of a special quality. Visual sense-organ, etc. perceive a substance together with a special quality inherent in it. Therefore, it is necessary that the organ of hearing also must perceive a quality which is sound. As sound is proved to be a quality and not a substance, the inference that 'sound is eternal, because it is a substance not having another substance as its material cause' stands refuted.

1.xvii. The Mīmāmsaka produces one more inference to prove the eternality of sound. 'Sound is eternal, because it is the quality of eternal ether, like the maximum grossness *(parama-mahat-parimāṇa)* abiding in ether'. Just as maximum grossness which is the quality of eternal ether is eternal, so also sound which is also the quality of eternal ether is eternal.

This inference is conditioned by an *upādhi*, viz. *akāryatva* (non-production). Wherever there is eternality, e.g. in the soul, etc., there is *akāryatva*. Thus, *akāryatva* is concomitant with the *sādhya*, viz. eternality. But wherever there is quality of eternal ether, e.g. the sound or maximum grossness, there is no *akāryatva*. Here *akāryatva* is not concomitant with the *hetu*,. viz. 'being the quality of eternal ether'. Since *akāryatva* is concomitant with the *sādhya* without being concomitant with the *hetu*, it is an *upādhi* of the *hetu*. Therefore, the inference being *sopādhika* involves the fallacy of *vyāpyatvāsiddhi*, and stands rejected. If the inference is accepted in spite of its being *sopādhika*, then the special quality of the soul, viz. knowledge, etc. also must be admitted as eternal, because they are also quality of the eternal soul, like the maximum grossness abiding in the soul. The Mīmāṁsaka of course would not subscribe to this position, because according to him knowledge is momentary.

The same *upādhi (akāryatva)* vitiates the following inference also. 'Sound is eternal, because it is apprehended by the sense of hearing, like soundness *(śabdatva)*'. Just as soundness which is apprehended by the sense of hearing is eternal, so also sound which is apprehended by the sense of hearing is eternal. That which is eternal is *akārya*, like soundness. Sound is not *akārya*, because it is produced; therefore, it cannot be eternal. If this is not admitted, then smell, colour, taste, and touch also will turn out to be eternal, like smellness, etc., because they are apprehended by the sense of smell, etc.

Therefore, it is established that the inference, viz. sound is not eternal, because it is produced, like a jar, is free from the fallacy of *bādha* and *satpratipakṣa*. Similarly, the inference is free from the fallacy of *virodha* and *vyabhicāra* also. The fallacy *of virodha* or *vyabhicāra* arises only when the *hetu* co-exists with the absence of *sādhya*. The *hetu* 'being produced' in this inference exists with the *sādhya* non-eternality, and not with the absence of *sādhya*, viz. absence of non-eternality. Therefore, the inference does not suffer from either *virodha* or *vyabhicāra*. The inference does not involve *asiddhi* also. As the *hetu* exists in the *pakṣa* sound, the fallacy of *hetvasiddhi* does not arise.

The fact that sound is produced may be established as follows. Sound is apprehended in various degrees as loud, louder, low, lower, etc., according to increase and decrease in the intensity of sound. These various degrees of sound depend on particular causes. Eternal entities that are uncaused have only maximum increase, like ether, etc., or only maximum decrease, like ultimate atom. Comparative degree of increase and decrease is not apprehended in uncaused eternal entities. If sound were to be eternal, it must have maximum increase or maximum decrease. But sound is not apprehended that way, it is apprehended as louder in comparison with a loud one and lower in comparison with a low one. The degrees of sound, viz. loud, louder, low, lower, etc., are determined by the degrees of the causal aggregate that produces the sound. Therefore, sound the origin of which is determined by a cause must be produced. Thus is established the non-eternality of sound by the *hetu* 'being produced'. The inference, therefore, cannot be charged with the fallacy of *asiddhi*.

Even though the inference mentioned above is free from *hetvābhāsa*, doubt may arise in the form 'let there be *hetu*, but let there be no *sādhya*', that is, let there be production, but let there be no non-eternality. This doubt may be removed by the following argument. There is a *vyāpti* that wherever there is production, there is proximate cause of destruction. Production is *vyāpya*, and proximate cause of destruction is *vyāpaka*. In respect of eternal ether, etc. there does not exist proximate cause of destruction. Hence, eternal ether, etc. are not subject to production. Proximate cause of destruction is present in sound, so sound must be subject to destruction. As sound is produced, the *vyāpaka,* viz. the proximate cause of destruction, must be present in the sound. Thus, sound is subject to destruction. And that which is subject to destruction is non-eternal. Therefore, there is no scope for such doubt, viz. let there be the *hetu* (production), but let there be no *sādhya* (non-eternality), to arise.

1.xviii. Objection may be raised that if sound were non-eternal, it will be a changing entity. Then it will not be possible to attain the knowledge of relation between word-sound and the meaning denoted by the changing word-sound. It may be argued that just as the knowledge of relation between eternal sound and non-eternal meaning denoted by the sound is possible, so also is possible the knowledge of the relation between changing sound and its signification.

The Mīmāṁsaka may say that according to him sound is eternal, and so is its signification eternal. Word-sound signifies *jāti* which is eternal; therefore, it is possible to attain the knowledge of relation between eternal word-sound and its eternal signification. With the Nyāya word-sound signifies individual *(vyakti)* which is non-eternal, and hence changing. Therefoe, it is not possible to have the knowledge of relation between changing word-sound and its changing signification.

The Nyāya does not admit this. *Jāti* cannot be the word-sense. Were *jāti* word-sense and not individual, then word would not give rise to the knowledge of individual.

It is not proper to maintain that words signify *jāti*, and individual is obtained through *ākṣepa. Ākṣepa* may be either inference or *arthāpatti.* It is not inference, for it is not possible to ascertain the *vyāpti* relation between *jāti* and the infinite number of individuals; hence, individual cannot be inferred from *jāti* which is denoted by the word. If, however, such *vyāpti* relation can be ascertained, then *vyāpti* can also be ascertained between words and infinite number of individuals signified by words. Thus, individual will be word-sense and not *jāti.*

Nor can it be argued that though *vyāpti* cannot be ascertained between *jāti* and infinite number of individuals, it can be established between *jāti* and individualness *(vyaktitva).*

For, even then only individualness may be inferred, but not individual.

Arthāpatti also does not give rise to the knowledge of individual. In *arthāpatti* some fact is presumed on the ground that some other fact cannot be explained without presuming this. For example, as fatness cannot be explained without presuming eating, eating is presumed on that ground. In the case under consideration it must be pointed out what remains inexplicable without presuming individual. It cannot be said that as *jāti* cannot be explained without presuming individual, individual is presumed on that ground. For, *jāti* exists even after the destruction of the individual and prior to the production of the individual. Hence, there is no reason to say that the presumption of individual is grounded in the *jāti's* being otherwise inexplicable.

The opponent may argue that the knowledge of *jāti* does not arise without the knowledge of individual. The knowledge of the *jāti* cowness denoted by the word 'cow' does not arise without the knowledge of the individual cow; hence, the knowledge of cowness presumes the knowledge of cow. As the knowledge of the individual is thus obtained through *ākṣepa*, the word signifies only *jāti* and not individual.

This position is untenable. Since the opponent maintains that word signifies only *jāti*, it stands out that word gives rise to knowledge of *jāti* only and not of individual. This being the case, he can no longer maintain that the knowledge of *jāti* does not arise without the knowledge of individual.

It may be said that the knowledge of *jāti* produced by the word 'cow' includes the individual also in its content. Thus, the knowledge about *jāti* does not arise without the individual as its content, and so individual is presumed on that ground.

Were this the case, then it may be admitted that the knowledge produced by the word 'cow' has as its content both the individual cow as well as the *jāti* cowness. Therefore, it may be maintained that the word 'cow' signifies both individual cow and the *jāti* cowness and not only the *jāti*.

It may be argued that the word 'cow' is related with the

jāti 'cowness' and word-*śakti* is inherent in the individual cow. The knowledge of the relation between the word and the *jāti* produces the knowledge of *jāti*. *Śakti* present *(svarūpasati)* in the individual gives rise to the knowledge of the individual. This way of the generation of the knowledge of individual by the word-*śakti* is what is meant by *ākṣepa*.

This argument stands refuted. The question arises whether the knowledge of *jāti* and that of individual arise in succession or simultaneously. They do not arise in succession, for there is no cause for their generation in succession. Nor can the succession be established on the ground that the knowledge of individual depends on the knowledge of *jāti*; so, the knowledge of individual succeeds the knowledge of *jāti*. Were it so, the presumption of word-*śakti* in the individual becomes useless, as in that case there would arise the knowledge of *jāti* first from the knowledge of the relation of the word with the *jāti*, and from this the knowledge of the individual would arise. Therefore, presumption of word-*śakti* in the individual to produce the knowledge of individual becomes needless.

Moreover, if the knowledge of individual were to succeed the knowledge of *jāti*, it would result that *jāti* is the *pramāṇa* for *vyakti* or individual. Then, there arises the contingency of postulating a seventh *pramāṇa*, viz. *jāti*, apart from the six admitted by the opponent Mīmāṁsaka.

Therefore, to obviate all these difficulties it may be admitted that after hearing the word 'cow' there arises the knowledge of both *jāti* and individual, and word-*śakti* inheres in the individual qualified by *jāti*. Just as the knowledge of word-*śakti* arises through non-eternal individual, so also arises the knowledge of relation between the non-eternal word and its signification. Therefore, for the sake of obtaining relation between the word and its signification, word-sound need not be postulated as eternal.

It has already been stated that the destruction of letter-sound is established by such perception as 'the sound *ga* heard before does not exist now'. Thus, letter-sound is non-eternal owing to its destruction. As a result word which consists of group of letters and sentence which is an assemblage of words arranged in a particular way are also non-eternal. This again results in the non-eternality of the Vedas which consist of sentences. Since Vedas are non-eternal, they are destroyed during the dissolution of the world and are created again during the creation of the world. And the creator of the Vedas is God.

1.xix. The opponent contends that even if letters were non-eternal, it is not proved that Vedas are created by God and are not eternal. The Vedas are independent and eternal in the sense that the course of the utterances of words that constitute the Vedas continues eternally from generation after generation depending on earlier and still earlier generation. This process does not stand in need of a person on whose excellence in the form of the possession of the valid knowledge of the Vedas depends the validity of the Vedas. Therefore, the Vedas are valid intrinsically and do not depend on God for their validity.

This contention is untenable. Everything is destroyed at the cycle of dissolution. The course of the study of the Vedas too must come to an end at that period. During creation there arises the necessity of creating those Vedas again, and it is God who creates these Vedas.

The opponent argues that there does not exist a period of dissolution, when everything simultaneously comes to an end. There is a *vypāti* that the present day and night are preceded by the previous day and night, and the previous day and night are preceded by still previous day and night, and so on. Following this *vypāti* day and night existing in the beginning of creation also must be preceded by the previous day and night. But were there cosmic dissolution, first day and night existing in the beginning of creation would not be preceded by the previous day and night, as such day and night would be absent owing to dissolution. As the postulation of dissolution violates this *vypāti*, dissolution cannot be accepted.

Again, if dissolution is admitted, it must be accepted that fructification of all virtuous and sinful deeds are suppressed at the same time. This is not true, because fruition of actions takes place at different times; fruits are not enjoyed at the same time, but they are enjoyed at different times owing to the different timings of fructification. Therefore, it is not proper to say that fructification of all the actions are suppressed during the period of dissolution. Further, the postulation of dissolution cannot account for the tradition of caste-system also. A person is called *brāhmaṇa* when he is born to *brāhmaṇa* parents and another a *kṣatriya* when he is born to *kṣatriya* parents. If dissolution takes place, at the beginning of creation there being absence of *brāhmaṇa* or *kṣatriya* parents, the first created person would not be called a *brāhmaṇa* or *kṣatriya*, etc. Thus, caste-system would be abolished.

Further, the signification of the word jar, etc. is ascertained on the ground of previous usage. During dissolution there being no such usage, at the beginning of creation signification of the word jar, etc. cannot be ascertained at all. Therefore, there is no reason to support the concept of dissolution. And in the absence of dissolution, it can very well be established that the study of the Vedas through the teacher-disciple tradition is existing eternally. The Vedas are not composed anew at any time.

वर्षादिवद्भवोपाधिर्वृत्तिरोधः सुषुप्तिवत्।
उद्भिद्वृश्चिकवद्वर्णा मायावत् समयादयः।।२।।

As in the case of the days of the rainy season the existence of the universe is the condition; there is suppression of action as in deep sleep; (origination of) the caste is as in the case of herbs, scorpions; significations (of words, etc.) are like jugglary.

2.i. All the arguments that are advanced to disprove dissolution, are examined and rejected. As regards the first objection it is stated that there is a *vyāpti* that day and night are preceded by the previous day and night. Even if dissolution is admitted, this *vyāpti* can be maintained. The last day and night existing prior to the dissolution may precede the first day and night existing in the beginning of creation. Just as the first day of a rainy season is preceded by the last day of the rainy season of the previous year and not by a day of the rainy season existing immediately before it, so also are the first day and night at the beginning of creation preceded by the last day and night existing prior to the dissolution and not by the day and night existing immediately before the first day and night at the beginning of creation.

The objection that as fructification of all the virtuous and sinful actions of human beings cannot be suppressed during dissolution so dissolution cannot be admitted, is unsound. During deep sleep, according to the Nyāya, fruition of action does not take place. Similarly, it may be admitted that during dissolution also actions are not fructified.

It may be argued that during deep sleep fruition of all the actions is not suppressed, as all the actions do not fructify uniformly at the same time; hence, fruits of some actions may be experienced even during deep sleep.

The argument advanced above does not stand to reason, because one who remains unconscious during deep sleep is not capable of enjoying any experience. The body is bereft of any use during deep sleep, yet it continues then in order to enjoy experience later on.

Objection may be raised that during deep sleep fruition of actions of one person may be suppressed, but the fructification of actions of all the people of the universe cannot remain suppressed during deep sleep owing to the reasons that actions are innumerable, actions do not fructify at the same time, and that by nature one action destroys another.

This objection is met on the ground that if suppression of the fruition of actions is possible during deep sleep in respect of one person, then there is no logic why it should not be possible for all the people, after all all the people do have deep sleep.

It does not stand to reason to argue that as night sets in at different time in different countries, deep sleep takes place in succession and not simultaneously. For, the succession of effects depends on the succession of their causes. Effects can take place simultaneously too when their causes occur simultaneously. To explain this point an example may be cited. Human body possesses three defects *(dosa),* viz. wind, bile and phlegm; the state of their equilibrium and lack of equilibrium determines the disease and absence of disease of the body respectively. There is nothing to determine the order of succession of their state of equilibrium, increase and abatement, but sometimes all the three may increase simultaneously destroying the body thereby. In the like manner when the fire and wind of final dissolution and the great ocean that consumes the universe combine together, there occurs the deluge of the whole universe. At that state of deluge human beings become devoid of experience simultaneously. Thus is established that simultaneous suppression of the fruition of actions is possible, even though actions fructify at different times.

Caste-system can be explained even if dissolution takes place. It is noticed that scorpions in general are born out of scorpions, but the first scorpion is born out of cowdung. So also the first *brāhmaṇa*, etc. at the beginning of creation are not born out of *brāhmaṇa* parents, etc., but they are born out of five elements in accordance with their respective *adṛṣṭa*. Therefore, caste-system would not be violated even if dissolution is admitted.

It is not proper to object that at the beginning of creation the signification of the word 'jar', etc. cannot be ascertained, as there is absence of the previous usage of such words. Just as a juggler having made a puppet pulled by strings, bids it bring a jar and the jar is brought, so also at the beginning of creation God after creating human beings and uttering the word 'jar', etc. asks them to bring it and thus teaches them the signification of the words. Therefore, the postulation of dissolution is not riddled with any contradiction.

2.ii. Objection may be raised that even though there is no *bādhaka-pramāṇa* against the theory of creation and dissolution, yet in order to establish creation and dissolution *sādhaka-pramāṇa* must be stated. To meet this objection two inferences are offered as *sādhaka-pramāṇa*, i.e. proofs that establish creation and dissolution.

The first is as follows. 'The series of effects constituting the world is produced by the material causes that are devoid of any visible causal series, because it is a causal series, like the series of fire generated by the *araṇi*-wood.' In this inference, the *sādhya* 'production by material causes' proves creation and 'devoid of any visible causal series' proves dissolution. Just as ultimate fire-atoms inherent in the *arani*-wood exist without producing effects before, but give rise to a series of fire when woods are put into friction, so also ultimate atoms during dissolution exist without producing the series of effect in the form of binary compound, etc., but later on produce the series of binary compound, etc. at the will of God. This is the purport of the inference mentioned above. That atoms exist without producing the series of effect signifies dissolution in the form of the destruction of effects. And the production of effects from the ultimate atoms later on signifies creation.

Following is the second inference. 'The ultimate atoms that constitute the present world produce the series of effect similar to the series of effect produced before, because being eternal they generate that series, like the ultimate atoms of a flame.' Just as the series of a flame in the form of binary compound, etc., produced by the ultimate atoms of flame, destroys the previous series and a similar series is produced, so also after the destruction of the series of cause and effect constituting the present world, a similar causal series is produced by the ultimate atoms. The destruction of one series establishes dissolution, and the production of a similar one establishes creation. Therefore, inference is the *sādhaka-pramāna* for the establishment of creation and dissolution. According to some the *Śruti*, viz. *dhātā yathā pūrvamakalpayat* (the creator made the creation as before), etc. is also the proof for creation and dissolution.

The inference advanced above may stand invalidated on the following ground. Creation and destruction of a thing may occur by adding and removing of parts to the thing, and yet at the same time there is no break in the series of creation, for at the time of the destrution of a thing another may be produced without causing any break in the series.

The ground set forth above to invalidate the inference establishing creation and dissolution is improper. Were this the case, then following the same logic there will be absence of break in the production of a cloth even when the cloth is destroyed. This, however, is contrary to our experience.

To argue that break in the production of a cloth is caused by the attainment of particular experiences of the cloth-maker and the purchaser, does not hold good. For, such experience being absent in the case of binary compound, there will be absence of break in the generation of binary compound. And in the absence of the withdrawal of its parts, binary compound will turn out to be eternal; in that case it will not be designated as binary compound at all, since a binary compound is produced out of two atoms. Therefore, it must be admitted that every effect is subject to origination and existence. Origination consists of relation of the effect with the first moment; existence consists of effect's relation with the continuing second moment, etc. If an effect depends for its existence on another, it does not exist in the absence of that another. If a thing has another thing as its cause, then the thing is produced only when that thing, i.e. its cause, is present. The cause of the existence of an effect cannot be eternal, because this leads to the eternal existence of an effect. If the cause of origination were eternal, then the effect will turn out to be eternal; and its nature of existing only for sometime *(kādācitkatva)* will be violated. Hence, both the cause of origination as well as that of existence are non-eternal. Thus, when the conjunction of part which is the cause of the existence of a part is removed, the part is removed; and in the absence of the introduction of the cause of the production of another part, that part is not produced. Thus, break in the series of creation is inevitable. The experience of the fruition of actions is also determined by the time factor that regulates the production of an effect only when its cause is present. Everything in the universe is subject to destruction. Therefore, there is established *vyāpti* relation between 'break', the probandum and 'series of effect', the probans. Thus is established creation and dissolution.

2.iii.. During dissolution the entire universe is reduced to ultimate atoms, and the ultimate atoms remain separately being devoid of causal efficiency. In that state, it may be enquired what happens to the hosts of living beings that live in the universe.

All living beings during dissolution are completely destroyed, what remain are the ultimate atoms. Just as the swarms of mosquitoes existing inside a berry that has fallen into the jaws of an infuriated monkey are destroyed, or just as the heaps of worms struggling in the inside of a tree that is being burnt down by wild fire are burnt down, or just as the crowds of passengers in a ship that has fallen into the *aurva* fire that is blazed up by the storms of deluge are destroyed, so also during dissolution along with the universe all living beings in the universe are destroyed.

Again, formerly purification was effected in the sacrificial grain, then in the wife and now it is done āfter the child is born.

Previously, all the Vedas were studied including the thousand subsidiaries, then as classified, then one of the branches with six subsidiaries. But now a *brāhmaṇa* seldom studies even a single branch.

Brāhmaṇas, formerly, used to live on grains collected from the field, then on alms, then on agriculture and commerce. Now they serve even the *śūdras* by accepting monthly salary.

Before *brāhmaṇas* used to take food at home, gradually they started eating at *kṣatriya* and *vaiśya* home, now they eat the food of even *śūdras*. Formerly, *dharma* consisted of penance, knowledge, sacrifice and charity. Gradually, the first three began to decay. Now only charity is existing — that too is found here and there as the real spirit of charity is deteriorating day by day owing to lack of faith, avarice, anger, etc.

Thus, on the evidence of the gradual deterioration in the field of study, profession, *dharma*, and of the gradual destruction of some of the branches of the Vedas that were studied before, even during the period of creation, it is quite reasonable to infer that during dissolution all the Vedas are destroyed, as there is none to study the Vedas.

Again, some of the *dharmas* mentioned in the Smrti are not found in any of the Vedas that are studied now. From this it may be concluded. that the branches of the Vedas that dealt with such *dharmas* are destroyed now. Similarly, certain *dharmas* existing in *śiṣṭācāra* (observances of representative men) are not found in the Vedas that are studied now. This also suggests that the branches of the Vedas concerned with these *dharmas* are destroyed now.

It is not proper to argue that there is no break in the Vedic tradition, as it is not desirable to assume more branches of the Vedas than what are being studied through the tradition of Smṛti and the observances of the representative men. For, Smṛtis and observances do not possess independent authority. Manu and others, the authors of Smṛti books, cannot be accepted as authority on the presumption that they were prompted to write Smṛtis after having direct knowledge of supernatural things. As we can not have knowledge of things that are supernatural, so also Manu and others lacked capacity of possessing knowledge of supernatural things. To say that Smṛtis are rooted in earlier observances and earlier observances are derived from still earlier Smrtis, thus establishing eternality of Vedic tradition, will lead to the admission of a blind series, as the tradition of the Vedas will not be based on a direct authority.

Destruction of the tradition of the Vedas cannot be avoided by saying that the Vedas are always obtained through inference. For, what has not been heard about from the beginning of this transmigratory world and are ever inferred, cannot be called the Vedas.

Moreover, the Vedas consist of letters; letters are to be defined in terms of their origination, cognition and intention, otherwise meaningful arrangement of the letters will not be possible. If the Vedas are to be ever inferred, the letters of the Vedas being not defined in terms of their origination, cognition and intention, meaningful arrangement of them will not be possible, and what is not meaningful will not consist of the Vedas.

2.iv. The following inferences may also be considered. 'This observance is what leads to that what is desired, because it is performed by the representative men', 'this observance is what ought to be done, because it is performed by the representative men' — these inferences establish that which is to be established by the Vedas. The means of attaining what is desired and what ought to be done are established by the Vedas; since inference itself establishes the means of attaining what is desired and what ought to be done, the inference of the Vedas is redundant.

It is not proper to argue that this inference itself is the Veda which is to be inferred, in so far as it establishes *dharma* pertaining to the means of what is desired and what ought to be done. This inference is established by perception; hence, it cannot be the Veda which is, according to the opponent, to be eternally inferred. Further, the term *veda* signifies words establishing *dharma;* hence, no inference other than words *(śabda)* can be the Veda.

One more inference may be set forth in order to establish authority for observance. The inference is as follows: 'This observance is based on authority, since it is performed by representative men' — this inference purports to infer the Vedas as the authority for the observance, thus establishing the inference of the Veda.

The inference advanced above establishes what is already established. Observances are amenable to perception; hence, they have perception as their authority. Observances as a means of attaining what is desired or as what ought to be done are not based on perception, but they have inference as their authority; hence, the inference advanced above does not prove anything other than what is already established.

It may be argued that *śabda* (Veda) is inferrred as the authority for observances, and *dharma* regarding what ought to be done is known through the Vedas.

This position is untenable. For, observances have perception as authority; hence, they require no other authority.

The opponent may rejoin that as *śabda* cannot be inferred from observances, the former may be inferred from *dharma* regarding what ought to be done. *Dharma* as to what ought to be done is not amenable to perception; it being unknown cannot give rise to the inference of *śabda* also; regarding *dharma* there is no perceptible scriputure too. Hence, *dharma* as to what ought to be done is to be inferred from the observance of representative men, and then *śabda* is to be inferred as the authority of *dharma* — from the *dharma* as to what ought to be done. The inferences are as follows: 'Observance has *śabda* as its authority, for it is what ought to be done', and '*dharma* as to what ought to be done has *śabda* as its authority, for it is observed as what ought to be done by the representative men'.

This position stands rejected, because *dharma* as to what ought to be done being established prior to the inference of *śabda,* the latter becomes futile. Knowledge of what ought to be done is derived from *śabda;* since that knowledge is established earlier than the inference of *śabda,* the inference of *śabda* is of no utility.

It may be argued that there is a *vyāpti* relation between *dharma* pertaining to what ought to be done and Āgama as the basis of *dharma,* from this it is inferred that observance too is derived from the Āgama.

Were it so, then on the observation that other *śabdas* are based on perception or inference, it may be argued that the Āgama too is based on perception or inference.

To argue that *śabdas* having beginning are based on perception or inference, but eternal *śabda* is not so based, will not save the position. For, in that case observances too will have to be considered as eternal, but eternal observances are not derived from the Āgama, as eternal observances may exist even without the Āgama.

2.v. The opponent (Prābhākara Mīmāṁsā) may rejoin that there is no harm in admitting observance and the inference of what ought to be done as eternal.

This position goes against two principles admitted by the opponent. In the first place, the stand that the Veda is to be eternally inferred stands rejected. Secondly, inference is also being admitted as authority it can no longer be maintained that in respect of *dharma* scriptural injunction is the only authority.

In fact, eternality of the observance of the Vedas cannot be established; it is inferred that like *rājasūya* sacrifice, etc. the Vedic observance is also subject to destruction. Even though *rājasūya*, *aśvamedha* sacrifices, etc. are prescribed by the Vedas, yet they are extinct now, because they are not performed. It is not that such sacrifices were never performed, because, were it so, such Vedic saying as 'the king desirous of kingdom performs *rājasūya* sacrifice' would turn out to be non-authoritative. It is also not that such scriptural sayings are non-authoritative, because the scriptural observance entailing new-moon sacrifice, etc. are performed now, and other scriptural observances too should be treated at the same level of authority.

It may be argued that *rājasūya*, *aśvamedha* sacrifices, etc. are not totally extinct, they may be performed in future by someone having eligibility to perform them. Similarly, other observances also get extinct for sometime, and then are performed at some other time. Observances of representative men are also to be treated in the similar manner; hence, there is no break in the observance. And in the absence of any other authority, observances are to be admitted as having the Āgama as their authority, just as *rājasūya* sacrifice, etc. are based on the authority of Āgama.

The argument mentioned above is untenable. After the destruction of observance when the creation starts, there being absence of observance, the inference of what ought to be done as well as the inference of *śabda* will be impossible. And in the absence of authority observance itself will not gain ground, and whatever observances are seen will turn out to be erroneous. Therefore, at the beginning of creation, perceptible scripture alone is the authority for observance of the Vedas. Those branches of the Vedas which are not studied now are destroyed, because they are not backed by any perceptible scripture.

It is not fair to say that the branches of the Vedas dealing with those *dharmas* that are metnioned in the Smṛti exist elsewhere. This means that those branches do not exist in India, but exist in some other country.

The question arises as to why those branches do not exist in India now. It may be due to one of the following reasons, viz that the family of the teachers must have gone elsewhere, or the family is extinguished, or there is break in the study of those branches.

The first two alternatives are untenable. For, had all the teachers gone elsewhere or extinguished, those branches being non-existent in India, *dharmas* mentioned by them would not have been observed at all in India.

It is not reasonable to argue that the teachers after coming back to India have re-introduced the observance of those *dharmas*. Were it so, they would have re-introduced the study of those branches of the Vedas also.

The third alternative also does not hold good. For, as the teachers and disciples existing in India study the other branches of the Vedas, they can as well study those branches on which is based the Smṛti, hence break in the study is not possible.

It is, on the other hand, quite reasonable to hold that owing to gradual day by day decay of the powers like longevity, strength, self-control, faith, diligence, study and retention of what is studied, as the time passed it was not possible to study all the branches of the Vedas. That is why only some of the branches are studied now. It is quite possible that those branches that are not studied now, are destroyed.

2.vi. The opponent may argue that the observances of those *dharmas* that are not found in the existing Vedas, are not authentic, because they do not derive their authority from the Vedas. Such observances are continuing blindly.

It is preposterous to presume that the branches that lend authority to such observances are destroyed. For, in that case there would result lack of faith in any rite whatsoever, because each rite could be accomplished only by the methods of rites found in many schools. For example, *agnihotra* sacrifice admitted by all the branches of the Vedas cannot be accomplished owing to the lack of knowledge of the methods mentioned in the branches that are destroyed now.

This position is not accepted. It is not proper to argue that the observances that are accepted by wise people are unauthentic. Were it so, it can equally be argued that like observances the Vedas are also unauthentic and they are continuing blindly.

In fact, the great seers anticipating the destruction of the branches of the Vedas in the course of time wrote down Kalpasūtras by compiling the methods found in the various branches. So, any rite can be accomplished by following the methods mentioned in the Kalpasūtras, even though certain branches are destroyed now. Thus, the charge that the destruction of the Vedic branches would obstruct the accomplishment of the rite stands rejected.

The observances accepted by the wise people are based on the Vedas. The branches of the Vedas that are not found now are destroyed.

The break in the Vedic tradition is not caused by the process of the rise of right knowledge at the time of final release. This break is caused by the process of the culmination of forgetfulness, intoxication, conceit, idleness and nihilistic tendency, etc. The break in the Vedic tradition is followed by its revival, and then there occurs break once again. Just as the current of the river Sarasvati gets hidden in between and then starts flowing, so also the Vedic tradition gets hidden for some time and then revives. If the revival of the broken tradition is not admitted, then the treatises which are introduced to establish *dharma* will stand rejected. Further, if gradually all the branches of the Vedas get destroyed, there arises the contingency of the destruction of the merit which has been attained by performing various observances. Therefore, the course of revival is established. To establish break and revival of the Vedic tradition the following inference is offered. 'This course of the Vedic tradition is preceded by break, because this is the course of the Vedic tradition, like the course of the future Vedic tradition.' In the *Gītā* Lord Kṛṣṇa says that whenever there is decline of *dharma* and ascendency of *adharma,* He sends forth Himself as an incarnation; in order to protect the virtuous, destroy the evil doers and to establish *dharma,* He takes birth in cycle after cycle. The proclamation of the Lord also proves creation and break. Therefore, as observance is subject to break, in the beginning of creation perceptible scripture is the only authority for observance.

If the observances were not based on the Vedas, authority must be pointed out on the basis of which the observances were introduced in the world. Indolence, etc. cannot be the base of the observances, because the observance of the rites that are of a fatiguing and painstaking nature cannot be grounded in indolence, etc. Nor can it be said that those who are ineligible to adhere to authentic Śāstra take resort to observance. For, those who are honoured by others, viz. Buddhists and others, are also not permitted to take to observance. Nor is it that those who cannot abide by the do's and don'ts regarding food and drink come for observance, because the observance of the Vedic *dharmas* has restriction in food and drink.

It is not that due to association with deceitful heretics people observe *dharmas,* because the observance is introduced through the order of father, grandfather, etc. Observance is not a means of livelihood also, for without achieving any visible fruit the wise people take all pains to observe *dharmas.*

In fact, wise people observe *dharmas* on the ground that the Vedas prescribe certain *karmas* to be performed. Therefore, observance derives the authority from the Vedas.

The grounds mentioned above are actually causes of the observance of Bauddha-dharma, and not that of the observance of Vaidika-dharma.

Thus, indolent people accept the Buddhist religion, because there is always a lightness of work in the observance of religious rites. Those who are dropped out from the Vedic religion get admission into the fold of Buddhist religion.

There is no regulation regarding food and drink also in the Buddhist religion. The Buddhist religion does not prescribe that after performing *pañca-mahāyajña*, etc. one should take food in the afternoon; with this religion one can eat at every seven *ghaṭikā* (each ghaṭikā is 24 minutes). Life is very easy going in the Buddhist religion. Hence, indolent, deceitful people observe the Bauddha-dharma and not the Vaidika-dharma that is observed by wise people.

The branches of the Vedas that are the authority of the observance of Vaidika-dharma gradually get destroyed. Therefore, the Vedas are non-eternal.

> जन्मसंस्कारविद्यादेः शक्तेः स्वाध्यायकर्मणोः।
> ह्रासदर्शनतो ह्रासः सम्प्रदायस्य मीयताम्।।३।।
> *The deterioration of the tradition (of the Vedas) may be inferred from the observation of the deterioration of birth, ceremonial purification, learning, etc. of the capacity of study and observance (of the Vedas).*

3.i. It is not only during dissolution that the Vedas are destroyed, even at the cycle of creation also a deteriorating tendency marks the capacity of the study and the observance of the Vedas in different era, and there is break in the tradition of the study of some of the Vedas. From this it may be concluded that during the period of creation also some of the branches of the Vedas gradually get destroyed.

Deterioration is observed not only in the field of study, it is experienced in the case of birth, purification *(saṁskāra)*, etc. also Previously children were born through mere will, then they were born through the institution of marriage. Marriage was also previously conducted under Śāstraic instruction, and child-birth was the sole purpose of marriage. But now marriage is not conducted in the way directed by the Śāstra, and it is solely motivated by enjoyment.

In the state of dissolution fructification of all actions is arrested, nothing is produced, nothing is destroyed. The universe stands like an empty space where only ultimate atoms exist. The question arises as to how the creation starts again.

Actions of living beings performed before dissolution exist

during dissolution, and the time delimited by such actions marks the beginning of the cycle of creation again.

So far as the duration of time is concerned, one night of the *caturmukha-brahmā* consisting of thousand of *yugas* is the duration of the period of dissolution. The cycle of creation starts again at the termination of this night.

At the period of dissolution all the Vedas being destroyed, neither the Vedas nor the wise people who accept the validity of the Vedas exist. Therefore, at the beginning of the creation the validity of the Vedas cannot be ascertained on the ground that they are accepted by the wise men. The validity of the Vedas, at the beginning of creation, must be ascertained on the ground that the Vedas are uttered by a reliable person, and that reliable person at that time can only be God. Therefore, God is established as the author of the Vedas.

3.iii The Sāṅkhya contends that the seer Kapila and others who have realized what is *dharma* and *adharma* possess eight *aiśvaryas* (super-natural powers) in the form of *animā, laghimā,* etc. by the practice of the Vedic rituals and the *yoga.* The Kapila and others, out of supreme mercy on living beings burnt down in the embers of the circumlocutory world may bless the suffering humanity through the teaching of the Vedas, telling thereby what is pleasing and what is conducive to happiness. Therefore, Kapila and others can be the author of the Vedas. There is no need of postulating a Supreme Lord for that purpose.

This argument does not hold good, because one cannot repose confidence in any being other than God. The Vedas are concerned with many things that are supra-sensible. One who has the direct knowledge of such supra-sensible things can only be the author of the Vedas. There is no proof to attest the fact that the seer Kapila and others had the direct knowledge of such things.

By meditation also the seer cannot have the direct knowledge of supra-sensible things, because meditation is not possible on things that are not known before. Therefore, as meditation of unknown things is not possible, it cannot be argued that supra-sensible things mentioned in the Vedas can be directly known by the power of meditation. Further, the supposition that *karma* and *yoga* are the means to attaining happiness cannot at first be known to Kapila and others; and without knowing that *karma* and *yoga* are the means to the desired end, they will not perform them too; and without the performance of *karma* and *yoga* Kapila and others cannot have direct knowledge of *dharma* and *adharma*. Just mere knowledge of *karma* and *yoga* will not activate Kapila and others to perform them, because without knowing means and end one does not engage in any activity. Nor can it be said that the knowledge of *anvaya-vyatireka* between performance of *karma-yoga* and attaining happiness inspires Kapila and others to perform *karma-yoga*. For, the result of *karma* being experienced in another life, the knowledge of *anvaya-vyatireka* between *karma-yoga* and its fruit, viz. happiness, cannot be gained in the same life. Knowledge of cause and effect cannot also be attained by the recollection that 'I who was the doer am the experiencer of the fruit also', because such recollection of two different bodies is not possible. Even if recollection is admitted on the ground that fruit does not accrue to one who is not the doer, yet such recollection as 'I experience this fruit of that particular earlier action' is not possible.

It may be rejoined that Kapila and others might have direct knowledge of the Vedas and their supra-sensible contents in the previous birth. After recollecting that knowledge in the next birth they could teach the Vedas, just as the teachers after studying the Vedas go to sleep and the next day after waking up recollect the Vedas that are studied the previous day and teach them. Therefore, there is nothing wrong if Kapila and others are postulated as the author of the Vedas.

This position is unsound. It may be possible to recollect a thing today that was experienced in some other day in the same birth. But it is not possible to recollect a thing in this birth that was experienced in some other birth. Therefore, the argument that Kapila and others after recollecting the direct knowledge of the Vedas that was attained in the previous birth teaches the Vedas in the next birth is untenable. God is omniscient. He can recollect the Vedas that existed in the previous cycle of creation and teach them at the beginning of the new cycle. Moreover, in the beginning of creation, if the Vedas were not admitted to be composed, or caste-system in the form of *brāhmaṇa, kṣatriya,* etc. was not supposed to be established by God, then it would not have been possible for Kapila and others to know among the first created subjects who was *brāhmaṇa* and who was *kṣatriya,* etc. It would have been impossible for them to establish the regulation of *dharma* also in accordance with the division of caste as *brāhmaṇa, kṣatriya,* etc.

It may be argued that prior to the creation of this universe *(brahmāṇḍa),* there existed other universe. Following the caste-system prevalent in that universe, Kapila and others established the caste-system at the beginning of the creation of this universe.

This argument does not stand to reason. Kapila and others who were born in this universe could not possibly have the knowledge of the caste-system prevalent in the other universe. It is indeed a difficult job to migrate from one part of this world to another part. It is not at all conceivable how one can migrate from one universe to another one. As it is not possible to visit the other universe, it is also not possible to possess the knowledge of the caste-system existing in that universe. In the absence of knowledge Kapila and others cannot establish the caste-system in this universe by following the caste-system existing in that universe.

It is the omniscient trustworthy God who has composed the Vedas at the beginning of creation. Omniscient Lord has set up all the Vedic traditions in the beginning of creation. These traditions are still existing through the order of teachers and disciples.

कारं कारमलौकिकाद्भुतमयं मायावशात् संहरन्
हारं हारमपीन्द्रजालमिव यः कुर्वन् जगत् क्रीडति।
तं देवं निरवग्रहस्फुरदभिध्यानानुभावं भवं।
विश्वासैकभुवं शिवं प्रति नमन् भूयासमन्तेष्वपि।।४।।

May I remain worshipful unto my end to that God Śiva, the Bhava, the sole ground of confidence, whose splendour of eternal knowledge shines forth unhindered, who in sport creates this world of supernatural and wonderful nature out of his māyā *and then destroys it, and having destroyed recreates it as a magic show.*

STABAKA - III

NO PROOF FOR NON-EXISTENCE OF GOD

The validity of the Vedas has been set forth in the second *stabaka* on the ground of their being composed by God, and God is established as the author of the Vedas. This position does not find favour with the opponent, namely, the Mīmāṁsaka. He raises a series of objections against this stand and makes an attempt to prove the non-existence of God through the six means of valid knowledge, viz. perception, inference, analogy, verbal testimony, presumption and non-cognition. These six are the *bādhakapramāṇas* (invalidating proofs) that prove the non-existence of God; hence, God cannot be established as the author of the Vedas. Udayana examines all the objections raised by the opponent and proves that none of these *bādhakapramāṇas* can establish the non-existence of God.

योग्यादृष्टिः कुतोऽयोग्ये प्रतिबन्धिः कुतस्तराम् ।
क्वायोग्यं बाध्यते शृङ्गं क्वानुमानमनाश्रयम् ।।१।।

There can be no legitimate non-cognition of God who is not fit to be perceived; the contradiction (pointed out by the opponent) also cannot be established, because the siddhāntin *does not negate horn which is not fit to be perceived; inference is also not possible without the subject.*

1.i. Since God does not come within the range of our ordinary perception, it may be said that through non-cognition His non-existence is proved. But an examination of the nature of non-cognition shows that non-cognition neither independently nor with the help of sense-organ can prove the non-existence of God.

The arguments set forth to prove the non-existence of God may be stated thus. (a) That which exists is perceived; God is not perceived, therefore, God does not exist. It is not proper to say that God, though exists, is not perceived, because He is not fit to be perceived, as in that case hare's horn too may be taken as existent entity on the ground that it is not perceived owing to its absence of fitness to be perceived. (b) Further, sentience in general is perceived as delimited by body which is fit to be perceived. God who is also a sentient being must possess a body. But in the case of God one does not perceive a body which is fit to be perceived; hence, God's body does not exist. As body in the case of God does not exist, God delimited by the body also does not exist. Here the non-existence of God is proved by legitimate non-cognition, because body that delimits a sentient being is fit to be perceived, but such body is not perceived in the case of God who is a sentient being. (c) Again, an agent takes to a work either for his own sake or for the sake of others, none does anything irrespective of his own interest or that of others. God is the creator of this world; hence, He must have some purpose behind this creation. But He does not create the world for His own sake, because He has no desire unfulfilled, all of His desires have ever been satisfied. Nor can He create the world for others, as in that case He must make all people either happy or miserable, not some people happy and others miserable. Were it the latter, He will be charged with partiality of making some happy and others miserable. As God does not create the world either for His own sake or for the sake of others, He cannot be established as the agent of the world.

All these objections stand refuted. (a) So far as the first objection is concerned, mere non-cognition cannot prove the non-existence of anything. Non-cognition can ascertain the non-existence of a thing only when that thing is fit to be perceived. When it is said that 'there is no jar on the floor' the non-existence of the jar on the floor is ascertained on the ground that had it been there, it would have been perceived; since it is not perceived, it does not exist. But in the case of God this *reductio ad absurdum* argument is not applicable. God is not fit to be perceived; hence,

His non-existence can never be proved by non-cognition.

The example of hare's horn put forth by the opponent is not at all apt here. Horn in general is fit to be perceived, because horn is perceived in other animals, viz. cow, etc. There is *vyāpti* (invariable concomitance) between horn in general *(śrṅgatva)* and perceptibility *(pratyakṣa-yogyatva)*. As horn in general is fit to be perceived, its existence in the hare can be denied by non-cognition. In other words, the non-existence of horn in the hare is proved by legitimate non-cognition. But non-existence of God cannot be proved by legitimate non-cognition; therefore, the example of hare's horn does not suit the context.

(b-c) The attempt to prove God's non-existence on the ground that He does not possess a body which is fit to be perceived and that He lacks any purpose behind the creation, is also unfounded. These arguments may be expressed in the following inferential forms.

'God does not exist, because we do not perceive a body in Him which is fit to be perceived and He does not create the world either for His own sake or for the sake of others'. Here, the subject of the arguments, viz. God, being Himself unestablished, two probans (*hetus*), viz. absence of body and that of purpose, remain unestablished in the subject. Hence, the arguments are fallacious.

It is not reasonable to deny the existence of God on the ground that it is not perceived, because even one's own soul though fit to be perceived from such cognitions as 'I am happy', 'I am miserable', etc. is not perceived, but this absence of perception in no way proves the absence of one's own soul. Similarly, absence of perception of the Supreme Soul which could not be effectually perceived cannot also prove the non-existence of the Supreme Soul.

During deep sleep the soul is not perceived. It is because of the absence of the causal aggregate necessary to perceive the soul. The nature of the soul is that it is cognized through its qualities, viz. *jñāna, kṛti* in the form 'I know', 'I do', etc. During deep sleep there does not exist such qualities, viz. *jñāna*, etc., hence, the soul qualified by such qualities is not cognized.

The reason why knowledge, etc. do not arise during deep sleep and how they arise in waking state consists in the following fact. Mind in contact with the sense-organ gives rise to knowledge, etc. During deep sleep mind does not come in contact with the sense-organ, as it rests in pericardium *(purītat),* that is why no knowledge, etc. are produced in that state. In the waking state mind leaves pericardium and comes in contact with the sense-organ and as a result knowledge, etc. are produced. Thus, the soul though fit to be perceived through knowledge, etc. is not perceived during deep sleep owing to the absence of knowledge, etc. This non-cognition, even though it is legitimate, does never prove the non-existence of the soul. God is not at all fit to be perceived; therefore, mere non-cognition cannot preclude the existence of God.

1.ii. The view that knowledge does not arise during deep sleep, owing to the absence of mind-sense-organ contact, is not acceptable to the Mīmāṁsaka. With the Mīmāṁsaka mind is an all-pervading entity. All-pervasion *(vibhutva)* means contact with all corporeal objects *(sarvamūrtadravya-saṁyogitva).* And, corporeality means the quality of having definite dimensions. If the mind were all-pervasive, it must be always in contact with the sense-organs. All-pervasiveness of the mind may be proved by the following inferential arguments.

(i) Mind is all-pervasive, because it is a substance always devoid of touch *(sarvadā sparśarahitadravyatvāt).* That which is a substance and is devoid of touch at all times is all-pervasive, e.g. *ākāśa* (ether). Mind is also a substance and is devoid of touch always; therefore, mind is all-pervasive. In this inference the epithet 'devoid of touch' is added to the *hetu,* because otherwise earth, etc. also will be all-pervasive, as earth, etc. are also substances. Everything that is produced is devoid of qualities at the moment of its origin; hence, the *hetu* 'substance devoid of touch' being applied to those things at the moment of their origin will turn them into all-pervasive substances. In order to obviate this difficulty the epithet 'always' is added to the *hetu*, though things are devoid of qualities at the moment of their origin, they are not so always.

(ii) Mind is all-pervasive, because it is a substance always bereft of specific quality *(sarvadā viśesaguṇaśūnyadravyatvāt)*. That which is a substance always bereft of specific quality is all-pervasive, like time, etc. Mind does not possess any specific quality; therefore, mind is all-pervasive.

(iii) Mind is all-pervasive, because being eternal it is a substance that does not give rise to any other substance *(nityatve satyanārambhakadravyatvāt)*, like soul. That which is not the inherent cause of any other thing is all-pervasive, like soul; mind is also not the inherent cause of any other thing; therefore, mind is all-pervasive.

(iv) Mind is all-pervasive, because it is the substratum of the contact of non-inherent cause of knowledge *(jñānāsamavāyikāraṇasaṁyogādhāratvāt)*, like soul. Just as soul which is the substratum of soul-mind contact that is the non-inherent cause of knowledge is all-pervasive, so also mind is all-pervasive, as mind is also the substratum of the soul-mind contact.

All the above inferential arguments according to Udayana cannot prove that mind is all-pervasive, because all involve the fallacy, called *svarūpāsiddhi*. This fallacy arises when *hetu* does not exist in *pakṣa*. In the inferential arguments mentioned above mind is *pakṣa*, and the *hetu* substancehood does not exist in mind, and hence, all the arguments are vitiated by the fallacy of *svarūpāsiddhi*. Arguments that may be advanced to prove mind a substance turn out to be futile. If mind is inferred as the *karaṇa* of the perception of pleasure, etc., even then mind cannot be proved to be a substance, as *karaṇa* need not necessarily be a substance. *Karaṇa* of inferential knowledge is *vyāpti-jñāna* which is not substance.

It may be said that only a sense-organ is the means of the direct apprehension of pleasure, etc., and mind may be inferred as the sense-organ through which pleasure, etc. are apprehended. As a sense-organ cannot but be a substance, mind too must be a substance. Thus is proved the substancehood of mind.

The above position does not hold good. An all-pervading substance becomes a sense-organ only when it is limited by a condition, for example, all-pervasive *ākāśa* becomes the sense of hearing only when it is delimited by ear-cavity. All-pervasive mind too must possess some delimiting condition in order to be a sense-organ. Just as *ākāśa* becomes the sense of hearing by being delimited by a particular part of the body, viz. ear-cavity, so also if mind were delimited by a part of the body, then the perception of pleasure, etc. would arise only when mind is delimited by that part of the body and cease to arise when that part becomes defective. In the case of *ākāśa* also sound is perceived only when *ākāśa* is delimited by ear-cavity, and when the limiting adjunct is destroyed, sound is no longer perceived. If the whole body is postulated as the limiting adjunct of the mind, then pain in the head and delight in the foot cannot be explained, because if one part of the body suffers from pain, the other part too must have the same experience, since that part is also the limiting adjunct.

It may be said that the effects brought about by all-pervasive substances, like sound, etc. are restricted in their activities to particular places in accordance with the non-inherent cause (*asamavāyikāraṇa*); and in the case of the perception of pleasure, etc., non-inherent cause, viz. contact of the mind delimited by the body, restricts the pleasure, etc. in particular places. But even according to this it is not possible to explain delight in the foot, because the non-inherent cause, viz. contact of the mind delimited by the body, can explain delight in the whole body and also not in part of it, as there is nothing to determine it.

To obviate this difficulty, if various delimiting adjuncts, viz. body, its parts, i.e. hands, etc., their parts, viz. atoms, etc. are ascribed to the mind, then the law of parsimony will be violated. Moreover, if all the parts of the body become the limiting adjuncts, such experience as 'delight in the foot' etc. also cannot be explained. Therefore, some subtle suprasensible entity different from body and its parts must be admitted as the limiting adjunct that delimits the mind as a sense-organ. If that were the case, instead of postulating an all-pervading substance delimited by a subtle suprasensible entity as the sense-organ, it is simpler to postulate that subtle entity itself as the sense-organ. Thus, the arguments that are advanced to prove mind to be a sense-organ prove also that mind is an atomic entity. This position goes against the fundamental stand of the opponent that mind is an all-pervading substance.

1.iii. In order to avoid the above difficulty the opponent attempts to infer the mind in a different way. Sense-organs, viz. eye, etc., give rise to knowledge successively and not simultaneously. The cause of this succession in knowledge may be traced back to the operation of mind thus. Only that sense-organ gives rise to knowledge with which mind comes in contact and not the other sense-organs with which mind does not come in contact. Thus, mind is the auxiliary cause of the knowledge produced by the sense-organ. The following inference may be cited as a proof for the mind: 'Knowledge produced by the sense-organ depends on the auxiliary causality of a sense-organ, because knowledge arises in succession, and this auxiliary cause is the mind'.

This inference is fallacious, because it proves mind to be atomic and not all-pervasive. Production of knowledge in succession is offered as a *hetu* to prove the mind, but if the mind were all-pervasive, it will come in contact with all the sense-organs at the same time. As a result there will be simultaneous production of all knowledge and not successive production. Thus, the fact of the successive rise of knowledge disproves the all-pervasiveness of the mind. Knowledge can arise in succession only when an atomic mind comes into contact with the sense-organ.

It may be said that though all-pervasive mind comes into contact with all the sense-organs simultaneously, yet knowledge arises in succession due to *adṛṣṭa*. A particular knowledge arises only when the *adṛṣṭa* giving rise to that knowledge is present, and in the absence of the latter former does not arise. Thus, succession in knowledge can very well be regulated by the *adṛṣṭa* even if the mind is not postulated to be atomic in size.

To this position it may be said that if that were the case, there is no need to postulate the mind also, as every other thing can be explained by *adṛṣṭa*. Thus, in the absence of the mind the argument that seeks to prove the mind as an all-pervading substance will suffer from the fallacy, called *āśrayāsiddhi*.

It may be argued that knowledge arises in succession by nature, even though the mind is all-pervasive. It is the nature of the *karaṇa* to produce one action at a time. Mind is the *karaṇa* of knowledge, and hence, it produces one knowledge at a time.

Adṛṣṭa becomes the cause of an effect in either of the following ways. Sometimes it produces an effect directly without the help of other visible causes, like the production of motion in the atoms by *adṛṣṭa* during the time of creation. Sometimes *adṛṣṭa* becomes the cause of an effect by bringing together the visible causes. For example, a jar is produced when the visible causes, viz. stick, wheel, etc., are brought together by *adṛṣṭa*. In the latter case, as soon as the causes are brought together, the effect is immediately produced, and here the function of *adṛṣṭa* is just to assemble the causes. In the former case, as *adṛṣṭa* produces the effect directly, the effect is not produced until and unless *adṛṣṭa* is present.

Adṛṣṭa is not the direct cause of knowledge. It just brings object, sense-organ and mind together, and as soon as these three are assembled knowledge is generated immediately, Though the mind being all-pervasive remains in contact with object, sense-organ and the soul always, yet being a *karaṇa* it produces one knowledge at a time, as it is the nature of the *karaṇa* to generate one action at a time. Thus, the succession in knowledge can be explained by admitting that mind is all-pervasive.

Succession in knowledge can take place only when mind is postulated as the *karaṇa*. The soul cannot be the *karaṇa*, because being knower it is supposed to be an agent and is devoid of the nature of the *karaṇa*. Sense-organs are many; hence, if they were *karaṇa*, simultaneous rise of knowledge by all the sense-organs cannot be prevented. Mind is only one. If mind were the *karaṇa*, it can regulate succession in knowledge.

1.iv. An objection may be raised. Though one knowledge is generated at a time owing to mind's intervention, yet there is possibility of the generation of one knowledge about many objects simultaneously. The same sense-organ gives rise to the knowledge of many objects, viz. jar, cloth, etc. Similarly, knowledge of many objects apprehended by various sense-organs may also be considered as one knowledge. For example, a person eating a piece of cake possesses knowledge comprising many objects, such as smell, taste, touch, etc., apprehended by various sense-organs.

The *siddhāntin* does not lend support to this view. The belief about the generation of one knowledge comprising many objects arises due to mal-observation. Apparently one may have the impression of having simultaneous perceptions of different objects through different senses. These are as a matter of fact different perceptions resulting from different contacts of the mind with the different senses taking place at different moments. They only appear to be simultaneous because of the extremely swift fluctuations of the atomic mind, as a result of which it comes in contact with the different senses in rapid succession. When mind is attentive, succession in the generation of knowledge is apprehended. And from this it can be safely inferred that at other times also the nature of the mind consists in operating successively.

Any attempt to explain succession in knowledge by desire, that is, knowledge arises only in respect of that object which the knower desires to know about and not about other objects, is untenable. Desire does not prevent the generation of knowledge about undesired object; if the causal aggregate of knowledge is present, knowledge arises about any object desired and undesired. But there is a difference, knowledge generated about undesired object does not leave any impression that gives rise to memory at a later time, whereas knowledge generated about desired object leaves that impression. If desire stands in the way of the generation of knowledge about object that is not desired even though the causal aggregate of the knowledge is present, then the eye opened for seeing the jar will not see the cloth at all.

To opine that the desire to know draws the mind away from other sense-organs and fixes it to a particular sense-organ that can apprehend the desired object and thus helps in the generation of knowledge in a particular order is also untenable. If the mind is held to be all-pervasive, it will be in contact with all the sense-organs even when there is desire to know a particular object, and this being the case it is beyond the capacity of the desire to draw the mind away from other sense-organs and to fix it to a particular sense-organ.

It may be argued that the mind does not generate knowledge directly, it does it so through its *vyāpāra,* and *vyāpāra* takes place in succession. Hence, knowledge too arises in succession. Thus, the succession in knowledge being based on the succession in *vyāpāra,* there is nothing wrong in holding mind to be all-pervasive.

The question arises as to the nature of this *vyāpāra* whether it is *saṁyoga* (contact) or something other than *saṁyoga.* If it is the first, then all-pervasive mind being in contact with all the sense-organs there arises the contingency of the generation of all knowledge at the same time. If the *vyāpāra* of the mind be something other than contact, it may be either action (*kriyā*) or quality *(guṇa).* It cannot be action, because all-pervasive entity is devoid of any action, like all-pervasive ether which does not have any action. If the *vyāpāra* be a quality, it may be either eternal or non-eternal. It cannot be the first, as an eternal quality cannot account for the sequence; and *vyāpāra* being eternal knowledge cannot arise in succession. If the *vyāpāra* were a non-eternal quality, then there arises the following difficulty. The generation of a non-eternal quality in an eternal all-prevading substance requires the contact with a non-all-pervading substance as its non-inherent cause. Production of sound in the all-pervading ether may serve as an illustration, where contact with the non-all-pervading substance, viz. contact between drum and stick, is the non-inherent cause of the production of sound in the ether. Similarly, non-eternal quality can be generated in the all-pervasive mind only when it has as its non-inherent cause the contact with a non-all-pervading substance. If a non-all-pervading substance is admitted, that can be taken as the mind. Instead of presuming an all-pervasive mind and one more non-all-pervading substance, it is simpler to presume mind as a non-all-pervading substance, and there is no harm in admitting mind to be atomic.

Therefore, succession in knowledge cannot be explained by the *vyāpāra* of the mind; it can be explained even if the mind is taken to be an atomic entity. Thus, only that sense-organ gives rise to knowledge which comes in contact with the atomic mind and not others which do not come in contact with the mind.

As the mind is atomic, during deep sleep it cannot come in contact with the sense-organs, because at that time it enters into pericardium *(purītat),* and consequently, no knowledge arises during deep sleep. In the absence of knowledge the soul qualified by the knowledge is also not apprehended at that time. The fact that the mind produces the knowledge of the soul only when the soul is qualified by its specific qualities, viz. knowledge, etc., are proved by *anvaya-vyatireka* relation. The mental perception of the soul in the form 'I know' is produced only when there is presence of knowledge in the soul. And, in the absence of knowledge such perception does not arise. This is not the peculiarity of the mind only, all the sense-organs that perceive substance possess this peculiarity. The visual sense-organ or the tactual sense-organ perceives only that substance which is qualified by its special quality, viz. colour or touch.

The question may arise as to how to explain the experience that arises in dream, since the mind does not come into contact with any sense-organ in dream. In reply it may be said that the experience one has in dream is erroneous, it is not knowledge proper. Due to the contact of the mind with the tactual sense-organ there arises cognition of the body-heat, etc. in dream, and this cognition gives rise to impression of the things experienced before. From this impression arises the remembrance of those things. Things remembered are imposed on some other things, and thus arise all delusions in dream. When the mind leaves its contact with the tactual sense-organ and enters into pericardium (*purītat*), there arises deep sleep. In the absence of mind-tactual-sense-contact there does not arise the cognition of body-heat, and as a result neither the impression of the things experienced before nor their remembrance arises. Thus, no knowledge, erroneous or valid, is generated during deep sleep.

The soul which is fit to be perceived is not perceived in deep sleep. Yet this fact does not negate the existence of the soul. The Supreme Soul is not fit to be perceived; therefore, there is no reason whatsoever to prove His non-existence through non-cognition. One perceives one's own soul, but does not perceive the soul belonging to others. The reason for not perceiving other's soul does not consist in the fact that other's soul lacks perceptibility, because in that case others also will not be able to perceive their own souls. Nor can it be said that the knower does not have capacity (*yogyatā*) to perceive other's soul. So far as knowledge is concerned, capacity means having inherent causality; the knower possesses that capacity, because it is the inherent cause of knowledge. It is also not the nature of the mind to perceive only one soul, and so the mind in contact with one soul does not perceive the other. The mind according to the opponent is all-pervasive; hence, it must have contact with all the souls. Therefore, the mind cannot be held responsible for not perceiving other soul.

If it is said that it is the nature of the mind to belong to one soul only, then at the time of liberation also the mind will belong to the soul, and consequently, there arises the contingency of the generation of knowledge at the time of liberation. If the contact of the mind with the soul is not admitted at the time of liberation, then the view that it is the nature of the mind to belong to one soul only cannot hold good. To say that it is the nature of the mind to belong to a particular soul means that it must belong to the soul throughout, at the time of liberation also; if it does not belong to the soul at the time of liberation, it means that this belonging is not its nature.

1.v. The *siddhāntin* solves the problem why the mind perceives one's own soul and not the soul belonging to others thus. It is *adṛṣṭa* that determines the perception of the soul; due to the intervention of *adṛṣṭa* the mind perceives only one's own soul and not other's. Again, owing to *adṛṣṭa* particular soul comes in contact with particular mind, body and sense-organs, and the mind perceives only that soul with which *adṛṣṭa* brings its contact and not any other soul. In the same way the mind, sense-organs, etc. generate knowledge inherent in that soul only, not in any other soul. Thus, the mind which is in contact with the individual soul lacks capacity to perceive the Supreme Soul. As the Supreme Soul is not fit to be perceived, He cannot be the subject of legitimate non-cognition; hence, His existence cannot be denied. And it is an accepted fact that mere non-cognition cannot negate the existence of anything. The counter-example of hare's horn advanced by the opponent is not proper. Had the existence of hare's horn been negated on the ground that it is not fit to be perceived as in the case of God,then the example would have been befitting. But hare's horn is fit to be perceived, that is, it can be the subject of legitimate non-cognition, which is not the case with God; hence, the former cannot be a counter-example of the latter.

It may be said that the purport of advancing the counter-example of hare's horn is as follows. If the hare's horn which is fit to be perceived is not perceived, its existence may be negated by the legitimate non-cognition. But if the hare's horn which is not fit to be perceived is not perceived, its existence cannot be denied, as the existence of God who is not fit to be perceived cannot be denied by mere non-cognition.

In reply to this objection it may be pointed out that hare's horn may be established only when there is proof for its existence. Legitimate non-cognition may not be a *bādhaka* for the existence of hare's horn which is not fit to be perceived. But mere absence of *bādhaka* does not guarantee the existence of a thing, there must be *sādhaka-pramāṇa* in support of its existence. There is no positive proof for the existence of hare's horn; therefore, hare's horn cannot be established.

The following inferential argument may be offered as a proof for the existence of hare's horn. 'The hare possesses horn, because it has animality' — but this argument does not go against the existence of God. Rather it goes in its favour, as the similar argument may be advanced in support of God's existence. The argument is as follows — 'The earth, etc. must have a creator, because they are effects'.

The opponent may argue that what he intends to place before is that if God whose existence cannot be denied by non-cognition is established by the inference, then the existence of hare's horn which cannot be disproved by non-cognition may also be established by the inference. Were it not so, that is, if the *hetu* 'animality' fails to prove 'possession of horn', then the *hetu* 'effecthood' too will not establish 'creatorship'; consequently, God cannot be established as the creator of the earth, etc.

The conclusion that if 'animality' does not prove 'possession of horn', then 'effecthood' also will not prove 'creatorship' stands rejected. If there is *vyāpya-vyāpaka* relation between 'effecthood proving creatorship' and 'animality proving possession of horn', then in the absence of the *vyāpaka*, viz. 'the failure of animality to prove possession of horn', absence of *vyāpya,* viz. 'the failure of effecthood to prove creatorship', can be inferred. But there is no such relation between the two. Hence, the absence of the 'effecthood proving creatorship' cannot be inferred from the absence of 'animality proving possession of horn.'

It may be further argued that just as the *sādhya*, viz. possession of horn, is unestablished because it is not perceived in the hare, so also God, the *sādhya* in the inference 'the earth, etc. must have a creator, because they are effects' is unestablished, since He is not perceived. This objection has already been replied to, as it is said that mere non-cognition cannot disprove the existence of anything.

If mere non-cognition, it may be argued, does not nullify the existence of God, then it cannot negate hare's horn also. Hence, the inference 'hare possesses horn, because it has animality' stands unrefuted.

To this it may be replied that the inference mentioned above cannot stand, because there is no *anukūla-tarka (reductio ad absurdum* argument) that proves that animality cannot exist without being invariably related to the possession of horn. Animality, the *hetu,* can very well exist without being connected with horn, the *sādhya,* for hare has animality but does not possess horn. Therefore, the *sādhya* remains unestablished. In the case of God, there is *anukūla-tarka* that definitely proves that there is invariable relation between effecthood and creatorship. The doubt that let there be effect, the *hetu*, without having a creator, the *sādhya*, can be dispelled by the *reductio ad absurdum* argument that if the earth, etc. do not have a creator, they cannot be effect, because an effect cannot take place without an agent or creator. Therefore, as the two inferences stand on different levels, the conclusion arrived at in respect of one cannot prove the other, that is, the conclusion that God must be the creator of the earth, etc. on the ground that the earth, etc. are effects, cannot prove that hare must have horn, because it possesses animality.

1.vi. Attempt may be made to prove God's non-existence from a different point of view. It is thus — from the absence of the *pramāṇa* that proves the existence of an object, the absence of that object is inferred. For example, visual sense-organ is inferred from the perception of colour as its *pramāṇa*, and the absence of the former is inferred from the absence of the latter. A blind man is inferred to be devoid of visual sense-organ, because he does not perceive colour. We do not have the perceptual knowledge of other's soul as we have of our own soul. Other soul is inferred on the basis of the *vyāpāra* of body, etc. belonging to that soul. Therefore, the *vyāpāra* of body, etc. is the *pramāṇa* for other soul, and from the absence of the former, absence of the latter is inferred. God is other than one's own soul, but we do not perceive the *vyāpāra* of body, etc. in Him. Hence, in the absence of *pramāṇa*, the absence of God is inferred.

The above position is untenable, the *vyāpāra* of body etc. is not the *pramāṇa* for agency, as during deep sleep there is absence of the *vyāpāra* of body, etc. but soul is not absent. Hence, such *vyāpāra* cannot be the *pramāṇa* of other soul.

The *vyāpāra* of vital force exists during deep sleep; therefore, it may be said that the *vyāpāra* of vital force is the *pramāṇa* for inferring other soul. Since this *vyāpāra* is absent in God, the absence of God may be inferred from the absence of the *vyāpāra* of the vital force.

This argument is unsound; during the time of *prāṇāyāma* when the process of respiration is controlled, there is absence of the *vyāpāra* of the vital force, but the soul exists. If the *vyāpāra* of the vital force were the *pramāṇa* of the soul, then during *prāṇāyāma* the soul would cease to exist. Since the soul exists during *prāṇāyāma*, the *vyāpāra* of the vital force cannot be the *pramāṇa* of other soul.

Further, since at the time of *prāṇāyāma* the particular form of the body exists, it cannot be said that the particular form of the body is the *pramāṇa* of the soul. That form of body is absent when a body falls down on account of fainting caused by consuming poison, but this in no way proves the absence of the soul at that time.

Nor can it be said that body-heat that is present in a faint body is the *pramāṇa* of the soul, because body-heat is absent when water is poured over the faint body, but the soul is present even at that time.

As all the factors mentioned as *pramāṇa* to prove the existence of other soul stand rejected, it may be concluded that particular sentient soul is inferred from the observation of the *vyāpāra* belonging to particular body. In the same way, the Supreme Soul, viz. God, may be inferred from the *vyāpāra* of the creation of the world.

Moreover, it can never be definitely said that the absence of *prameya* is inferred from the absence of *pramāṇa*. The absence of *prameya* can be inferred from the absence of *pramāṇa* only when there is *vyāpti* relation between *prameyatva* and *pramāṇa-gamyatva*, that is, the *vyāpti* that wherever there is a *prameya* it must be known by *pramāṇa*. Such *vyāpti* relation is not possible. It may be asked whether this *pramāṇa-gamyatva* includes all *pramāṇas* or some *pramāṇas*. If it is the first, then a *prameya* must be known by all the *pramāṇas*. This is not true, heaven is known only by *śabda-pramāṇa*, as there is no other *pramāṇa* to prove its existence. The second alternative is also not tenable, because according to it a *prameya* must be known by some *pramāṇa*, and in that case that which is not known by any *pramāṇa* will turn out to be non-*prameya*.

It may be said that that which is a *prameya* must be known by a certain person with the help of some *pramāṇa*. This is not tenable, because a *prameya* which is not known by a certain person may be known by some other person with the help of some *pramāṇa*. But this does not prove that the *prameya* is non-existent, because it is not known to a certain person.

Nor can it be said that that which is not known by all the *pramāṇas* is not *prameya*, because the fact that a particular object is not known by all the *pramāṇas* at all times cannot definitely be ascertained.

It may be asked if the absence of *prameya* is not determined by the absence of *pramāṇa*, then how the absence of eye is determined in the case of a blind person. In fact, the absence of eye is not ascertained from the absence of *pramāṇa*, viz. perception of colour. Perception of colour is *vyāpaka*, not a *pramāṇa*. There is a *vyāpti* relation between eye and perception of colour in the form 'wherever there is eye, there is perception of colour'. Here eye is *vyāpya*, perception of colour is *vyāpaka*. Whenever there is absence of *vyāpaka*, viz. perception of colour, the absence of *vyāpya*, viz. eye, is inferred. Thus, the absence of eye in the blind is determined by the absence of *vyāpaka* and not by the absence of *pramāṇa*.

व्यावर्त्याभाववत्तैव भाविकी हि विशेष्यता।
अभावविरहात्मत्वं वस्तुनः प्रतियोगिता।।२।।

Only a real thing becomes the substratum by possessing absence of that which is to be rejected; counter-correlateness which consists of the absence of absence also belongs to a real thing.

2.1. The opponent may argue that there is *vyāpya-vyāpaka* relation between agency and motive; if God were the agent of the world, He must have motive. Since God does not have any motive behind the creation, He does not exist. Here the absence of God's agency, the *vyāpya,* is inferred from the absence of motive, the *vyāpaka.*

The opponent tries to negate the existence of God from the following inference — 'God is not the agent, because there is absence of motive in Him'. This inference is vitiated by the fallacy of *āśrayāsiddhi,* since God, the *pakṣa* of the inference, is yet to be established.

2.ii. The opponent may defend that he seeks to establish absence of agency in God who is sought to be established by the Naiyāyika through erroneous reason. This argument stands refuted. If the existence of God is pseudo and not real, such God cannot be the subject of any inference. Inference arises only when there is doubt about the existence of something in the substratum. For example, in the inference 'there is fire in the hill, because there is smoke' doubt arises in two ways — (i) is there fire in the hill?, or (ii) is there absence of fire in the hill? The hill becomes the subject *(pakṣa)* of the inference when the presence of fire is established by ascertaining that there is absence of absence of fire in the hill. Thus, the subject-hood *(pakṣatva)* of the hill consists in its being the substratum of the absence of absence of fire. This being the case, only a real object can be the substratum and not an unreal object. The opponent does not admit the reality of God; hence, God cannot be the subject of inference. Similarly, an unreal object cannot be the counter-correlate *(pratiyogi)* too. Counter-correlateness is in the form of the absence of absence, e.g. the counter-correlate of the absence of jar is the absence of the absence of jar, viz. the jar. There is no absence of the absence of hare's horn; hence, hare's horn cannot be the counter-correlate of the absence of hare's horn. Since hare's horn cannot be the counter-correlate of its absence, its absence cannot be asserted.

दुष्टोपलम्भसामग्री शशशृङ्गादियोग्यता।
न तस्यां नोपलम्भोऽस्ति नास्ति साऽनुपलम्भने।।३।।

Legitimacy in the case of the perception of hare's horn consists of defective aggregate of perception; were this aggregate present, perception would definitely follow; if perception does not arise, there does not exist such an aggregate.

3. Mere non-cognition cannot negate the existence of anything, because in that case the existence of all super-sensuous things would remain unestablished. It is the legitimate non-cognition which can negate the existence of anything. Legitimacy of non-cognition consists in the presence of all the aggregate of perception other than the counter-correlate. In the case of the perception of hare's horn, legitimacy consists of defective aggregate of perception. Otherwise, if the aggregate were perfect, the hare's horn would have been really perceived. In the presence of the aggregate of perception, non-cognition would definitely follow. If, however, hare's horn is not perceived, it means that causal aggregate of the perception of hare's horn does not exist. Thus, it is legitimate non-cognition which can establish negation.

Negation consists in the knowledge of absence *(abhāva)*; valid knowledge of absence arises only when the counter-correlate the absence of which is asserted is real. Hare's horn is not real; hence, the knowledge of the absence of hare's horn also lacks validity. When it is said 'hare's horn does not exist', it means that horn does not exist in hare, its substratum. Horn is fit to be perceived; therefore, its existence can be denied in its substratum, viz. hare. That which is not fit to be perceived cannot be denied by mere non-cognition.

इष्टसिद्धिः प्रसिद्धेंऽशे हेत्वसिद्धिरगोचरे।
नान्या सामान्यतः सिद्धिर्जातावपि तथैव सा।।४।।

If the established part (individual soul) were the subject, then it establishes what is already established; if the unestablished part (Supreme Soul) were the subject, then the probans would be unestablished (in the substratum); if, however, the soul-in-general were the subject, then also there arises the same difficulty.

4. It is true that God who is not established cannot be the subject of any inference. But individual souls are subject to

perception; there is no doubt about the existence of individual souls. And it can be definitely proved that individual souls are devoid of omniscience and an agency of the earth, etc. The inference is as follows: 'souls other than my own are not omniscient, because they are sentient beings like me'. Similarly, 'they are not creators of the earth, etc., because they are individual beings like me'. These inferences prove that souls other than one's own are neither omniscient nor the creators of the earth, etc. Since God comes within the category of the souls other than one's own soul, it follows that He lacks omniscience and creatorship of the earth, etc. Hence, God cannot be established as omniscient creator of the world.

In both these inferences the subject is 'souls other than one's own'. If by the subject is meant individual soul *(jīvātmā)*, then it establishes what is already established, because individual souls are never admitted to be either omniscient or the agent of the world. If, on the other hand, 'souls other than one's own' refer to God, God being unestablished there arises the fallacy of *āśrayāsiddhi*. If by the subject is meant soul-in-general, then also the problem cannot be avoided, because even then it must be pointed out whether soul-in-general refers to God or individual soul other than God. If the first, the subject remains unestablished; if the second, it establishes what is already established.

One more inference may be advanced to prove absence of omniscience and agency in God. Soul-in-general *(ātmatva)* has no relation with a person who is omniscient and agent of everything, because it is a *jāti* like cowness. Just as the *jāti* cowness does not inhere in an individual who is omniscient and agent of everything, so also *(ātmatva) jāti* does not inhere in omniscient agent. This is to prove that there does not exist a soul as omniscient agent.

The *sādhya* in this inference, viz. a person who is omniscient and agent of everything, is unestablished. Therefore, the inference suffers from the fallacy of *sādhyāprasiddhi*. A fallacious inference cannot prove absence of omniscience and agency in God.

आगमादेः प्रमाणत्वे बाधनादनिषेधनम्।
आभासत्वे तु सैव स्यादाश्रयासिद्धिरुद्धता।।५।।

If validity of the Vedas is admitted, then the existence of God, being established by the Vedas, cannot be disproved by inference; if, however, the Vedas are admitted as invalid, then the fallacy of unestablished substratum cannot be avoided.

5.i. The opponent says that God who is established by the Vedas can be the subject of inference, and thus the fallacy of *āśrayāsiddhi* can be evaded. God is established by the Vedas, so the inference that proves absence of omniscience and agency in God cannot be charged with the fallacy of *āśrayāsiddhi.*

The question arises whether the opponent admits the validity of the Vedas or not. If he admits the validity of the Vedas, then God established by the Vedas also must be valid; so, His existence cannot be disproved by the inference. If he does not admit the validity of the Vedas, then God established by the invalid Vedas too will fall short of reality. And it has already been said that an unreal object cannot be the subject of any inference.

5.ii. The Cārvāka points out that one need not depend on legitimate non-cognition to negate the existence of anything. Non-cognition is enough to prove the non-existence of any object. That which is not perceived does not exist, and that which is perceived exists. God is not perceived, therefore, He does not exist.

If all that is not perceived is taken as non-existent, all inference will cease to exist, because the function of inference is to prove something that is not perceived. This position goes in favour with the Cārvāka, as he does not accept the validity of any *pramāṇa* other than *pratyakṣa* (perception).

In fact, people do put forth effort to have the knowledge of things, e.g. fire, through inference. The Cārvāka says that one need not take the help of inference to have the knowledge of fire, one can attain it through doubt. When one observes smoke in the hill, doubt arises whether fire exists in the hill or not.

This doubt can be dispelled by successful activity, that is, finding fire in the hill. Thus, doubt can serve the purpose of inference, and there is no need to postulate inference.

> दृष्ट्यदृष्टयोः क्व सन्देहो भावाभावविनिश्चयात्।
> अदृष्टिबाधिते हेतौ प्रत्यक्षमपि दुर्लभम्।।६।।
>
> *Doubt does not arise in perception as well as non-perception, because perception ascertains existence and non-perception ascertains non-existence; (in that case) even perception will be abolished, because the cause of perception (viz. eye, etc. not being amenable to perception) is precluded by non-perception.*

6.1. This view does not stand to reason. If that which is perceived exists and that which is not perceived does not exist, then doubt itself will not arise. Moreover, if the absence of perception determines the non-existence of a thing, then a Cārvāka who has gone out must be taken as non-existent by his family, because his family does not perceive him. Again, the Cārvāka also must admit that his family is non-existent, as he does not perceive his family. It cannot be maintained that after his returning home people of the family are produced anew, because the causes that produce them are non-existent owing to the reason that the causes are not perceived. And in the absence of the causes the effect does not arise. Further, if absence of perception were the determinant of non-existence, then sense-organs like eye, etc. will turn out to be non-existent, because they are not perceived. Were the eye, the cause of perception, non-existent, perception, the only *pramāṇa* accepted by the Cārvāka, will not take place. Therefore, the view that non-existence is determined by mere non-cognition stands refuted. By this, the view of the atheists who deny the existence of atom, colour of atom, sense-organ, heaven, *adṛṣṭa*, God, etc. by mere non-cognition is also refuted. The Mīmāṁsaka admits heaven and *apūrva*. So far as this point is concerned the *siddhāntin* does not have any quarrel with him. The Mīmāṁsaka does not believe in the existence of God and embodied deities, while the *siddhāntin* believes in them.

The opponent argues that if the existence of an unseen object cannot be denied by mere non-cognition, then the very inference through which the *siddhāntin* tries to prove God's existence will be unestablished. For, there is a possibility of every inference being conditioned by some unseen factor (*upādhi*), as a result the *vyāpti* relation which is the basis of inference will be unestablished. And in the absence of *vyāpti* inference will lack validity.

6.ii. Some section of the Naiyāyika maintains that even in the presence of some unseen *upādhi*, *vyāpti* relation can be established. *Vyāpti* is established provided four or five of the following conditions are fulfilled — (i) *pakṣasattva*, (ii) *sapakṣasattva*, (iii) *vipakṣa-asattva*, (iv) *abādhitattva*, (v) *asat-pratipakṣattva*. These conditions are fulfilled when the *hetu* is present in *sapakṣa* and absent in *vipakṣa*. Since *vyāpti* can be established provided four or five of these factors are present, there is no need to negate some unseen *upādhi*.

The opponent objects that if *vyāpti* is established even in the presence of *upādhi*, then *aprayojaka-hetu* too will be turned out as valid one. In reply to this objection, first it must be pointed out what is meant by *aprayojaka-hetu*. The author of the *Nyāya-sūtra* has mentioned five kinds of *hetvābhāsas* (fallacious reasons) in which five conditions of *vyāpti* mentioned earlier are absent. If *aprayojaka-hetu* comes under these *hetvābhāsas*, it cannot be a valid *hetu*. If it does not come under *hetvābhāsas*, its definition must be stated.

Aprayojaka-hetu cannot be defined as that which is neither *karaṇa* nor *kārya* of a *sādhya*. In *sāmānyatodṛṣṭa* inference the *hetu* is neither *karaṇa* nor *kārya* of the *sādhya*, but it is considered as a valid *hetu*.

Nor can *aprayojaka-hetu* be defined as a part of the cause from which the *sādhya* is inferred. In *pūrvavat* inference the effect is inferred from a part of the cause, e.g. a jar is inferred from the conjunction of two pot-sherds. But such cause is a valid one and not *aprayojaka*.

Vyabhicārī-hetu also cannot be called as *aprayojaka*. If the *hetu* does not exist in *vipakṣa*, it is not *vyabhicārī;* if it exists in *vipakṣa*, it is *vyabhicārī*. If a *hetu* is called *aprayojaka* owing to its *vyabhicāra*, then it may as well come under the group of *hetvābhāsa* called *vyabhicāra*, the *hetu* being *vyabhicārī*. Nor by *aprayojakatva* is meant a *hetu* about which there is doubt regarding *vyabhicāra*, though there is no cause of *vyabhicāra*. For, such doubt may exist in all inferences, and consequently, all valid *hetu* will turn out to be *aprayojaka*.

It is also not that if a *hetu* has *vyāpti* relation with a *sādhya* along with other *hetu* having the same *vyāpti* relation, the former becomes *aprayojaka*. If a *hetu* exists in a *pakṣa* having a *vyāpti* relation with a *sādhya*, that *hetu* is valid even though many other *hetus* may have the same *vyāpti* relation with that *sādhya*. The fire may be produced from various causes, but all the causes are valid ones.

To define *aprayojaka-hetu* as that which exists in lesser place than the *sādhya* is also untenable. Smoke which is the cause of inferring fire exists in a few places, whereas fire exists in more places, but smoke is not *aprayojaka-hetu*. Nor can it be said that if a *hetu* is presumed to prove a *sādhya* in deference with one which is generally admitted to prove that *sādhya*, the former becomes *aprayojaka*. Though *kāryatva* is generally admitted to prove destruction by the inference 'this is subject to destruction, because this is an effect', *sāvayavatva* too may be presumed to prove the same *sādhya* by the following inference 'this is subject to destruction, because it is made up of parts.' Here the latter *hetu* is not *aprayojaka*.

Some maintain that *aprayojaka-hetu* cannot be defined. To this it is rejoined that an *aprayojaka-hetu* is one about which there is doubt of *vyabhicāra*.

It has already been objected that such doubt lingers in all inferences even in the absence of any positive cause for it, and so no inference will function. In reply it may be said that as there is no criterion to determine whether a *hetu* is *vyabhicārī* or not, one cannot ascertain which *hetu* is *vyabhicārī* and which one is valid. In the absence of such criterion what induces doubt about *vyabhicāritva* is the perception of concomitance between *hetu* and *sādhya (hetu-sādhya-sāhacarya)*. Just as seeing a pillar from a distance doubt arises whether it is a man or pillar, since one perceives characteristics, viz. height, etc. common to both man as well as pillar, similarly seeing concomitance between *sādhya* and *hetu* both *vyabhicārī* as well as *avyabhicārī*, doubt arises whether the *hetu* is *vyabhicārī* or not. What causes doubt here is the perception of concomitance between *hetu* and *sādhya*. Thus, it must not be maintained that doubt arises uncaused.

6.iii. As doubt about the validity of *hetu* exists in all inferences, attempt may be taken to sort out some criterion on the basis of which validity of the *hetu* can be ascertained.

Svabhāva (nature) cannot be that criterion. It is not reasonable to say that it is the nature of some *hetu* to be *vyabhicārī* and that of others not to be so. For, there must be some reason why it is the nature of some *hetu* to be *vyabhicārī*, and not that of others. Repeated observation of the concomitance between *hetu* and *sādhya* also cannot be the criterion, because at times repeated observation also fails to ascertain the validity of a *hetu*. Thus, though it is repeatedly observed that wherever there is *pārthivatva* (earthhood) there is *lauhalekhyatva* (piercing the earth by iron axe), *lauhalekhyatva* is absent in *vajramaṇi* (a kind of gem which cannot be pierced through by iron axe). Therefore, *svabhāva* cannot be the determinant of *vyabhicāritva* and *avyabhicāritva*.

To opine that the *hetu* which is vitiated by *upādhi* is *vyabhicārī* and that which is not so vitiated is *avyabhicārī* is untenable. For, it is difficult to ascertain the absence of *upādhi*, as it has already been objected that doubt about the possibility of some unseen *upādhi* is present in all inferences.

It may be argued that if an object is not cognized by any of the *pramāṇas* having a particular characteristic, that object is devoid of that characteristic. For example, a crane is never known to be black by any of the *pramāṇas*; hence, it is proved that crane is not black. Similarly, in the inference 'there is fire in the hill, because there is smoke', smoke is never proved by any *pramāṇa* being conditioned by any *upādhi;* hence, it is proved that smoke is devoid of any *upādhi.*

This position is untenable. In order to establish the position mentioned above the opponent has to prove the following inference. 'Smoke is devoid of *upādhi',* because it is not cognized by anybody through any *pramāṇa* as conditioned by an *upādhi.* This inference remains unproved. As every inference is subject to the doubt of being conditioned by an unseen *upādhi,* this inference is also no exception to it. So, one more inference is needed to ascertain the absence of *upādhi,* and so on *ad infinitum.*

Moreover, it is not proper to maintain that the *upādhi* is not known to anybody, a thing not known to one may be known to others. Similarly, it cannot be maintained that *upādhi* is not perceived at any time, because there is probability of the *upādhi* which is not perceived now, being perceived at a later time.

Therefore, as it is not possible to ascertain the absence of *upādhi,* all inference being subject to the doubt of being vitiated by some unseen *upādhi,* inference itself remains unestablished.

6.iv. The Bauddha says that the validity of a *hetu* is determined by two factors — (i) either the *hetu* must be identical with the *sādhya (sādhyābhinna),* or (ii) it must be produced from the *sādhya (sādhyāt utpatti).* A *hetu* is concomitant with the *sādhya* only when it is identical with the *sādhya* or is produced from the *sādhya.* For example, in the inference 'this is a tree, because it is *śiṁśapā'* the *hetu* is *śiṁśapā* which is not different from the *sādhya,* viz. tree. Again, in the inference 'there is fire in the hill, because there is smoke' the *hetu* 'smoke' is produced from the *sādhya* 'fire'.

In the former case, the *hetu śiṁśapā* is established on the ground of *vipakṣa-bādhaka-tarka.* The *tarka* is as follows. If this were not *śiṁśapā,* this will not be a tree. If this were not tree, this will not be produced from the causal aggregate of the tree. Since *śiṁśapā* is produced from the causal aggregate of the tree, this must be a tree. Thus, identity (*tādātmya*) is established between the *hetu*, viz *śiṁśapā*, and the *sādhya*, viz. tree, by *reductio ad absurdum* argument, and by this is established the *vyāpti* relation between them.

In the inference 'there is fire in the hill, because there is smoke', the fact that smoke is produced from fire is established by *anvaya-vyatireka* relation. Wherever there is smoke, there is fire *(anvaya)*, wherever there is absence of fire, there is absence of smoke (*vyatireka*). The Bauddha argues that since *vyāpti* relation can be established by *tādātmya* and *tadutpatti*, doubt of *vyabhicāra* does not arise in respect of inference.

The Bauddha position is not accepted on the ground that sometimes '*tādātmya* being the determinant of the validity of *hetu* and sometimes *tadutpatti*', there is no single criterion that can determine the validity of *hetu* in all inferences. Moreover, an inference cannot be free from the doubt of *vyabhicāra* even though the *hetu* were identical with the *sādhya* or is produced from it. The facts that *śiṁśapā* is concomitant with tree and smoke is concomitant with fire are established by the following inferences — '*śiṁśapā* is concomitant with tree, because it is identical with tree'; 'smoke is concomitant with fire, because it is produced from fire'. Since these inferences are also not free from the doubt of *vyabhicāra*, *vyāpti* relation established by these inferences is subject to question.

Therefore, as the Bauddha cannot supply with a criterion that will remove the doubt of *vyabhicāra* from all inferences, the Cārvāka view that inference is not a *pramāṇa* at all remains unrefuted.

शङ्का चेदनुमाऽस्त्येव न चेच्छङ्का ततस्तराम्।
व्याघातावधिराशङ्का तर्कः शङ्काऽवधिर्मतः।।७।।

If there is doubt (about some unseen condition), there is inference; if there is no doubt, inference is there; doubt is removed by reductio ad absurdum *argument* (tarka) *which is employed so long as it does not contradict one's own action.*

7.1. The Naiyāyika refutes this objection in the following manner. The reason on which the Cārvāka disproves the validity of inference is that the *hetu*, even if it is not conditioned by an *upādhi* at present, may be so conditioned in future. With the *siddhāntin* future is not perceived now, hence, the knowledge of future must be attained through inference. The inference is as follows. 'Present is succeeded by future, because it is time, like past.' Thus, doubt of *vyabhicāra* may arise only when future is established by inference. If the validity of inference is not accepted, doubt of *vyabhicāra* will not hold good. And if the doubt of *vyabhicāra* is absent, validity of inference cannot be refuted.

With the Naiyāyika the doubt that the *hetu* may be *vyabhicārī* in future is removed by *tarka*. The *tarka* is as follows. In the inference, 'there is fire in the hill, because there is smoke', the question arises 'let there be smoke in the hill, but let there be no fire in it'. Then arises the doubt whether the *hetu* is conditioned by an *upādhi (sopādhika)* or is not concomitant with the *sādhya (vyabhicārī)*. This doubt is removed by the following *tarka*. If smoke were *sopādhika* or *vyabhicārī*, then it will not be produced from fire. In actual practice smoke is seen to be produced from fire. Thus, to maintain that smoke is *sopādhika* or *vyabhicārī* is against actual practice. Since this position contradicts experience, the first question that 'let there be smoke in the hill, but let there be no fire' cannot hold good. Afterwards repeated observation of the concomitance between smoke and fire proves smoke the *hetu* of fire, and from this is inferred fire in the hill. Therefore, it is *tarka* that dispels the doubt of *sopādhikatva* and *vyabhicāritva*.

The opponent raises an objection that the *tarka*, viz. if there is no fire, there is no smoke, is based on the *vyāpti* relation between absence of fire and absence of smoke. Doubt may arise about the validity of this *vyāpti* too. So, in order to remove this doubt one more *tarka* must be offered, and so on *ad infinitum*.

This objection is unfounded, doubt is employed so long as it does not contradict one's own action pertaining to body, mind and speech *(svakriyā-virodha)*. The moment it contradicts one's action or experience and leads to some undesirable conclusion, it must be rejected then and there. Thus, if smoke exists without fire, then smoke will not be produced from fire. Were it so, it must be admitted that effect takes place even in the absence of its cause. If this position is accepted, it may be aruged that hunger is satisfied even without eating; therefore, one will not desire to eat in order to satisfy one's hunger. As this conclusion goes against actual practice, the conclusion that effect takes place even in the absence of its cause contradicts everyday experience. Obviously, no effect is produced in the absence of its cause. Moreover, doubt regarding *vyabhicāra* is raised in order to disprove *vyāpti-jñāna*. If it is admitted that effect is produced even in the absence of its cause, then *vyāpti-jñāna* too may be disproved in the absence of doubt, and hence, the very purpose of doubt will be thwarted. Thus is removed doubt by *tarka*. Therefore, the doubt that 'let there be smoke, but let there be no fire' must not be entertained.

The purpose of *tarka* is to point out the absurdity involved in the opposite stand-point. The charge of infinite regress cannot be levelled against *tarka*, because *tarka* continues till it does not contradict accepted empirical usage, it comes to a termination when it goes against such usage.

Objection has been raised that there is doubt about the *hetu* being conditioned by some unseen *upādhi* in any inference. The Naiyāyika may be asked to single out a *pramāṇa* that removes this doubt. In fact, the Naiyāyika would say, if any unseen or supra-sensuous *upādhi* exists, its absence will be determined by the same *pramāṇa* that proves its existence, there is no need to find out any other *pramāṇa* to determine its absence. Doubt arises in general whether there is any factor other than the actual *hetu* that is concomitant with the *sādhya*. Such doubt is removed by *tarka* which has already been mentioned.

7.ii. It is established that *tarka* helps to ascertain the validity of inference. The author of the *Nyāya-sūtra* also has accepted *tarka* as a part of the *nyāya*. A *hetu* is *aprayojaka* when there is no *anukūla-tarka* to remove the doubt of its *vyabhicāritva*. *Aprayojaka-hetu* is of two kinds — (i) *śaṅkitopādhika*, (ii) *niścitopādhika.* If there is doubt about a *hetu* being conditioned by an *upādhi* or if it is definitely known that the *hetu* has an *upādhi,* that *hetu* is *aprayojaka. Upādhi* has been defined as that which is concomitant with the *sādhya* without being concomitant with the *hetu.* The ground for the use of the term *upādhi* is that, that which transfers its character to something existing near it is called *upādhi.* For example, hibiscus flower that transfers its red colour to a crystal existing near it is a *upādhi.* Similaly, in the inference 'there is smoke in the hill, because there is fire' wet-fuel is the *upādhi,* because wet-fuel is concomitant with smoke and not the *hetu* fire. The *vyāpti* relation between wet-fuel and smoke is mistakenly imposed on fire.

Thus, the validity of a *hetu* in an inference is ascertained on repeated observation of the concomitance between *sādhya* and *hetu* and with the help of *tarka* that removes the doubt about the presence of *sādhya* in *vipakṣa.* And the *hetu* may be in the form of *kārya, karaṇa,* etc. Even if *vyāpti* is ascertained on the grounds of *tādātmya* and *tadutpatti* as the Bauddha maintains, the help of *tarka* is needed. The question why should the effect smoke be concomitant with the cause fire and *śiṁśapā* with tree can be answered only with the help of *tarka,* viz. if smoke and *śiṁśapā* were not concomitant with fire and tree respectively, then smoke would not have been the effect of fire, and *śiṁśapā* identical with tree. Therefore, *tarka* determines *vyāpti* in all inferences by negating the possibility of the presence of *sādhya* in *vipakṣa.* The following is the example of an *apravojaka-hetu.* He is dark, because he is the son of Mitrā. A woman by name Mitrā had sons of dark complexion, the cause of which was eating of greens. After some time another son was born to her while she was not eating greens. Hearing this, one infers, 'that son is dark, because he is the son of Mitrā, like the other sons'. In this inference 'eating of greens' is *upādhi.* 'Eating of greens' is concomitant with the *sādhya,* viz. dark complexion, because wherever there is dark complexion, it is accompanied by eating of greens. This *upādhi* is not concomitant with the *hetu,* viz. being son of Mitrā, because the birth of a fair son of Mitrā is not accompanied by eating of greens. Therefore, the *hetu* in this inference, viz. being son of Mitrā, is *aprayojaka* as it is conditioned by a *upādhi*, viz. eating of greens.

It may be asked whether *aprayojaka-hetu* is included within any of the accepted *hetvābhāsas*. According to some, just as *siddhasādhana* is not included within the five *hetvābhāsas*, so also *aprayojaka-hetu* is not included within them. Thus, *siddhasādhana* is not included in the fallacy of *bādhita*, because unlike *bādhita* the *sādhya* here is already established. Nor can it be said that since *sādhya* is already established, *pakṣatā*, i.e. doubt whether the *sādhya* exists in the *pakṣa*, being absent, *siddhasādhana* may be grouped under the fallacy of *asiddha*. Were it so, the fallacy of *bādhita* too will come under *asiddha*, because absence of *sādhya* in the *pakṣa* being proved doubt regarding presence of *sādhya* in the *pakṣa* is absent in it. *Siddhasādhana* cannot be included in *anaikāntika-hetvābhāsa*, as the *hetu* is not *vyabhicārī* in the *sādhya*. Therefore, as *siddhasādhana* cannot be included in *bādhita*, *asiddha*, etc., it is established that *siddhasādhana* is a separate *hetvābhāsa*.

According to some. like *siddhasādhana*, *aprayojaka-hetu* is also a separate *hetvābhāsa*.

This view is not accepted. The author of the *Nyāya-sūtra* has divided *hetvābhāsa* into five kinds, no other *hetvābhāsa* is admitted. Both *siddhasādhana* and *aprayojaka* come under *asiddha*.

Knowledge of *pakṣadharmatā*, i.e. the presence of *hetu* in the *pakṣa*, accompanied by *vyāpti* relation, i.e. *vyāpti* between *hetu* and *sādhya*, is *siddhi*; absence of that knowledge leads to *asiddhi*. *Asiddhi* arises in the absence of the knowledge of *vyāpti*, or of *pakṣa*, or of *hetu*, which are respectively called *anyathāsiddhi*, *āśrayāsiddhi* and *svarūpāsiddhi*.

Āśrayāsiddhi results either from the absence of knowledge of *āśraya* or that of *pakṣatva* which qualifies the *āśraya*. In *siddhasādhana*, *sādhya* being already established *pakṣatva*, i.e. doubt regarding the presence of *sādhya* in the *pakṣa*, is absent, and in the absence of *pakṣatva*, *pakṣa* qualified by *pakṣatva* is also not known; thus arises the fallacy of *āśrayāsiddhi*. In *bādhita*, *vyāpti* is negated, not *pakṣatva*. Therefore, *siddhasādhana* comes under *āśrayāsiddhi* and *aprayojaka* under *anyathāsiddhi*, that is, in the latter *hetu* is established in the *pakṣa* in the absence of the *sādhya*.

Some opine that *aprayojaka-hetu* is liable to be *vyabhicārī*; so it comes under *anaikāntika*. This is not tenable. If the doubt of *vyabhicāra* were due to absence of the knowledge of *vyāpti*, then it must come under *asiddha*, as it has already been mentioned that absence of the knowledge of *vyāpti* gives rise to *anyathāsiddhi*. If the doubt of *vyabhicāra* arises from the observation of concomitance between *sādhya* and *hetu* both *vyabhicārī* and *avyabhicārī*, then such doubt may be removed by *tarka*. With the help of *tarka* validity of the *hetu* can be ascertained, and when the doubt of *vyabhicāra* is removed, there is no reason to group *aprayojaka* under *anaikāntika*.

परस्परविरोधे हि न प्रकारान्तरस्थितिः।
नैकताऽपि विरुद्धानामुक्तिमात्रविरोधतः।।८।।

*(*Bhāva *and* abhāva*) being mutually contradictory, there can be no category which is other than* bhāva *or* abhāva; *it cannot also be the same (pertaining to the nature of both* bhāva *and* abhāva*), because their contradiction is apparent from the statement itself.*

8. So far it has been established that perception and inference cannot be *bādhaka-pramāṇa* for the existence of God. Presently, it will be proved that analogy *(upamāna)* is also not a *bādhaka* for God's existence.

According to Vaiśeṣika *upamāna* is not a separate *pramāṇa*, because the object of *upamāna* is not something different from that of perception, etc. Therefore, *upamāna* cannot be a *bādhaka-pramāṇa* for God's existence. Prābhākara Mīmāṁsaka maintains that *upamāna* deals with similarity *(sādṛśya)*. Similarity is not a separate category. Similarity in order to be a category must be either a *bhāva* or an *abhāva*. Other than *bhāva* and *abhāva* no third category is an established fact. Similarity cannot be admitted as possessing the characteristics of both *bhāva* and *abhāva* too, because mutually contradictory entities cannot reside in the same substratum.

Category *(padārtha)* is of two kinds — *bhāva* and *abhāva*. The former is of two kinds — category possessing quality *(guṇavat)* and category devoid of quality *(nirguṇa)*. Category possessing quality is substance *(dravya)*. The latter consists of two kinds — having substratum *(āśrita)* and without substratum *(anāśrita)*. The latter is *samvāya*. The former is of two kinds — having *sāmānya* and devoid of *sāmānya*. In this division also the former consists of two kinds — *spanda* and *aspanda*, *spanda* is termed *karma*, and *aspanda* is *guṇa*. The latter consists of *ekāśrita* (having one substratum) and *anekāśrita* (having many substrata). *Ekāśrita* is *viśeṣa* and *anekāśrita* is *sāmānya*.

These are the divisions of category. Similarity must be included in any of these divisions; if it is not included, it cannot be a category. Thus, including *abhāva* categories are seven. Similarity may come under any of these categories, it is not a separate category. Therefore, there is no need to postulate a separate *pramāṇa*, namely, *upamāna* to cognize similarity.

It may be said that similarity is not a distinct category, it comes under *sāmanyā*, and that is the object of *upamāna*. Thus, the knowledge, namely, that *'gavaya* is similar to cow', arises when there is a sense-contact with *gavaya*. This knowledge does not arise merely by sense-contact, because mere visual sense-contact cannot give rise to the knowledge of similarity. This is produced by *upamāna-pramāṇa*. This position is not accepted. Sense-contact gives rise to the knowledge, viz. that *'gavaya* is similar to cow' only when there arises the memory of cow which was perceived before. In the absence of this memory such knowledge is not produced. Therefore, this knowledge which is produced by sense-contact with the help of memory is perceptual in character. Just as *pratyabhijñā* in the form 'this is that Devadatta' arises after having sense-contact with Devadatta in a different place accompanied by the memory of knowing him in certain other place, so also is this knowledge *'gavaya* is similar to cow'. Therefore, knowledge produced by *upamāna* is perceptual knowledge.

It may be said that when sense-organ comes in contact with *gavaya*, the knowledge in the form 'this is similar to cow' arises first—this is perception. After that arises the knowledge 'my cow is like this'—this is also perception. This knowledge is not memory because, even if cow was perceived before, similarity between cow and *gavaya* was not perceived before; hence, memory in the form 'cow is like this' cannot take place. Nor is it that just as the *pratyabhijñā*, viz. 'this is that Devadatta' that follows the previous *pratyabhijñā* 'that is this Devadatta', is perception, so also the knowledge 'cow is like this' that follows the knowledge 'this is like cow' is perception. In the former case *viśeṣya* Devadatta being same in both *pratyabhijñā* and there being sense-contact with Devadatta both knowledge, viz. 'this is that Devadatta' and 'that is this Devadatta', may be perceptual. In the latter case, cow and *gavaya* are different. And, while there is sense-contact with *gavaya*, there is absence of such contact with cow. Hence, the knowledge 'cow is like this' cannot be perceptual. This knowledge is based on the knowledge of similarity between *gavaya* and cow which was perceived before and remembered at the time of the perception of *gavaya*. Hence, this is produced by a spearate means of knowledge, namely, *upamāna*.

साधर्म्यमिव वैधर्म्यं मानमेवं प्रसज्यते।
अर्थापत्तिरसौ व्यक्तमिति चेत्प्रकृतं न किम्? ।।६।।

Like similarity, dissimilarity too should be known by a separate means of knowledge; if presumption is admitted to be the means of knowing dissimilarity, then why not similarity too be known by the same presumption?

9. If *upamāna* is accepted as a separate *pramāṇa* to give rise to the knowledge of similarity, then one more *pramāṇa* must be admitted to generate the knowledge of dissimilarity. After knowing that 'camel is dissimilar to cow' knowledge may arise in the form 'cow is dissimilar to camel'. This is not a perceptual

knowledge, because there is no sense-contact with 'cow'. Since this knowledge is based on the knowledge of dissimilarity between camel and cow, one distinct *prāmaṇa* is required to cognize this dissimilarity.

The opponent may argue that there is no need to admit a separate *pramāṇa* to cognize dissimilarity, knowledge of dissimilarity arises from the process, called *arthāpatti* (presumption). *Arthāpatti* is the assumption of something to account for what apparently clashes with what is seen, heard, or proved. The knowledge, viz. 'camel is dissimilar to cow', necessarily implies 'cow is dissimilar to camel', because one cannot perceive in camel dissimilarity to cow unless in cow there is dissimilarity to camel. But the same logic holds good in the case of similarity also. The knowledge 'cow is similar to *gavaya*' may be implied by the knowledge '*gavaya* is similar to cow', since one cannot perceive in *gavaya* similarity to cow, unless in cow there is similarity to *gavaya*. Therefore, *upamāna* too may be included in *arthāpatti*. And in the absence of *upamāna* as a distinct *pramāṇa*, the question of its being a *bādhaka* to God's existence does not arise at all.

सम्बन्धस्य परिच्छेदः संज्ञायाः संज्ञिना सह।
प्रत्यक्षादेरसाध्यत्वादुपमानफलं विदुः।।१०।।

The knowledge of the relation of a name with the thing named is the result of comparison, as it cannot be produced by perception, etc.

10. The Naiyāyika subscribes to the view that *upamāna* is a distinct means of knowledge, as the knowledge produced by *upamāna* cannot be produced by any other means of knowledge. The knowledge of the connection of a name with the thing named is the result of *upamāna*; and this cannot arise from perception, etc. Thus, the knowledge or ascertainment of the connection, i.e. power of meaning , of the name as *gavaya*, with the thing named, i.e. the animal distinguished by the species *gavaya*, can be attained only through *upamāna*. Hearing the statement that

'*gavaya* is like a cow' a villager may come across a cow-like animal in the forest. Then he recalls the meaning of the statement '*gavaya* is like a cow' and concludes that this is what is meant by the word *gavaya*. This knowledge does not arise merely from the statement, because in that case knowledge may arise even when *gavaya* is not perceived. Nor does the knowledge arise from the mere perception of *gavaya*. If it were so, a person who has not heard the statement that '*gavaya* is like a cow' would also come to know what is meant by the word *gavaya* when he perceives a *gavaya*. Nor can it be said that this knowledge results from the statement '*gavaya* is like a cow' together with the perception of *gavaya*. The time of hearing the statement and the time of perceiving *gavaya* being different, both cannot take place together. Again, the meaning of the statement may be remembered even when the statement is forgotten. It is not proper to say that the knowledge 'this is what is meant by the word *gavaya*' arises from the perception of *gavaya* aided by the remembrance of the meaning of the statement. For, the knowledge does not arise even when *gavaya* is perceived unless one perceives in *gavaya* similarity with a cow. Therefore, *upamāna* must be admitted in order to give rise to the knowledge of similarity on which is based the knowledge, viz. 'this is what is meant by the word *gavaya*'.

The Vaiśeṣika does not admit *upamāna* a separate *pramāṇa*; with him the knowledge 'this is what is meant by the word *gavaya*' is a *pratyabhijñā* perception. The meaning of the word *gavaya* is known from the statement that 'the word *gavaya* signifies a cow-like animal'. When one perceives a *gavaya* and attains the knowledge in the form 'this is what is meant by the word *gavaya*', then there results what is known as *pratyabhijñā*.

The above knowledge can be attained through inference also. The *vyāpti* is as follows. The word *gavaya* stands for that which is like a cow, because the learned people call it so. When the learned people use some word in some specific meaning and there is absence of other secondary meaning, that word signifies that meaning, just as the word 'cow' stands for cow. Similarly, as the word *gavaya* is used to refer to a cow-like animal, the word stands for a cow-like animal. Therefore, as the knowledge of *upamāna* can be attained either in the form of *pratyabhijñā* or inference, there is no need of postulating *upamāna* to give rise to that knowledge.

सादृश्यस्यानिमित्तत्वान्निमित्तस्याप्रतीतितः।
समयो दुर्ग्रहः पूर्वं शब्देनानुमयाऽपि वा।।११।।

Likeness is not being the ground (for the use of the term gavaya*), the ground being not known, the relation between the name and the thing named cannot be known earlier by the statement or by inference.*

11. This position is not accepted. That the word *gavaya* stands for a cow-like animal cannot be known from the statement. The ground for the use of the term *gavaya* may be determined first. It may be either likeness to cow *(gosādrśya)* or *gavayatva*. It is not the first, because one who does not know a cow, also uses the word *gavaya*. If likeness to cow were the ground for the use of the term *gavaya*, then he cannot use the word *gavaya* as he lacks the knowledge that *gavaya* is like a cow.

The second alternative also is untenable, because a villager who has not seen a *gavaya* cannot learn immediately after hearing the statement that *gavayatva* is the ground for the use of the term *gavaya*. Nor can the ground be both likeness to a cow as well as *gavayatva*, for the presumption of two grounds for the use of one term violates the law of parsimony. Therefore, as the relation between the name *gavaya* and the thing named cannot be known from the statement, it is not proper to maintain that the knowledge 'this is what is meant by the word *gavaya*' that arises after perceiving *gavaya* is *pratyabhijñā* perception.

श्रुतान्वयादनाकाड्.क्षं न वाक्यं ह्यन्यदिच्छति।
पदार्थान्वयवैधुर्यात्तदाक्षिप्तेन सड.गतिः।।१२।।

The statement that is heard, already having connection does not have expectancy for any other sentence (for its comprehension); when the relation of the meanings of the words is incomplete, then only its connection with an implied meaning is needed.

12.i. It may be rejoined that at the time of the study of the Veda the meaning is not understood; afterwards when grammar, etc. are studied, the meaning of the Veda which has already been studied is understood. Similarly, a villager does not understand what the statement that '*gavaya* is like a cow' means when he hears it first, as he lacks the knowledge of *gavayatva* which is the ground for the use of the term *gavaya*. Afterwards when he perceives a *gavaya* in the forest, he remembers the statement heard before and attains the knowledge that *gavaya* stands for the word *gavaya*. This knowledge is generated from the statement heard before and remembered at the time of perceiving *gavaya*. In the absence of the perception of *gavaya* one cannot understand the meaning of the statement '*gavaya* is like a cow' owing to the lack of definite knowledge as to whether 'likeness to cow' is the ground for the use of the term *gavaya* or is the *upalakṣaṇa* (implied meaning) for *gavaya*. After one perceives a *gavaya*, following the logic of parsimony one comes to the conclusion that 'likeness to cow' is a *upalakṣaṇa* for *gavaya* and thus understands the meaning of the statement '*gavaya* is like a cow'. This view is untenable. The fact that the statement '*gavaya* is like a cow' does not give rise to knowledge when it is heard first is not accepted, because it cannot be proved. When the statement is heard, its meaning is also comprehended. Hence, there is no scope of its being understood with the help of an implied meaning. The Veda is not understood only when it is studied by an ignorant person. But the person who hears the statement '*gavaya* is like a cow' is not ignorant. Hence, there is no reason why he should not understand the meaning.

Therefore, it may be concluded that the knowledge in general that the word *gavaya* stands for an animal that is like a cow arises earlier from hearing the statement. Only the desire to know specifically which cow-like animal is signified by the word *gavaya* is satisfied by such knowledge as '*gavaya* is signified by the word *gavaya*'. This knowledge is generated by *upamāna*, it does not result from the statement. Again, the knowledge that *gavaya* stands for the word *gavaya* cannot be attained by inference. Inference may give rise to the knowledge that the word *gavaya* signifies a cow-like animal, but the knowledge that *gavaya* is signified by the word *gavaya* cannot be produced by inference. Therefore, *upamāna* must be admitted as a distinct *pramāṇa* to give rise to this knowledge.

The knowledge that *gavaya* is signified by the word *gavaya* may be attained by the following inference. 'The word *gavaya* stands for *gavaya*, because in the absence of any secondary meaning of the word the learned people use the word *gavaya* for the object *gavaya*, just as the word cow stands for the object cow.'

This inference is fallacious. The *hetu* in this inference is 'the use of the word *gavaya* for the object *gavaya* in the absence of any secondary meaning of the word'. The *hetu* consists of two parts — (a) use of the word *gavaya* for the object *gavaya* and (b) in the absence of any secondary meaning of the word, both are unestablished. Since *gavaya* is not perceived before, the fact that the word *gavaya* is used for the object *gavaya* cannot be known.

Again, as it is not known prior to this inference that the word *gavaya* stands for the object *gavaya*, it cannot be ascertained whether the word has any secondary meaning. It is not proper to maintain that the knowledge that the word *gavaya* stands for the object *gavaya* arises from the statement, viz. that *gavaya* is like a cow. For, *gavaya* being not known before and likeness to a cow being not accepted as the ground for the use of the term *gavaya*, the statement cannot give rise to this knowledge.

It may be argued that the statement gives rise to the following *vyāpti* relation. 'That which is like a cow is *gavaya*.' From this *Vyāpti* is inferred 'this is *gavaya*, because this is like a cow'. The knowledge that the word *gavaya* stands for the object *gavaya* arises from the *hetu* 'likeness to cow' — hence, this knowledge is the result of inference and not that of *upamāna*.

This position stands refuted. If at all any *vyāpti* is derived from the statement, it must be in the following form — 'that which is *gavaya*, is like a cow'. The statement '*gavaya* is like a cow' is offered as an answer to the question, namely, what is a *gavaya*. Having the knowledge in general about *gavaya*, the knower desires to have specific knowledge of *gavaya*. In order to satisfy this desire it is stated that *gavaya* is like a cow. From this there follows the *vyāpti* 'that which is *gavaya* is like a cow'. But this *vyāpti* gives rise to the contrary inference, viz. 'this is like a cow, because this is *gavaya*', and not 'this is *gavaya*, because this is like a cow'.

It may be said that the question 'what is *gavaya*' is put in order to know the *lakṣaṇa* of *gavaya*. And the answer '*gavaya* is like a cow' purports to establish that likeness to cow is the *lakṣaṇa* of *gavaya*. *Lakṣaṇa* takes the role of *hetu*, e.g. smellness is the *lakṣaṇa* of the earth, and earth is inferred from smellness. Similarly, knowing from the statement that likeness to cow is the *lakṣaṇa* of *gavaya*, one may infer the following fact — 'this is *gavaya*, because this is like a cow'.

The inference of the *sādhya* from *lakṣaṇa* is always of the *vyatireka* form, because the purpose of *lakṣaṇa* is to distinguish a thing from the things other than it and to specify its use, which is possible only by *vyatireka* inference. The inference mentioned above may have the following *vyatireka* form. 'This is *gavaya*, because this is like a cow; that which is not a *gavaya* is not like a cow, e.g. an elephant.' The example in this inference is elephant. But elephant cannot be an example here, because the fact that elephant does not stand for the word *gavaya* cannot be proved. Some people may use the word *gavaya* for an elephant, as sometimes a single word is rendered with various meanings. Therefore, the example being fallacious, the *vyatireka* inference stands rejected. Consequently, *gavaya* cannot be inferred from likeness to cow.

One may argue that the question 'what is a *gavaya*' is asked in order to know the *liṅga* (mark) that gives rise to the knowledge of what is meant by the word *gavaya*. In answer to this question 'likeness to cow' is offered as the *liṅga*; since 'likeness to cow' is the *liṅga*, *anvaya-vyāpti* is possible between this *liṅga* and *gavaya*. It is only in respect of *lakṣaṇa* that the problem of *vyatireka-vyapti* arises. This position is also untenable. One can desire to know *liṅga* only when one has knowledge of *liṅga* in general. The opponent has no knowledge of what is meant by the word *gavaya,* and he is also not aware of the fact that *gavaya* may be known through inference. This being the situation, it is not reasonable to draw the conclusion that the knower enquires about the *liṅga*.

The question as to what is a *gavaya* purports to know what is the ground for the use of the term *gavaya*. The speaker being unable to point out the ground directly states that *gavaya* is like a cow. Likeness to cow is *jñāpaka* that helps the knower to find out what is a *gavaya*. When the knower comes across a cow-like animal and knows what is a *gavaya,* he decides about the ground for the use of the term *gavaya* through *tarka* which is as follows. If both likenss to cow and *gavayatva* were the grounds, then that violate the law of parsimony. Hence, the ground for the use of the term *gavaya* must be *gavayatva*. The knowledge of similarity between a cow and *gavaya,* mentioned by the statement, must be attained through *upamāna-pramāṇa*. And the knowledge that *gavaya* stands for the word *gavaya* also is generated by the same *pramāṇa*. Therefore, it is proved that *upamāna* is a distinct *pramāṇa*.

It is not proper to argue that like similarity dissimilarity too must be cognized by a distinct *pramāṇa*. *Upamāna* is competent to give rise to the knowledge of both similarity as well as dissimilarity. *Upamāna-pramāṇa* is that which gives rise to the knowledge of the meaning of a word not known before with the help of the meaning of certain other words which are used along with the word in order to clarify its meaning. For example, the meaning of the word *gavaya* is understood with the help of the meaning of such words as 'similarity with cow', 'dissimilarity with horse', uttered along with the word *gavaya* — that is, the knowledge of *gavaya* arises when it is found to have similarity with cow or dissimilarity with horse.

The Naiyāyika gives *upamāna* the separate status of *pramāṇa* but concludes that since *upamāna* ascertains the direct significance of a word and has nothing to do with establishing the objects, existence or non-existence, it cannot preclude the existence of God.

12.ii. We now proceed to examine *śabda-pramāṇa* (verbal testimony). According to the Vaiśeṣika *śabda* cannot disprove the existence of God, as it is not a separate *pramāna.* It comes under inference. *Śabda* gives rise to *śabdabodha,* i.e. the knowledge of logical connection between word-meanings. When a sentence is heard, words that constitute a sentence give rise to the recollection of their meanings. The mutual relation between word-meanings consists of sentence-meaning.

The Vaiśeṣika maintains that the knowledge of mutual connection between word-meanings, i.e. sentence-meaning which is the result of *śabda,* can be accomplished by inference. Therefore, *śabda* may be included within inference. The inference that produces the knowledge of sentence-meaning may take two forms — in one 'word-meanings' being the *pakṣa,* and in the other 'word'.

(a) These word-meanings are mutually connected on the ground that they are brought to recollection by the aid of words which possess expectancy, compatibility and juxtaposition, like word-meanings 'bring the cow'. Here the *pakṣa* is 'word-meanings'. In the instance 'bring the cow' word-meanings are recollected by the aid of words, and they are also mutually connected. Similarly, in the instance 'Devadatta goes to the village' word-meanings are established to be mutually connected.

In this connection it may be noted that expectancy (*ākāṅkṣā*) means a word's incapacity to convey a complete meaning without some other word to complete the construction. Compatibility *(yogyatā)* consists in this that a word should not have an incompatible meaning with the meaning of other words in the sentence. Juxtaposition *(āsatti)* consists in the enunciation of the words without a long pause between them.

The other form of inference is as follows:

(b) These words must have been preceded by the speaker's knowledge of the connection between the several meanings which these words respectively call to recollection on the ground that they are words possessing expectancy, etc. Here the *pakṣa* is 'words'. When one says 'Devadatta is going to the village', this saying is preceded by one's knowledge of the connection between the word-meanings. This inference establishes knowledge of connection between word-meanings, and connection is also established as adjectival *(viśeṣaṇa)* to knowledge. Therefore, as the knowledge of the connection between word-meanings can be attained by inference, the Vaiśeṣika asserts that *śabda* is not different from inference.

अनैकान्तः परिच्छेदे सम्भवे च न निश्चयः।
आकाङ्क्षा सत्तया हेतुर्योग्यासत्तिरबन्धना।।१३।।

If the probandum in the inference were 'the relation between word-meanings', then the inference involves the fallacy of inconstancy; if the prbandum were 'the possibility of relation between word-meanings', then the relation between word-meanings cannot be ascertained; expectancy by itself being present becomes the hetu *of the knowledge of sentence-meaning; mere juxtaposition accompanied by compatibility cannot establish the probandum.*

13.i. In the former inference where the *pakṣa* is 'word-meanings', the *sādhya* may be either 'relation between word-meanings' or 'possibility of the connection between word-meanings', i.e. word-meanings possessing capacity to be connected *(sambhāvitasaṁsarga)*.

If the first, there may arise the knowledge of connection between word-meanings uttered by unreliable person, e.g. 'there are fruits on the bank of the river', though in fact there are no fruits on the bank. Since there is absence of *sādhya,* i.e. connection between word-meanings, in the words uttered by unreliable person, the *hetu,* viz. recollection of meanings by the aid of words that possess expectancy, etc., is fallacious.

To avoid this difficulty, the *hetu* may be qualified by *āptatva,* i.e. the *hetu* is recollection of meanings by the aid of words possessing expectancy, etc. that are uttered by a reliable person. The question arises as to what is meant by the word 'reliable'. It may mean absence of deceiving or possession of valid knowledge. It is not the first, because one who has no desire to deceive may erroneously utter a sentence which lacks connection between word-meanings.

If the second alternative, it may be asked whether a reliable person is one who has valid knowledge of all things at all times, or one who possesses valid knowledge of some of the things. The first alternative is untenable, for, no person other than God can possess valid knowledge of all things at all times.

The second alternative is rejected, because in that case an unreliable person too could be trusted as reliable, because he may possess valid knowledge of some of the things.

It may be said that a reliable person is one who possesses valid knowledge of the connection between word-meanings constituting the sentence under consideration. Then the inference will be as follows. 'The word-meanings are mutually connected, because meanings are recollected by the aid of words uttered by a reliable person who possesses valid knowledge of the connection between word-meanings.'

Here, the *sādhya* is the mutual connection between word-meanings. This is unknown prior to inference. Hence, the fact that reliable person possesses the valid knowledge of connection between word-meanings remains unestablished. Reliability being unestablished, the *hetu* that is qualified by reliability also is not established. And such *hetu* cannot give rise to inferential knowledge.

Therefore, the connection between word-meanings cannot be the *sādhya* in the inference. If 'possibility of connection between word-meanings', i.e. word-meanings possessing capacity to be connected, were the *sādhya,* then even if possibility is established, connection may not be established. Ascertainment of the connection between word-meanings is the result of *śabda-pramāṇa*, this cannot be established by inference; therefore, *śabda* cannot be included under inference.

Further, 'possibility of connection between word-meanings' cannot be *sādhya*, because it is already included in the alleged *hetu*, viz. on the ground that they are brought to recollection by the aid of words which possess compatibility, etc. — 'compatibility' mentioned there only means that words indirectly possess a character which necessitates a logical connection between their meanings.

If the epithet 'compatibility' is removed from the *hetu*, then the *hetu* stands thus: 'on the ground that they are brought to recollection by the aid of words which possess expectancy and juxtaposition'. This *hetu* exists in such instance as 'he sprinkles with fire' — but the *sādhya*, viz. 'possibility of connection between word-meanings' is absent here, as fire lacks compatibility to be sprinkled with. Hence, the *hetu* is fallacious.

The other form of inference is — 'the words must have been preceded by the speaker's knowledge of the connection between the several meanings which these words respectively call to recollection on the ground that they are words possessing expectancy, etc.' The epithet 'expectancy' added to the *hetu* must be known first. Expectancy consists in the mind's inquiry after certain additional meanings which are supplied by words suggested by the construction. Expectancy by its very presence, whether it is definitely known or not, is therefore a cause of verbal knowledge. But if it were to be included in the *hetu* of an inference, it must be actually known in order to be so included.

Expectancy does not mean the relation of *viśeṣaṇa* and *viśeṣya* between two word-meanings. If it were the relation of *viśeṣaṇa* and *viśesya*, then being a relation between word-meanings it constitutes the sentence-meaning. Hence, it must be inferred as a *sādhya* and cannot be included in the *hetu*. It is also not that the possible relation of *viśeṣaṇa-viśeṣya* between word-meanings is expectancy, the epithet 'compatibility' included in the *hetu* already suggests this possibility.

The inseparable relation between two word-meanings is also not what is meant by expectancy, because in the instance 'blue lotus' there arises knowledge of the sentence-meaning even in the absence of any inseparable relation between 'blue' and 'lotus' — as there exists lotus without being blue and there is blue flower without being lotus. Nor can it be said that inseparable relation between quality and substance, 'blue' being quality and 'lotus' being substance, is expectancy, as quality does not exist without substance and substance does not exist without having quality. If it were so, two sentences uttered in succession, namely, 'how clean the water of the river is', and 'on the bank of the river is grazing the bull', will turn into one sentence. For, there being inseparable relation between 'river' and 'bank', one has expectancy for the other. In fact, 'river' has relation with 'water' and not with 'bank' — and the latter sentence is not the continuation of the former one, both are different. Therefore, expectancy does not stand for inseparable relation.

By expectancy is not meant the desire to know about the meaning also. For, in the case 'there is a cloth' there being desire to know what kind of cloth it is and in the absence of any adjective, viz. red colour, etc. to satisfy that desire, sentence-meaning will not be comprehended.

Moreover, desire to know being itself present becomes the *hetu* of the knowledge of the sentence-meaning, and not being known. But in the inference *liṅga* becomes the *hetu* of the knowledge by being known — that is, in the inference the knowledge of *hetu* gives rise to knowledge, whereas in the case of desire the very presence of desire gives rise to knowledge. Therefore, desire to know, the presence of which produces knowledge, cannot be included as adjectival to the *hetu* the knowledge of which produces knowledge. And the *hetu* which includes only juxtaposition accompanied by compatibility cannot establish the *sādhya,* viz. the mutual connection between word-meanings.

It is not proper to say that it is the knowledge of the desire to know that produces the knowledge of sentence-meaning, and not mere desire to know. For, comprehension of sentence-meaning arises provided there is juxtaposition, compatibility and desire to know even in the absence of the knowledge that one desires to know. For example, seeing a thing that possesses height, from a distance one desires to know whether it is a man or a pillar, and going nearby one knows that that is a pillar. This knowledge arises even in the absence of the knowledge that one desires to know it. Similarly, sentence-meaning is comprehended provided there is desire to know even in the absence of the knowledge of that desire. Therefore, desire becomes the *hetu* of knowledge by being present and not by being known.

Admitting for the sake of argument, if the knowledge of the desire to know were presumed to be the *hetu* of knowledge, then also the inference mentioned earlier will not be free from fallacy. Thus, someone may utter 'this is the son of the king — get the man out'; and, an inattentive person may not hear the first part, viz. 'this is the son' *(ayameti putro),* and may hear 'get the man of the king out' *(rājñaḥ puruṣaḥ apasāryatām).* Here, the *hetu,* viz. words having expectancy, compatibility, etc., exists, but the *sādhya,* viz. mutual connection between word-meanings, is absent, because the word 'king' is uttered in order to show its relation with 'son' and not with 'man'. Hence, the *hetu* being present in the absence of the *sādhya,* the inference suffers from the fallacy, viz. *vyabhicāra.*

It may be argued that the *hetu* requires inclusion of all the words uttered by the speaker, possessing expectancy, etc. In the instance mentioned above the knowledge of the connection between word-meanings does not arise, because the listener misses some of the words uttered by the speaker. Thus, the *hetu,* viz. inclusion of all the words uttered by the speaker, being absent, the *sādhya,* viz. mutual connection between word-meanings, is also absent; hence, the inference cannot be charged with the fallacy of *vyabhicāra.*

This is not tenable. For, in that case comprehension of the sentence-meaning that arises after hearing some of the words will not take place. If it is said that such knowledge is erroneous, then a *hetu* must be pointed out to give rise to this erroneous knowledge. Defect in the sense-organ is not the *hetu*, because defective sense-organ can be the *hetu* of that erroneous knowledge which is perceptual in character; erroneous knowledge of the sentence-meaning is not perceptual, and hence, defective sense-organ cannot be the *hetu*.

Therefore, it may be concluded that erroneous knowledge of the sentence-meaning arises due to the defect of hearing of some of the words and not hearing some others. This leads to the further conclusion that erroneous sentence—erroneous due to inclusion of some of the words, gives rise to the erroneous knowledge of the sentence-meaning without taking into consideration the knowlege of *vyāpti*, etc. Similarly, it must be admitted that a valid sentence produces the valid knowledge of the sentence-meaning without the help of the knowledge of *vyāpti*, etc. Thus, the view that sentence-meaning depends on inference for its comprehension stands refuted.

It may be rejoined that mutual connection between word-meanings is not comprehended in the case where a part of the sentence is heard, what is comprehended is the absence of knowledge of the absence of connection *(asaṁsargasya ajñāna)*. In the instance 'this is the son of the king, get the man out' — the word 'king' is related with 'son' and not with 'man'. But the absence of relation between 'king' and 'man' is not known to the listener owing to his not hearing the first part of the sentence, viz. 'this is the son'. Thus arises the absence of knowledge of the absence of connection in the case where the full sentence is not heard.

Even then the inference is not free from the charge of *vyabhicāra*. In the case of a sentence uttered by an unreliable person, mutual connection between word-meanings being absent, there arises the contingency of the *sādhya*, viz. precedence by the knowledge of connection being absent, while the *hetu*, viz. words possessing expectancy, etc. is present. The presence of *hetu* in the absence of *sādhya* involves *vyabhicāra*.

To obviate this difficulty, if the *sādhya* is changed as 'precedence by the absence of knowledge of the absence of connection' instead of 'precedence by the knowledge of connection', then the supposed inference will prove absence of knowledge of the absence of connection and not the mutual connection between word-meanings, which consists of sentence-meaning.

Therefore, it stands out that inference is not competent to give rise to the knowledge of the mutual connection between word-meanings. *Śabda* must be admitted as a separate *pramāṇa* to generate this knowledge.

13.ii. So far the inferential argument that has been advanced by the Vaiśeṣika to produce the verbal knowledge has been refuted taking expectancy in the sense of desire to know. In fact, expectancy does not stand for desire to know; sentence-meaning is comprehended, when it is heard even in the absence of any desire to comprehend. Expectancy means compatibility (*yogyatā*) towards the desire to know. Expectancy has been defined as the prior non-existence of the knowledge of connection produced by the inseparable relation existing between recollected word-meanings or meanings implied by the recollected word-meanings. For example, in the expression 'the earth possesses smell' — there is inseparable relation between 'smell' and 'earth' recollected by the two words 'smell' and 'earth', and this knowledge is preceded by the prior non-existence of the knowledge of connection between 'smell' and 'earth' — this is what is meant by expectancy between 'smell' and 'earth'.

In the expression 'blue lotus' also there is expectancy between two words 'blue' and 'lotus', as there is inseparable relation between quality and substance which are implied by the two word-meanings 'blue' and 'lotus', 'blue' implying quality and 'lotus' substance.

Expectancy in this sense becomes the cause of knowledge by being present *(svarūpa-sati)* and not by being known. Hence, it cannot be included in the *hetu* of an inference which becomes the cause of knowledge by being known. Thus, it is proved that the second form of inference where the *pakṣa* is 'words' is unestablished.

> निर्णीतशक्तेर्वाक्याद्धि प्रागेवार्थस्य निर्णये।
> व्याप्तिस्मृतिविलम्बेन लिङ्गस्यैवानुवादिता।।१४।।
>
> *As in the Vedas words have capacity to produce sentence-meaning, so also secular words produce sentence-meaning prior to the application of inference;* liṅga *in the inference depends on the knowledge of* vyāpti *which arises after the sentence-meaning has already been produced by the words; hence, it is a repetition.*

14. The Prābhākara who advocates the theory of *anvitābhidhānavāda* maintains that *śabda* produces the verbal knowledge of the Vedas, and secular verbal knowledge is generated by the inference. Words possess two kinds of *śakti* (power), *smāraka* and *anubhāvaka*. *Smāraka-śakti* in words gives rise to the recollection of word-meanings first, then *anubhāvaka-śakti* produces the knowledge of word-meaning in relation with other word-meanings, provided words possess expectancy, compatibility, etc. The theory of *anvitābhidhāna* consists in the knowledge of individual word-meanings in relation with other word-meanings, generated by the *śakti* inherent in words.

The Vedas, according to the Mīmāṁsā, is not composed by any person, so the Vedic sentences cannot be preceded by the speaker's knowledge of them. This being the case, the inference, viz. 'these words must be preceded by the speaker's knowledge of them', cannot be applied in the case of the Vedas. Therefore, it must be admitted that verbal knowledge of the Vedas is produced by the Vedic words that are heard. Secular words can be preceded by the speaker's knowledge of them, because they are uttered by human beings. Hence, verbal knowledge in secular matters is attained through inference — this is the view advocated by the Prābhākara Mīmāṁsā. This view does not stand to reason. Whatever causal aggregate, viz. expectancy, compatibility, etc., is present in the Vedic sentence, the same is present in secular sentence also. Therefore, if verbal knowledge of the Vedas arises without the help of inference, the same must be the case in the secular verbal knowledge also.

First there arises the sentence-meaning which is produced by the words themselves. Later on inferential knowledge of the sentence-meaning takes place. Inferential knowledge arises through the knowledge of *liṅga*. The knowlege of *liṅga* in the case of secular verbal knowledge is the knowledge of the meanings of words possessing expectancy, etc. which are the causal aggregate for the rise of verbal knowledge. After the rise of the knowledge of *liṅga*, the inferential knowledge has to depend on the knowledge of *vyāpti* between *hetu* and *sādhya*. Owing to the dependence on the knowledge of *vyāpti* the inferential knowledge of the sentence-meaning is generated later than the generation of sentence-meaning by the words themselves. Since the knowledge of the relation of word-meanings arises earlier than the inferential knowledge, the latter is a repetition as it cognizes what has already been cognized. Therefore, as in the Vedas, so in the secular sentence too *śabda* produces verbal knowledge and not inference.

It may be said that on the secular level the reliability of the speaker needs to be ascertained, and for that purpose inference is required. In the Vedas the question of reliability of the speaker does not arise; hence, inference is not required to give rise to the verbal knowledge of the Vedas. It is not so, if verbal knowledge of the Vedas arises without depending on the ascertainment of the reliability, then secular verbal knowledge too may arise in the same manner. Vedic sentences are in no way different from secular sentences. Therefore, if the reliability of the speaker needs to be examined in the latter case, it should be done in the former also. If the verbal knowledge of the Vedas does not depend on the reliability of the speaker for its validity, secular verbal knowledge too need not depend on it.

It is seen that the knowledge of the connection between word-meanings does not arise in connection with a sentence uttered by an unreliable person. Doubt arises even when a sentence is uttered by a reliable person and verbal knowledge does not arise owing to that doubt. For this reason it is necessary to ascertain the reliability of speaker.

The question arises as to what is it that is doubted here. Expectancy, compatibility, etc. are not doubted here, because, as in the Vedas, they are not subject to doubt in the secular verbal usage also. Nor can the doubt be in regard to the validity of the knowledge which is to be produced from the sentence. For, there can be doubt about the validity of knowledge only when knowledge is produced from the sentence and not before that. Since doubt does not arise before the production of the verbal knowledge, objection that verbal knowledge does not arise owing to the doubt makes no sense.

Doubt regarding the reliability of the speaker also cannot prevent the generation of verbal knowledge. Were the ascertainment of the speaker's reliability the cause of verbal knowledge, it could be said that verbal knowledge would not arise if there is doubt about the reliability of the speaker. In fact, ascertainment of the speaker's reliability is not the cause of verbal knowledge, verbal knowlege of the Vedas is produced even in the absence of the speaker's reliability, as the Vedas are *apauruṣeya,* i.e. not composed by any personal being. Not only the Vedic, secular verbal knowledge too may arise irrespective of the question whether the speaker is reliable or not. Thus, one attains the verbal knowledge of the works composed by Manu while one is under the impression that like the Vedas these compositions are also *apauruṣeya,* because they are read out in the same manner as the Vedas are read out. Here one attains the verbal knowledge without ascertaining whether the speaker is reliable or not.

Therefore, it may be concluded that like the Vedas secular verbal knowledge also arises without depending on inference.

It may be argued that the cause of verbal knowledge does not consist in ascertaining the reliability of the speaker. The cause is, on the other hand, absence of doubt of unreliability, i.e. absence of the doubt that certain sentence is uttered by an unreliable person. Such doubt does not arise in respect of the Vedas, when the Vedas are established as *apauruṣeya*. That is, when the Vedas are known as *apauruṣeya,* the question does not arise whether they are composed by reliable person or not. In secular verbal usage this doubt is removed, when it is ascertained that the sentence is uttered by a reliable person.

This position is untenable. It may be asked whether the verbal knowledge of the Vedas arises in respect of a person who does not possess the knowledge that the Vedas are *apauruṣeya.*

According to the opponent, doubt regarding unreliability of the speaker is not removed unless there arises the knowlege that the Vedas are *apauruṣeya.* Therefore, with him verbal knowledge of the Vedas is not produced in the absence of the knowledge that the Vedas are *apauruṣeya.* This conclusion is unfounded, because there is no reason to substantiate this conclusion.

व्यस्तपुंदूषणाशङ्कै स्मारितत्वात्पदैरमी।
अन्विता इति निर्णीते वेदस्यापि न तत् कुतः।।१५।।

These Vedic sentence-meanings are mutually connected, because they are brought to recollection by words which are free from any imputation of defects incidental in the case of a human being; is not the verbal knowledge of the Vedas too a repetition (first the knowledge of the Vedic sentence-meaning being produced by inference and then by śabda*)?*

15.i. It may be said that the verbal knowledge of the Vedas arises even if the Vedas are not known as *apauruṣeya,* but there remains doubt about the validity of that knowledge. This position is not accepted, because doubt can be removed by successful activity, etc. that arise after the production of the verbal knowledge. If this is not accepted, then it must be admitted that like secular verbal knowledge, verbal knowledge of the Vedas also is produced by inference. The inference runs thus — 'these Vedic sentence-meanings are mutually connected, because they are brought to recollection by words which are free from any imputation of defects incidental in the case of a human being'. Is not the verbal knowledge of the Vedas too a repetition (first the knowledge of the Vedic sentence-meaning being produced by inference and then by *śabda)?* As the knowledge of word-meaning being connected with several other word-meanings can be attained by the inference mentioned above, *śakti* may be admitted as inherent in the connection of words, and thus, the theory of *anvitābhidhāna* which admits *śakti* in words does not hold good.

We shall now discuss the question: What does the theory of *anvitābhidhāna* mean? If by *anvitābhidhāna* is meant the generation of the verbal knowledge of word-meaning being connected with other word-meanings, the Naiyāyika has no objection to it, because he too admits it. If it is said that *anvitābhidhāna* means that the words convey their isolated meanings with a view to signifying the relation among those meanings, then also the Naiyāyika does not object to it, as he admits it.

The Naiyāyika would only object to the contention that *anvitābhidhāna* means the generation of the knowledge of sentence-meaning by *śakti* inherent in words. *Anvitābhidhāna* does not mean that the power that rests in a word gives rise to the knowledge of the sentence-meaning. Although the word-meanings are known earlier, the sentence-meaning which is only the relation among the word-meanings is not known earlier, so there is no possibility of the generation of any knowledge in the form that the words are significative of the sentence-meaning.

Nor can *anvitābhidhāna* mean that the word-*śakti* that inheres in word-meanings gives rise to the knowledge of sentence-meaning. Word-*śakti* that exists in word-meanings ceases to function after producing the knowledge of word-meanings, and so it cannot produce the knowledge of sentence-meaning. The same word-*śakti* does not produce the knowledge of both word-meanings as well as sentence-meaning If it produces so, the knowledge of word-meanings as well as that of sentence-meaning will be produced simultaneously, and the relation of antecedence and subsequence that exists between the knowledge of word-meanings and that of sentence-meaning will be violated.

If both the knowledge of word-meanings and of sentence-meaning arise simultaneously, then just as knowledge of word-meanings arises without depending on expectancy, compatibility, etc. so also the knowledge of sentence-meaning will arise without depending on expectancy, etc. This, however, is not accepted.

It may be argued that word-*śakti* produces the knowledge of word-meanings without depending on anything else, but it produces the knowledge of sentence-meaning through the *vyāpāra* of the knowledge of word-meanings. Thus arises the sequence between the knowledge of word-meaning and that of sentence-meaning.

In reply to this it may be said that were the *śakti* the *karaṇa* of the knowledge of sentence-meaning, it could produce the knowledge of sentence-meaning through the *vyāpāra* of the knowledge of word-meanings. But *śakti* is not *karaṇa*. Words possessing *śakti* are the *karaṇa* of the knowledge of sentence-meaning. Hence, the position mentioned above stands rejected.

It may be rejoined that what the theory of *anvitābhidhāna* advocates is that words possessing *śakti* give rise to the knowledge of word-meanings which favours the production of the knowledge of sentence-meaning. If *anvitābhidhāna* means only this much, the Naiyāyika has nothing against it. For him word-*śakti* produces the knowledge of word-meanings, and the knowledge of sentence-meaning arises on the strength of expectancy, etc.; word-*śakti* has nothing to do with the production of the knowledge of sentence-meaning. If this is what is meant by *anvitābhidhāna-vāda,* the Naiyāyika admits it.

The opponent may define the *anvitābhidhāna-vāda* in the following way. The *śakti* of the word *kāraka* (case) inheres in the *kāraka* which is related with the *kriyā* (verb) recollected by the word *kriyā.* The *śakti* of the word *kriyā* inheres in the *kriyā* which is related with the *kāraka* that is recollected by the word *kāraka.* When the knowledge of *kāraka* arises as being related to *kriyā*, the *kāraka* is the *viśeṣya* and the *kriyā* is *viśeṣaṇa.*

Again, when the knowledge of *kriyā* arises as being related to *kāraka*, the *kriyā* is the *viśeṣya* and the *kāraka* is *viśeṣaṇa*. By *anvitābhidhāna* is meant inherence of word's *śakti* in that which is related *(anvite padānāṁ śaktireva anvitābhidhānam)*. This definition needs further explanation. If by this is meant the relation of a certain meaning implied by a certain word-*śakti* with other meanings, this goes in favour with the Naiyāyika. The Naiyāyika admits that the object-cow implied by the word 'cow' is related to some *kriyā*. The *śakti* of the word 'cow' inheres in 'cow' which is related and not in *kriyā* with which 'cow' is related.

The opponent may say that *anvitābhidhāna* consists in the knowledge of inherence of the word-*śakti* in related word-meaning. Therefore, word-*śakti* must be admitted to inhere in relation *(anvaya)* also.

The Naiyāyika replies that there is no proof to the effect that there exists the significative relation between words on the one hand and the meanings as related to each other on the other hand. When the word 'cow' is heard, there arises the knowledge that the word 'cow' stands for the object-cow. But no knowledge arises as to the fact that the word 'cow' stands for cow which is related with something else. Therefore, it is admitted that after word-meanings are recollected by the aid of words, the knowledge of the mutual relation between word-meanings arises with the help of expectancy. Consequently, there is no proof to support the fact that word-*śakti* inheres in relation *(anvaya)*.

If the knowledge of the mutual connection between word-meanings arises without the help of word-*śakti*, then that knowledge may arise even when meanings are not recollected by the aid of words, but are recollected through some other means. To avoid this, word-*śakti* must be admitted in relation also.

This objection is baseless. There is no hard and fast rule that knowledge of the connection between word-meanings arises only when meanings are recollected by the aid of words. Were this the case, nobody would have understood the significance of a piece of poetry where metaphorical usages are much in vogue.

A poet may come to know the connection between word-meanings that are recollected through reflection and not by the aid of words, and convey that connection through the poetry. Therefore, it is not necessary that one must postulate word-*śakti* in relation also in order to have the knowledge of the mutual connection between word-meanings.

A piece of poetry may be composed either about a word-meaning which has connection with other meanings or about a word-meaning which has no connection with other meanings, but on which is imposed the connection. In the latter case where the connection is imposed on the meanings that are not connected, what is established is not the knowledge of the connection between word-meanings, but is the absence of the knowledge of the absence of connection. In the former case where the composition is about word-meaning mutually connected with other meanings, what is established is the knowledge of connection.

This position is rejected on the ground that there is no reason to prove the fact that sometimes words give rise to the knowledge of connection and sometimes, absence of knowledge of the absence of connection.

The opponent points out that words uttered by a reliable person give rise to the knowledge of connection between word-meanings and that uttered by an unreliable person give rise to the absence of knowledge of the absence of connection. Thus, reliability and unreliability of the speaker determine the knowledge of connection and the absence of knowledge of the absence of connection. But the difficulty is that it is not possible to ascertain the difference between the word 'cow' uttered by a reliable speaker and that uttered by an unreliable speaker. The observation made by the opponent may be analysed in the following way. By this the Prābhākara may mean that the reliable speaker produces some *śakti* in the word uttered by him and that *śakti* gives rise to the knowledge of mutual connection between word-meanings. Or he may mean that words qualified by reliable utterance produces the knowledge of connection between word-meanings. Here reliability of the speaker stands as a *viśeṣaṇa* to words.

If it is the first, the Prābhākara will have to postulate three *śaktis*.

In this context it may be said that the Bhāṭṭa Mīmāṁsaka holds the *anvitābhidhāna-vāda* theory of meaning. According to this theory word-*śakti* produces the knowledge of word-meanings; then there arises the knowledge of the mutual connection between these meanings. This theory postulates two *śaktis*, one inherent in words that produces the knowledge of meanings and the other inherent in meanings that produces the knowledge of their mutual connection, that is the knowledge of sentence-meaning. The Prābhākara says that since according to this theory meanings that produce the sentence-meaning are derived from words, the Bhāṭṭa must postulate one more *śakti* in words that produces sentence-meaning generating *śakti* in word-meanings. Postulation of three *śaktis* goes against the law of parsimony. But the same difficulty arises, as it has already been mentioned, in respect of the theory advocated by Prābhākara also. Thus, words possess a *śakti* that produces the recollection of word-meanings; one more *śakti* is postulated to produce the knowledge of connection between recollected word-meanings. Since word-meanings uttered by a reliable speaker are supposed to produce the knowledge of connection between meanings, a third *śakti* must be postulated in the reliable speaker which produces the connection generating *śakti* in words. Thus, the postulation of three *śaktis* in the Prābhākara theory also violates the law of parsimony.

If the second alternative is accepted, that is, if it is maintained that words qualified by reliable utterance produce the knowledge of connection between word-meanings, then the Bhāṭṭa too may hold that meanings qualified by their being derived from words produce the knowledge of connection between word-meanings. Thus, there is no need of postulating a third *śakti* in words to produce the sentence-meaning generating *śakti* in meanings.

Therefore, it is not proper on the part of the Prābhākara to find fault with the Bhāṭṭa theory of *anvitābhidhāna-vāda*.

15.iii. The Prābhākara may ask the Naiyāyika to point out exactly what is wrong in holding the view that words uttered by a reliable speaker produce the knowledge of connection between word-meanings, and there arises the absence of knowledge of the absence of connection when words are not uttered by a reliable person. The Naiyāyika replies that what he intends to hint at is that if the knowledge of connection between word-meanings arises due to a reliable speaker's uttering the words and not due to word-*śakti*, then the Prābhākara can no longer maintain that words possess *śakti* to produce the knowledge of the connection between meanings. According to the Naiyāyika, as it has already been mentioned, the knowledge of connection between meanings that are recollected by the aid of words is produced with the help of expectancy, etc.

The difference between the Nyāya theory of meaning and the Bhāṭṭa theory of meaning is this — according to the Bhāṭṭa, meanings, the knowledge of which is produced by words, are the *karaṇa* of the knowledge of connection between meanings, whereas according to the Nyāya words are the *karaṇa* of the knowledge of connection and the recollection of meanings is the *vyāpāra*. The Bhāṭṭa view that word-meaning is the *karaṇa* of the knowledge of connection does not hold good. For, the meanings that apply to past and future are absent at the time when the verbal knowledge is produced; and, hence, they cannot be *karaṇa*. Therefore, it is reasonable to hold the Nyāya view that words are the *karaṇa*. Since the verbal knowledge arises after hearing the words, the charge that *karaṇa* is absent at the time of the generation of knowledge cannot be levelled against the Nyāya.

Thus, it is established that *śabda* is not included in inference, it is a distinct *pramāṇa*. *Śabda* cannot preclude the existence of God also The opponent cites the *Bhagavad-gītā* to prove that *śabda* is a *bādhaka* for the existence of God. One couplet in the *Gītā* says that "though actions are carried out by the qualities of *prakṛti*, the soul blinded by egoism thinks 'I am the agent'." That is to say the soul is in reality devoid of agency, the agency that exists in the qualities of *prakrti* is mistakenly imposed on the soul. The Naiyāyika admits God, the Supreme Soul as the agent of the world. Since this couplet proclaims that the soul in reality is devoid of agency, the agency in God who is the Supreme soul is also nullified.

न प्रमाणमनाप्तोक्तिर्नादृष्टे क्वचिदाप्तता।
अदृश्यदृष्टौ सर्वज्ञो न च नित्यागमः क्षमः।।१६।।

If the couplet cited (for establishing absence of agency in God) is the testimony of an unreliable person, it is bereft of authority; if it is the testimony of a reliable person, he must possess the knowledge of such suprasensible fact (conveyed by the couplet), otherwise he will not be admitted as a reliable person and such a person must be the omniscient God; the Vedas cannot be eternal (hence, must have an author, viz. God).

16. In reply to this objection it may be said that if the couplet which the opponent brings forward to establish absence of agency in God is the testimony of an unreliable person, it has no authority. If it is the testimony of a reliable person, then he must possess the knowledge of such transcendental fact conveyed by the couplet, otherwise he cannot give such testimony. And such a person must be the omniscient God. Thus, God is established as an omniscient author of the *Bhagavad-gītā*. Further, since the Vedas are not eternal, they must have an author, and God is established as that author.

न चासौ क्वचिदेकान्तः सत्त्वस्यापि प्रवेदनात्।
निरञ्जनावबोधार्थो न च सन्नपि तत्परः।।१७।।

The scriptural statement referring to the non-existence of God is not the determinant, because the scriptures do state about the existence of God; the verse (cited from the Bhagavad-gītā*) aims at establishing the fact that the soul is devoid of all the qualities (viz.* sattva, rajas, *etc.*

of prakṛti*), it does not purport to establish the non-existence of God.*

17. Moreover, *śabda* establishes the existence of God. The Vedas mention the existence of God in various contexts. The verse cited by the opponent from the *Gītā*, viz. 'though actions are ever done by the qualities of *prakṛti* ..... etc.' does not purport to prove the non-existence of God. It aims at establishing the soul which is devoid of all the qualities, viz. *sattva*, etc. of *prakṛti*. The portions of the Vedas, the *Gītā*, etc. that testify to the existence of God, do not imply any secondary meaning. (This will be discussed in the fifth chapter). Thus, it is established that *śabda* is not a *bādhaka-pramāṇa* for the existence of God.

हेत्वभावे फलाभावात्प्रमाणेऽसति न प्रमा।
तदभावात्प्रवृत्तिर्नो कर्मवादेऽप्ययं विधिः।।१८।।

As in the absence of cause the effect does not arise, so also in the absence of the śabda-pramāṇa *(Vedas) there will not arise valid knowledge (of the Vedic instructions regarding sacrifice, etc.) and in the absence of knowledge there will not arise action (towards sacrifice, etc.); the same is the case with the theory of* karma *too.*

18. Having shown that *śabda* cannot preclude the existence of God, the *siddhāntin* proceeds to examine *arthāpatti* (presumption) and proves that *arthāpatti* also cannot prove the non-existence of god. *Arthāpatti* consists in presuming a fact on the basis of another fact. When certain fact already asserted cannot be explained without presuming another fact, we have the means of cognition called *arthāpatti*. For example, when it is said 'fat Devadatta does not eat in the day time', the fact of Devadatta's being fat cannot be explained unless it is presumed that he eats during night. Thus, the fact of Devadatta's eating during night is known by *arthāpatti*. Similarly, instructions given by the author of the Vedas cannot be explained unless God is presumed to be non-existent. Hence, non-existence of God is proved by

arthāpatti. The author of the Vedas gives instructions in the Vedas in order to inspire people to perform the Vedic rituals. Since God is admitted to be omniscient, He could as well inspire people to follow the Vedic rituals through some other device instead of giving instructions. Thus, had there been an omniscient God, He would have made people perform the Vedic rituals through some other means even without giving instructions. Hence, the very fact that the author of the Vedas gives instructions presumptively implies that that author is not God. Thus, there does not exist an omniscient God. This is the objection raised by the opponent to prove God's non-existence by *arthāpatti.*

The objection finds the following reply. It is an accepted fact that the effect does not take place in the absence of cause. If the Vedas, the *śabda-pramāṇa*, do not exist, the knowledge of *dharma* that arises from the knowledge of the Vedas too will not arise. In the absence of the knowledge of *dharma,* people will not have *pravrtti* towards *dharma.* The cause of *pravṛtti* is the knowledge of the Vedas; therefore, to account for *pravrtti* the Vedic instructions must be admitted. And these instructions cannot come from anybody other than God. Hence, non-existence of God cannot be proved by *arthāpatti* on the ground that the Vedic instructions cannot be explained unless God is presumed to be non-existent. On the contrary, the instructions can be explained only when God is admitted as their instructor. As the instructions can be explained without presuming the non-existence of God, God cannot be disproved by *arthāpatti.*

If this is not admitted, the Pūrva Mīmāṁsaka too will face the same difficulty. According to them *adṛṣṭa* inspires human beings to take to their respective *karma.* If *adṛṣṭa* were the cause of *pravṛtti* and set people into action, the Vedic instructions stand useless. In the absence of *adṛṣṭa* the people will not take to action in spite of the instructions given in the Vedas. Therefore, to account for the Vedic instructions, the Mīmāṁsaka must agree with the Naiyāyika that the knowledge of the Vedas is the cause of *pravṛtti* towards *dharma,* and that knowledge arises from the Vedic instructions, God being the instructor.

अनियम्यस्य नायुक्तिर्नानियन्तोपपादकः ।
न मानयोर्विरोधोऽस्ति प्रसिद्धे वाऽप्यसौ समः ।।१६।।

Unpervaded cannot be that which cannot be explained without presuming some fact (i.e. that which cannot be explained without presuming some fact is the vyāpya*); non-pervader cannot be that which is presumed in order to explain some fact (i.e. that which is presumed in order to explain some fact is* vyāpaka*); there is no contradiction between two means of knowledge; where inference is admitted, it may as well be the presumption.*

19. Moreover, *arthāpatti* is not a distinct *pramāṇa*, it comes under inference. The following is the stock example of *arthāpatti*. Hearing that living Devadatta is not at home, there arises the knowledge that he is away, because the fact of living Devadatta's not being at home cannot be explained unless it is presumed that he is away. That which cannot be explained without presuming some fact is *vyāpya*, and that which is presumed in order to explain some fact is *vyāpaka*. That which cannot exist without something else existing is the *vyāpya*. For example, smoke which does not exist without fire is the *vyāpya* of fire. Similarly, the fact of living Devadatta's absence at home that seems implausible without his staying out is the *vyāpya* of his staying out. The fact of Devadatta's staying out is the *vyāpaka*. The knowledge of *vyāpaka* with the help of *vyāpya* is attained by inference. If staying out were not the *vyāpaka*, the absence at home cannot be explained. The nature of staying out consists in being absent at home. There is *anvaya-vyāpti* between absence at home and staying out. Thus, *arthāpatti* comes under *anvayi-anumāna*. This would be Udayana's reply to those who maintain that there is no *anvaya-vyāpti* in *arthāpatti*. When *arthāpatti* is defined as that wherein certain fact is presumed on the ground that certain other fact cannot be explained without presuming it, this is *vyatireka-vyāpti* expressed in different words. *Vyatireka-vyāpti* consists of the absence of *hetu* in the absence of *sādhya*. Therefore, *arthāpatti* comes under *kevala-vyatireki anumāna*.

It may be objected that in *arthāpatti* the *vyāpti* in the form of inseparable relation *(avinābhāva)* exists, yet it does not depend on the knowledge of the *vyāpti.* Inferential knowledge depends on the knowledge of the *vyāpti.* Due to this difference between *arthāpatti* and inference the former cannot be grouped under the latter.

It may be asked whether *arthāpatti* depends on the knowledge of the fact that *hetu* cannot be explained without the *sādhya.* If it does not depend, then also knowledge arises even though the knowledge of the fact of *hetu* being not explained without *sādhya* is absent. If *arthāpatti* depends on the knowledge, then it is established that *arthāpatti* depends on the knowledge of *vyāpti.* The knowledge of *vyāpti* consists in the knowledge that the *hetu* cannot be explained without the *sādhya.* Therefore, *arthāpatti* comes under inference.

It may be said that *vyāpti* cannot be established between the fact of Devadatta's absence at home and the fact of his staying out, because there is no example in support of this *vyāpti.*

In fact, experience shows that when one is at home, one does not stay out and when one stays out of home, one is absent at home. Following one's own experience one can very well come to the conclusion that when living Devadatta is not at home, he must be out of home. Thus, everybody can know the *vyāpti* between the fact of living Devadatta's absence at home and the fact of his staying out following one s own experience. Therefore, there is no opportunity to doubt that *vyāpti* cannot be established in *arthāpatti* due to lack of example.

Some object that in *arthāpatti, hetu* is not present in *pakṣa.* In inference both the knowledge of *vyāpti* as well as that of *pakṣa-dharmatā* (presence of the *hetu* in *pakṣa*) are required. In the instance 'Devadatta is out of home, because there is absence of him at home' the *hetu* is absence of Devadatta at home. But this *hetu* is not present in the *pakṣa,* viz. Devadatta. Therefore, *arthāpatti* cannot be an inference as in it the *hetu* does not exist in *pakṣa.*

This argument is untenable. If it is said that in *arthāpatti* the knowledge of the *vyāpti* gives rise to the knowledge even in the absence of the *hetu* in the *pakṣa* then following the same logic it may be argued that in inference also knowledge arises even in the absence of the *hetu* in *pakṣa*. In that case *pakṣa-dharmatā* will not constitute inference. Moreover, during the generation of the knowledge of the *vyāpti* between *hetu* and *sādhya*, there arises the knowledge of the presence of *hetu* in the *pakṣa* also. The knowledge of the *vyāpti* arises in threefold ways — (i) that which possesses the *hetu*, possesses the *sādhya* also;(ii) wherever there is *hetu*, there is the *sādhya* also; (iii) whenever there is *hetu*, there is the *sādhya*. All these three imply the cognition of the copresence of the *hetu* and the *sādhya* in the same substratum. Therefore, the inference should be formulated in such a way that the *hetu* and the *sādhya* can exist in the same substratum. The instance cited in *arthāpatti* may take the following inferential form — 'Devadatta is out of home, because there is absence of him at home'. Here, the *hetu* 'absence of him at home' is present in the *pakṣa* 'Devadatta'; thus, *pakṣa-dharmatā* is present in *arthāpatti*.

The opponent may say that *arthāpatti* comes under inference only when by *arthāpatti* is meant the presumption of staying out on the ground of living Devadatta's not being at home. But the opponent does not take *arthāpatti* in that sense. When there arises contradiction between two *pramāṇas*, resolving of this contradiction is what is meant by *arthāpatti*. This is not what happens in inference. In the instance 'living Davadatta is not at home' — the word 'living' implies 'existing alive'. 'Existing alive' is possible only when one exists somewhere, thus the word 'living' implies 'existing somewhere alive'. The sentence 'living Devadatta is not at home' means that 'Devadatta who is existing somewhere is not at home'. The word 'somewhere' may mean home, the sentence being 'Devadatta who is existing at home, is not home' — this is a contradiction. *Arthāpatti* resolves this contradiction by presuming that 'somewhere' implies 'out of home', and thus, there arises the knowledge that 'Devadatta who is not at home is out of home'. Since knowledge does not arise in this manner in inference, *arthāpatti* cannot be included in inference.

This position is untenable. If there is contradiction between two *pramāṇas,* one turns out to be invalid. For example, when the knowledge of snake in a rope is contradicted by the subsequent knowledge 'this is not a snake', the former turns out invalid. Both the knowledges cannot be valid. If both were valid, there will be no contradiction, e.g. there is no contradiction between two knowledges 'this is a jar' and 'this is a cloth', because both are valid. In the instance under consideration 'living Devadatta is not at home' implies that 'living Devadatta exists somewhere' — even then no contradiction can be pointed out. Just as there is no contradiction between two statements 'exists somewhere' and 'does not exist elsewhere', so also there is no contradiction between the statements 'exists somewhere' and 'does not exist at home'. If there is contradiction between the two, one would be invalid.

It is not proper to say that *arthāpatti* determines both the knowledge as valid, and so one does not turn out invalid. *Arthāpatti* is applied only when something cannot be explained. It must be found out what exactly here is inexplicable. If the apparent contradiction between the two statements is what cannot be explained, then it may be concluded that both relate to different objects *(viṣaya).* For, contradiction arises only when both relate to the same obejct; contradiction resolves when both statements relate to different objects. Thus, one person cannot be at two places at the same time. In the statement 'exists somewhere' the word 'somewhere' implies a place other than home. The *vyāpti* is between 'the knowledge of absence at home' and 'that of being at a place other than home'. *Arthāpatti,* thus, consists in relating the statements with different objects.

As *arthāpatti* depends on the knowledge of the *vyāpti* which consists in the instance under consideration between 'the knowledge of absence at home' and 'that of presence at a place other than home', it comes under inference. Threfore, it is settled that without inference *arthāpatti* has no existence.

If this is not accepted, then inferential knowledge too may be attained through *arthāpatti,* and inference may be abolished as a *pramāṇa.* Thus, in the instance 'there is fire in the hill, because there is smoke', on the observation of smoke the knowledge of fire may be gained through *arthāpatti* on the ground that smoke cannot be explained without fire; just as seeing fat Devadatta who does not eat in the day time, his eating during night is presumed on the ground that fatness cannot be explained without presuming eating during night.

The opponent may admit that *arthāpatti* is what the Naiyāyika calls *kevala-vyatireki* inference. It is because that in *arthāpatti* there is no *anvaya-vyāpti* in the form wherever there is *hetu*, there is *sādhya.* In that case *arthāpatti* must be called *kevala-vyatireki* inference, and so it cannot be a distinct *pramāṇa.* The ground (*hetu*) for the use of the term inference for *arthāpatti* is the inseparable relation of the *hetu* with the *sādhya* (*sādhya-avinābhūtaliṅga).* That is, the knowledge of *arthāpatti* arises on the ground of the inseparable relation of the *hetu* with the *sādhya.* Inseparable relation consists in the *hetu's* not being present without the *sādhya.* Such relation is common to both *anvayi* as well as *vyatireki* inference. Therefore, *arthāpatti* is another name for inference or one of its forms, viz. *kevala-vyatireki* and not a distinct *pramāṇa.*

प्रतिपत्तेरपारोक्ष्यादिन्द्रियस्यानुपक्षयात्।
अज्ञातकरणत्वाच्च भावावेशाच्च चेतसः।।२०।।

(Non-cognition is included in perception, etc., because) the knowledge of abhāva *is perceptual, the capacity of the sense-organ is not exhausted, instrumental cause is not cognized, the mind produces knowledge with the help of a positive cause.*

20. That non-cognition, which the Vedāntin and the Pūrvamīmāṁsaka add as a sixth *pramāṇa*, cannot prove non-existence of God, has already been established. Now it will be shown that non-cognition is not a separate *pramāṇa*, but is included in perception, etc.

Non-cognition sometimes comes under perception,

sometimes under inference. The reasons for including non-cognition under perception are four.

The first is *pratipatterapārokṣyāt,* that is, the knowledge of *abhāva* being perceptual. The knowledge which is perceptual in character is produced by sense-organ, e.g. the knowledge of colour. The knowledge of colour which arises through perception is produced by sense-organ. Similarly the knowledge, viz. 'there is no jar on the floor' is perceptual, hence it is produced by sense-organ. Non-cognition assists the sense-organ in producing the knowledge. Thus, non-cognition is the auxiliary cause of the knowledge of *abhāva.*

That the knowledge of *abhāva* arises through perception can be proved from the following fact. The knowledge is mediate when it is produced by the knowledge of *karaṇa*, e.g. inferential; verbal knowledge, etc. are mediate, because they are produced by the knowledge of *liṅga* and that of *śabda.* Perceptual knowledge is produced by the *karaṇa* itself and not by its knowledge. For example, perceptual knowledge of colour arises by the visual sense-organ itself and not by the knowledge of the visual sense-organ.

The knowledge of *abhāva,* viz. 'there is no jar on the floor', is produced by the non-perception of jar, it is not produced by the knowledge of the non-perception of jar.

Since the knowledge of the absence of jar is produced by the non-perception of the jar, which is its *karaṇa,* it is a perceptual knowledge. Therefore, the *karaṇa* of the knowledge of *abhāva* must be sense-organ, as perception gives rise to knowledge through the instrumentality of sense-organ. Non-cognition is the accessory cause towards the knowledge of *abhāva.* Thus, the knowledge of *abhāva,* which is produced by the non-perception and not by the knowledge of it, must be included in perception.

Sometimes the knowledge of *abhāva* is produced by the knowledge of non-perception. In that case non-perception produces the knowledge through inference. For example, when a servant ordered by his master to fetch Caitra from the temple replies to his master that in spite of his best efforts he could not perceive Caitra in the temple, and hearing this the master infers that Caitra was absent in the temple at the time when the servant entered the temple — this knowledge of *abhāva* is based on the knowledge of non-perception. Here, the inferential knowledge of the absence of Caitra in the temple is produced by the knowledge of non-perception of Caitra in the temple which is imparted to the master by the servant. This kind of non-cognition comes under inference.

Similarly, the knowledge of *abhāva* produced by the knowledge of previous non-perception comes under inference. Thus, going to a temple in the morning to worship God and after coming back one may ask oneself in the afternoon whether Caitra was there in the morning. Then one may infer— 'Caitra was not present there in the morning, since I did not perceive him in the temple precinct, which I have searched out properly'. Here, the inferential knowledge of the absence of Caitra is based on the knowledge, that arises in the afternoon, of the absence of Caitra in the morning.

The second reason for the inclusion of the knowledge of *abhāva* in perception is *indriyasya anupakṣayāt*. Knowledge that arises after sense-contact, owing to latter's incapability to come in contact with further object, is perception. For example, when visual sense-organ comes in contact with colour, owing to its incapability to come in contact with any object other than colour, there arises the perceptual knowledge of colour. Similarly, the knowledge of the absence of jar on the floor that arises after visual sense-contact is perception.

It may be argued that sense-contact gives rise to the knowledge of the substratum, viz. the floor, and not the knowledge of the absence of jar. The knowledge of the absence of jar is produced by the non-perception of jar. Therefore, the knowledge of the absence of jar cannot be perception, as it is not produced by sense-contact.

This is not accepted. In that case, even a blind will possess the knowledge of the absence of colour in jar, because he can have tactual perception of jar, and there is non-perception of the colour of the jar also.

The Naiyāyika too may be charged with the same defect as they maintain that the same sense-organ that perceives the substratum, perceives the absence also. Thus, when a jar is perceived by the tactual sense-organ, there arises the contingency of perceiving the absence of a particular colour in jar.

This charge cannot be levelled against the Naiyāyika. The Naiyāyika subscribes to the view that the sense-organ that perceives the counter-correlate *(pratiyogi)* perceives the absence also. The counter-correlate of the absence of colour is colour which is perceived by the visual sense-organ. The same sense-organ perceives the absence of colour also. The blind does not possess visual sense-organ. Hence, he cannot have the knowledge of the absence of colour, when he perceives a jar through tactual sense-organ.

According to the Mīmāṁsaka non-cognition produces the knowledge of *abhāva* after the substratum is perceived by the same sense-organ that perceives the counter-correlate. The blind perceives the jar through tactual sense-organ, not by the visual sense-organ which perceives the counter-correlate of the absence of colour, viz. colour. For this reason, the blind does not possess the knowledge of the absence of colour, even though non-perception of colour exists in the substratum jar.

If this view is accepted, then the knowledge of the absence of colour in air which is perceived by the tactual sense-organ will not arise. For, the visual sense-organ that perceives colour, the counter-correlate of the absence of colour, does not perceive the substratum air.

It may be said that the substratum need not be perceived by the same sense-organ that perceives the counter-correlate; for the production of the knowledge of *abhāva* non-cognition produces the knowledge of *abhāva,* when the substratum comes in contact with the sense-organ that perceives the counter-correlate. Air comes in contact with the visual sense-organ that perceives colour, and then the non-perception of colour produces the knowledge of the absence of colour.

This being the case, it stands out that the visual sense-organ which air comes in contact with, has no relevance towards the knowledge of air, that is, it cannot perceive air. It perceives only the absence of colour. Therefore, as the visual sense-organ is incapable to perceive anything other than the absence of colour, the knowledge of the absence of colour that arises after the visual sense-contact is perception. Perceptual knowledge has only sense-organ as its *karaṇa* and not non-cognition.

It may be argued that in the instance 'there is no jar on the floor' visual sense-contact has relevance neither to the perception of substratum nor to that of *abhāva*. It has relevance only to the perception of some object. So, the knowledge of *abhāva* is not produced by the sense-contact. This is not true, sense-contact with some object is not relevant to the knowledge of *abhāva*. If the knowledge of *abhāva* is produced by the perception of some object which is in contact with the sense-organ, then there will not arise the knowledge, viz. 'there does not exist any object possessing touch', that arises in the darkness after one knows the absence of the object possessing touch by stretching the hand. The knowledge does not arise, because the cause of the knowledge, viz. the perception of some object, is absent as the tactual sense-organ does not perceive any object which is in contact with it.

Again, a person whose eyes are focussed on an object in front will possess the knowledge of the absence of blackness in a white jar existing behind. For, eyes perceive an object in front, and this perception is the cause of the knowledge of *abhāva*.

It may be said that the cause of the knowledge of *abhāva* consists also in the forward existence of the object in which is perceived the absence of colour. The jar that exists behind lacks this forward existence, so the knowledge of the absence of colour is not produced in it. This amounts to saying that the knowledge of *abhāva* arises only when the visual sense-organ comes in contact with the substratum. That means the knowledge of *abhāva* is perception, as it is produced by the sense-contact. Further, visual sense-organ produces perception only when it comes in conjunction with light, conjunction with light being the auxiliary cause of the knowledge of *abhāva*. Thus, the visual sense-organ gives rise to the perception of *abhāva* being assisted by the auxiliary cause, viz. conjunction with light.

It may be argued that the visual sense-organ has relevance to legitimacy (*yogyatā*), even if it has no relevance to other things. Thus, the cause of the knowledge of *abhāva* is not mere non-cognition, but legitimate non-cognition. Such legitimacy arises only when the sense-organ comes in contact with the substratum. Thus, the sense-organ is relevant to legitimacy and not to the perception of *abhāva*. Hence, the knowledge of *abhāva* is not perception.

It may be replied that legitimacy consists in the presence of all the causal aggregate that are required to produce the knowledge. The contact between the eyes and the substratum is also one of such causes. Therefore, it is quite in the fitness of things to maintain that the knowledge of *abhāva*, which is produced by such contact, is perception.

The opponent may say that the visual sense-contact does not produce the knowledge of *abhāva* directly, but it produces the knowledge of an object that is present in the substratum in which is perceived the absence. Thus, after the production of the knowledge of some object existing on the floor through the visual sense-organ, there arises the knowledge of the absence of jar on the floor through the non-cognition of the jar. The cause of this knowledge of the absence of jar is non-cognition and not the sense-organ. It is true that in this case perception of the absence of colour in air is not possible, because as neither air nor any object existing in air is amenable to visual sense-perception, it is not possible for the sense-organ to produce the knowledge of any object existing in air. But this defect does not affect the opponent, because he does not admit that the knowledge of the absence of colour in air is a perceptual knowledge. This knowledge arises through the inference — 'air does not possess colour, because it is never perceived as possessed of colour'.

This inference is not quite a valid one. The *hetu* in the

inference is absence of perception. It may be asked to point out the *hetu* that produces the knowledge of the absence of perception. The opponent advocates that the knowledge of *abhāva* is produced by non-cognition. Then it may be admitted that the knowledge of the absence of perception is produced by another non-cognition. That non-cognition being a *abhāva* may be known by another non-cognition, and so on *ad infinitum*. Thus, the knowledge of the *hetu* in the inference being unestablished, the *hetu* is unestablished.

The opponent has a further say in this matter. He says that the knowledge of the *hetu*, viz. the absence of perception, is not produced by non-cognition, it is inferred from the absence of *jñātatā* (knownness). If an object possesses *jñātatā*, there arises the knowledge of that object also, because *jñātatā* is not produced in the absence of the knowledge of that object. Thus, knowledge is inferred from *jñātatā*, absence of *jñātatā* guarantees the absence of knowledge also. The knowledge of the absence of perception is inferred on the ground of the absence of *jñātatā*. And the absence of colour in air is inferred by the absence of perception of colour.

This explanation is not free from defect. If the absence of *jñātatā* of colour is proved in air, then only absence of perception of colour can establish the absence of colour in air through inference. But such absence of *jñātatā* is not a proved fact, hence the *hetu*, absence of perception, cannot be known from the absence of *jñātatā*. Thus, the inference advanced to prove the absence of colour in air stands rejected.

The opponent says that like the Naiyāyika he too admits perception. According to him the knowledge of colour is a perception, what non-cognition produces is the knowledge of the absence of the knowledge of colour. From this is inferred the absence of colour in air.

It may be replied that if the opponent subscribes to the view of the Naiyāyika, then he must admit that the knowledge of the absolute non-existence of colour in air arises through perception, just as the knowledge of the prior as well as posterior non-existence of sound in ether is perception. Therefore, the attempt to infer the absence of colour in air should be given up.

Further, the opponent holds that non-cognition produces the knowledge of *abhāva* directly, and the knowledge of the substratum or the property of substratum *(adhikaraṇadharma)* produced by the sense-organ that perceives the counter-correlate is the auxiliary cause.

The question arises whether the knowledge of the property of substratum is admitted as the auxiliary cause owing to the reason that it is the cause of the knowledge of *abhāva*, or that there is *vyāpti* between the knowledge of *abhāva* and the knowledge of the property of substratum. It is not the first, because the knowledge of *abhāva* arises even in the absence of the knowledge of any property of the substratum. Therefore, the effect being present in the absence of the cause, the knowledge of the property of substratum cannot be the cause of the knowledge of *abhāva*.

If the second alternative is accepted, there being *vyāpti* between the knowledge of *abhāva* and that of the property of substratum, when the knowledge of the absence of colour in air is produced, the knowledge of the property of the substratum, viz. air,is also required.

The opponent maintains that non-cognition produces the knowledge of *abhāva* in two ways — either with the help of *liṅga* or without *liṅga*. He cites an example to clarify this idea. The knowledge of *brāhmaṇahood* may be produced by *liṅga* or without *liṅga*. 'He is a *brāhmaṇa*, because he was born by *brāhmaṇa* parents' — here *brāhmaṇahood* is known by *liṅga*, viz. giving birth by *brāhmaṇa* parents. In this case sense-contact is not necessary. *Brāhmaṇahood* is known by perception when the knowledge of giving birth by *brāhmaṇa* parents arises through perception — knowledge of this kind requires sense-contact. Similarly, when non-cognition produces the knowledge of *abhāva* through *liṅga*, e.g. air does not possess colour, because it is not perceived as possessed of colour, the knowledge of the property which is present in the substratum is not necessary. But if non-cognition produces the knowledge of *abhāva* without *liṅga*, e.g. there is no jar on the floor, the knowledge of the property present in the substratum is necessary. In either way sense-contact has no part to play in the knowledge of *abhāva*.

This position is not accepted. The view that sometimes the knowledge of the property that is present in the substratum is required for the knowledge of *abhāva* produced by non-cognition and sometimes not is not supported by any reason. This being the case, it is reasonable to suppose that everywhere non-cognition produces the knowledge of *abhāva* without the knowledge of any property present in the substratum. On this view since the sense-contact can produce neither the knowledge of substratum nor that of its property, it may be concluded that it produces the knowledge of *abhāva*. Thus, the knowledge of *abhāva* that arises after the sense-contact which is incapable to perceive anything other than *abhāva*, is perception. Non-cognition assists the sense-organ in producing the knowledge.

The third reason for including the knowledge of *abhāva* in perception is *ajñātakaraṇatvāt* (instrumental cause not being cognized). The knowledge which is not produced by a cognized *karaṇa* is directly produced by the sense-organ. For example, the perception of colour is produced by the visual sense-organ which is not cognized before. Similarly, the knowledge of *abhāva*, viz. 'there is no jar on the floor', arises through uncognized instrumental cause; hence, it is produced by the sense-organ.

The opponent says that a knowledge is called perception only when its subject is positive entity and is produced by uncognized *karaṇa*. The knowledge of *abhāva* is not produced by the sense-organ, even if it is produced by an uncognized *karaṇa*. Thus, the *vyāpti* is as follows. The knowledge that has positive entity as its object and is produced by uncognized *karaṇa*, arises through the sense-organ. The knowledge of *abhāva* does not arise through the sense-organ, because it does not have positive entity as object. This *vyāpti* narrows down the previous *vyāpti*, viz. knowledge which is produced by uncognized *karaṇa*, is produced by the sense-organ.

In reply it may be pointed out that previous *vyāpti* cannot be rejected, because there is no argument to invalidate it. If it is rejected for the reason mentioned by the opponent, then it may be equally argued that inference is possible only about positive objects and not about *abhāva*. Thus, no knowledge of *abhāva* can be attained through inference.

The opponent may further argue that let the knowledge of *abhāva* be produced by uncognized *karaṇa*, but let it not be produced by the sense-organ. There is no argument to invalidate the absence of causality of the sense-organ towards the knowledge of *abhāva*.

It is replied that observation testifies to the fact that wherever knowledge arises through uncognized *karaṇa*, it is produced by the sense-organ. If this is not admitted in the case of the knowledge of *abhāva*, there is no reason why this should be admitted in other cases. Moreover, there is established causal relation between knowledge produced by uncognized *karaṇa* and the causality of the sense-organ. If this is not accepted, then it may be concluded that effect arises even in the absence of cause. Thus, like the knowledge of *abhāva* the knowledge of colour, etc. also will not be produced by the sense-organ.

The causal relation between the knowledge produced by uncognized *karaṇa* and the causality of the sense-organ is established in the following way. In the case of inference, etc. it is observed that knowledge arises through the knowledge of *liṅga*, etc.which are the *karaṇa*. In the case of the knowledge of colour, etc. knowledge arises without the knowledge of *karaṇa*. Thus, it is concluded that the knowledge of colour, etc. is produced by uncognized *karaṇa*. And this *karaṇa* is the sense-organ. Following this, a general rule is formulated that sense-organ is the cause of the knowledge-in-general which is produced by uncognized *karaṇa*, that is, which is produced without the knowledge of *karaṇa*. Since the knowledge of *abhāva* arises without the knowledge of *karaṇa*, it must be produced by the sense-organ.

The fourth reason for the inclusion of the knowledge of

abhāva in perception is as follows. It is an acknowledged fact that the mind produces the knowledge of external object with the assistance of *pramāṇa* which is a positive entity. For Example, the mind produces the knowledge of fire being assisted by a positive *pramāṇa*, viz *liṅga*. Similarly, in order to produce the knowledge of *abhāva* which is an external entity, the mind depends on the assistance of a positive *pramāṇa*. This *pramāṇa* must be the sense-organ, not non-cognition because non-cognition is not a positive entity. Therefore, the mind does not produce the knowledge of *abhāva* merely with the help of non-cognition. It depends on the assistance of something other than the non-cognition, viz. the sense-organ, to produce the knowledge of *abhāva*. Thus, the knowledge of *abhāva* produced by the mind with the assistance of the sense-organ cannot but be a perception.

All these four reasons discussed above establish that the knowledge of *abhāva* is perception. So, non-cognition which is a cause of the knowledge of *abhāva* is not a distinct *pramāṇa*, it only assists the sense-organ which is the *pratyakṣa-pramāṇa*.

प्रतियोगिनि सामर्थ्याद्व्यापाराव्यवधानतः ।
अक्षाश्रयत्वाद्दोषाणामिन्द्रियाणि विकल्पनात् ।।२१।।

The sense-organ having the power of perceiving the counter-correlate; (the cause of) operation being inseparable (from the cause of the perception of non-existence); defects (of the erroneous perception of non-existence) residing in the sense-organ, perception of non-existence being a determinate knowledge.

21. The *siddhāntin* adduces four more reasons to establish the fact that sense-organ is the *karaṇa* for the knowledge of *abhāva*.

A. *Pratiyogini sāmarthyāt* is the first reason. It is a general rule that the knowledge of the non-existence of a thing is produced by the same *pramāṇa* that produces the knowledge of the counter-correlate. For example, just as the *liṅga* or the sense-organ produces the knowledge of the counter-correlate, viz. fire or jar, similarly they produce the knowledge of the absence of fire or the absence of jar also. The visual sense-organ that perceives the counter-correlate jar, perceives the absence of the jar also. Therefore, the knowledge, viz. 'there is no jar on the floor' is produced by the sense-organ. This is what is meant by *pratiyogini sāmarthyāt.*

B. The second reason is *vyāpāra-avyavadhānataḥ.* This may be explained in the following manner. The knowledge of the absence of jar produced after the production of the knowledge of substratum, viz. floor, by the sense-organ, is caused by the sense-organ. For, the knowledge of the substratum is the *vyāpāra* of the sense-organ, which is followed by the production of knowledge. This may be explained by an example. Stick is the cause of the operation of wheel in the production of a jar. But it is the cause of the jar also which is produced after the operation of wheel, because operation of wheel is the *vyāpāra.* Here stick becomes the cause of the jar through the *vyāpāra* of the operation of wheel. Similarly, the sense-organ also becomes the cause of the knowledge of *abhāva* through the *vyāpāra* in the form of the knowledge of the substratum. Otherwise, if the cause of a *vyāpāra* is not admitted as the cause of the effect produced after the *vyāpāra,* the *savikalpaka* (determinate) knowledge cannot be established as produced by the sense-organ. First there arises the *nirvikalpaka* (indeterminate) knowledge through the sense-organ, after that arises the *savikalpaka* knowledge. If the sense-organ that gives rise to *nirvikalpaka* knowledge does not produce the *savikalpaka* knowledge, then the latter cannot be ranked as perceptual knowledge, as it is not produced by the sense-organ.

It is objected that if the cause of a *vyāpāra* becomes the cause of the effect that follows the *vyāpāra,* then inferential knowledge too may turn out a perceptual one. The knowledge of fire in the hill arises after smoke is perceived by the sense-organ. Thus, the sense-organ may turn out to be the cause of the knowledge of fire through the *vyāpāra* in the form of the perception of smoke. The knowledge of fire then no longer will be a inferential one, but a perceptual one.

This objection arises due to mistaken idea of what is meant by *vyāpāra.*

When a thing cannot be a *karaṇa* without giving rise to a particular action, that action is the *vyāpāra* of that *karaṇa*. For example, a stick cannot be the *karaṇa* towards a jar without being a operating cause of the wheel. Thus, operation of the wheel is the *vyāpāra* of the stick which is the *karaṇa* of the jar.

If the visual sense-organ cannot produce the knowledge of fire without producing the knowledge of smoke, then the knowledge of smoke can be the *vyāpāra* towards the knowledge of fire. But this is not the case. Sometimes the knowledge of fire may arise by the visual sense-contact without the knowledge of smoke. Therefore, the knowledge of smoke is not the *vyāpāra* of the visual sense-organ, and so the knowledge of fire which is produced after perceiving smoke is not produced by the sense-organ.

C. *Akṣāśrayatvāt doṣāṇām* is one more reason that points out that sense-organ is the *karaṇa* of the knowledge of *abhāva*. Experience attests that defective sense-organ produces erroneous knowledge. For instance, defective eyes perceive a shell as yellow; this is an erroneous perception. But defectless eyes perceive the shell as white; this is a valid perception. From this it is established that the same *karaṇa* that produces erroneous knowledge owing to its defects produces the valid knowledge also being free from the defects. Thus, the knowledge of *abhāva* arises in respect of a positive thing also, e.g. there arises the perception of absence of white colour in a shell. This is an erroneous knowledge of *abhāva*. The cause of this error must be pointed out. The error cannot be due to the defect in the knowledge of the substratum, or in the remembrance of the counter-correlate, or in the non-cognition of the counter-correlate, because all these three are free from defects. Therefore, it must be the defect of the *karaṇa*, viz. the sense-organ, which is the cause of the erroneous knowledge of *abhāva*. Since the erroneous knowledge of *abhāva* arises due to defect in the instrumental causality of the sense-organ, the valid knowledge of *abhāva* too must arise through the defectless instrumental causality of the sense-organ. This establishes that the valid knowledge of *abhāva* is a perceptual knowledge. The following inference is advanced as a proof — 'sense-organ is the *karaṇa* of the valid knowledge of *abhāva*, because it is the *karaṇa* of the erroneous knowledge of *abhāva;* that which is the *karaṇa* of erroneous knowledge is the *karaṇa* of the valid knowledge too, e.g. sense-organ, *linga*, etc.'

D. The last reason is *vikalpanāt*. 'The floor has non-existence of jar' is a *savikalpaka* knowledge; hence, it must be produced by the sense-organ. The knowledge, viz. 'the floor has non-existence of jar', arises after the sense-contact. Here floor is the subject of the non-existence of jar. According to the opponent the sense-organ does not apprehend *abhāva,* because it gets diminished after perceiving the floor. Were it so, it may be argued that the non-cognition also gets diminished after apprehending *abhāva*, hence, it will not perceive the floor. Thus, the sense-organ will produce only the knowledge of floor, and the non-cognition will produce only the knowledge of the non-existence of jar. The knowledge in the form 'the floor has non-existence of jar, which comprises both floor as well as non-existence of jar', will be produced neither by the sense-organ nor by the non-cognition. Therefore, the opponent has to formulate one more *pramāṇa* apart from the six *pramāṇa*s to give rise to this knowledge.

It cannot be said that the sense-organ and the non-cognition produce the knowledge of floor and that of non-existence of jar individually, and the knowledge comprising both arises by combining the sense-organ with the non-cognition. Both cannot be combined as their subject-matter is different. Just as *kulāla,* the agent of a jar, cannot be combined with *kuvinda,* the agent of a cloth or as stick, wheel, etc. which produce a jar, cannot be combined with yarn, shuttle, etc. which produce a cloth, so also the sense-organ that produces the knowledge of floor cannot be combined with the non-cognition which produces the knowledge of the non-existence of jar.

It may be argued that *kulāla,* etc. cannot be combined with *kuvinda,* etc., because no effect is seen to be produced from their combination. But the sense-organ can be combined with the non-cognition, because the knowledge, viz. 'the floor has non-existence of jar', arises from their combination.

This argument is rejected on the following ground. The knowledge of the floor is immediate perceptual knowledge, because it is produced by the sense-organ. The knowledge of the non-existence of jar is again mediate knowledge, because it is produced by the non-cognition. Thus, if the knowledge 'the floor has non-existence of jar' arises from the combination of the sense-organ with the non-cognition, then the same knowledge will entail two different kinds of knowledge. This is not accepted.

Therefore, it must be admitted that the knowledge of the non-existence of jar arises in the same way as the knowledge of fragrant sandalwood arises. In the latter case, the sense-organ perceives only the sandalwood, but it produces the knowledge of fragrant sandalwood with the help of the remembrance of fragrance which is perceived before by the sense-organ of smell. Similarly, the sense-organ produces the knowledge of the non-existence of jar on the floor with the help of the non-cognition.

The opponent argues that non-cognition must be admitted as a separate proof. He says that *viśiṣṭa-jñāna* does not arise in the absence of the knowledge of *viśeṣaṇa*. So, prior to the generation of the *viśiṣṭa-jñāna,* viz. the floor has the non-existence of jar, the knowledge of *viśeṣaṇa,* viz. the non-existence of jar, is required. The knowledge of *viśeṣaṇa* arises from the non-cognition. The visual sense-organ may give rise to the *viśiṣṭa-jñāna*, but it cannot produce the knowledge of *viśeṣaṇa*. Just as the visual sense-organ cannot produce the knowledge of *viśeṣaṇa*, viz. fragrance, even though it produces the *viśiṣṭa-jñāna,* viz. sandalwood is fragrant, so also the sense-organ cannot produce the knowledge of *abhāva*, i.e. the knowledge of *viśeṣaṇa*. Hence, non-cognition must be admitted as a distinct *pramāṇa* in order to produce the knowledge of *viśeṣaṇa* in the form of *abhāva*.

Further, the sense-organ produces the *savikalpaka* knowledge of an object only when that object is known earlier by *nirvikalpaka* perception. Non-existence is not known earlier by *nirvikalpaka* perception; hence, *savikalpaka* knowledge of an object qualified by that non-existence is not possible. Therefore, a separate *pramāṇa*, viz. non-cognition, must be admitted in order to produce that knowledge. Just as the inference, etc. produce the *viśiṣṭa-jñāna* of an object even in the absence of the *nirvikalpaka* knowledge of that object, so also non-cognition produces the *viśiṣṭa-jñāna* of *abhāva* even in the absence of the *nirvikalpaka* knowledge of that *abhāva*.

One more objection is that the sense-organ has no contact with the *abhāva*; hence, the knowledge of *abhāva* cannot be produced by the sense-organ. The contact between the sense-organ and the *abhāva* cannot be conjunction, etc., because the sense-organ is not directly conjoined with the *abhāva*. It is not the contact of *viśeṣaṇatva* also. The relation of *viśeṣaṇatva* is preceded by other relation, e.g. the jar on a floor is the *viśeṣaṇa* of the floor, but this relation of *viśeṣaṇatā* is preceded by the relation of conjunction *(saṁyoga)* as the jar is conjoined with the floor. The non-existence of jar is not conjoined with the floor; hence, the relation between the non-existence of jar and the floor cannot be *viśeṣaṇatva*. Therefore, non-cognition must be admitted as a *pramāṇa* to produce the knowledge of *abhāva*.

Moreover, those who advocate that the knowledge of *abhāva* is produced by the sense-organ must admit non-cognition. Surely the sense-organ that produces the knowledge of a jar does not produce the knowledge of the non-existence of jar.

All these four objections stand rejected. The *viśiṣṭa-jñāna*, viz. the floor has the non-existence of jar, need not be preceded by the knowledge of *viśeṣaṇa*, viz. the non existence of jar. Therefore, there is no necessity to postulate non-cognition to produce the knowledge of *viśeṣaṇa*.

अवच्छेदग्रहध्रौव्यादध्रौव्ये सिद्धसाधनात्।
प्राप्त्यन्तरेऽनवस्थानान्न चेदन्योऽपि दुर्घटः।।२२।।

In the perception of non-existence substratum and counter-correlate being cognized (hence, the perception of non-existence is determinate); if substratum and counter-correlate are not cognized. then establishing what is already established (perception of non-existence being indeterminate will be cognized by the sense-organ); non-existence being related with the substratum through self-same relation, because acceptance of any other relation will lead to infinite regress; if this is not accepted, any other alternative will be untenable.

22. Again, the knowledge of *abhāva* does not arise without the substratum and the counter-correlate. Thus, *abhāva* cannot be known by *nirvikalpaka* perception also. But the general rule that the *savikalpaka* knowledge of an object is possible only when that object is known by *nirvikalpaka* perception applies only to positive entities and not to negative entities. Therefore, the second objection that as the *savikalpaka* knowledge of *abhāva* is not possible without the *nirvikalpaka* knowledge of *abhāva*, non-cognition must be admitted as a *pramāṇa* to produce the knowledge of *abhāva*, stands refuted.

The contact between the sense-organ and the *abhāva* is that of *viśeṣaṇatva* which is a *svarūpa-sambandha*. *Abhāva* abides in the substratum by the *svarūpa-sambandha* which is not preceded by any other relation. The visual sense-organ comes in contact with *abhāva* through the contact between the substratum, viz. the floor, and the non-existence of jar, the *viśeṣaṇa* of the substratum. Therefore, the third objection that the sense-organ does not produce the knowledge of *abhāva*, because it has no contact with *abhāva*, also stands rejected.

To deny *svarūpa-sambandha* between the substratum and *abhāva* and to argue that all the relations are different from the things related would lead to infinite regress, for, in that case the relation between the substratum and *abhāva* will depend on another relation, and so on *ad infinitum*.

To admit non-cognition as the means of perceiving *abhāva* is also not free from difficulties. Thus, all the knowledge produced by different *pramāṇas* are preceded by *nirvikalpaka* knowledge. The inference of fire in the hill is preceded by the perception of fire in the kitchen. Perceptual knowledge depends on the *nirvikalpaka* knowledge directly. Inferential knowledge, being dependent on perception, depends on *nirvikalpaka* knowledge indirectly. Hence, non-cognition too cannot give rise to the *savikalpaka* knowledge of *abhāva* without producing the *nirvikalapaka* knowledge of *abhāva*. As in the perception of *abhāva*, *nirvikalpaka* knowledge of *abhāva* does not arise, non-cognition cannot be the *pramāṇa* for the perception of *abhāva*. Regarding relation too, if *svarūpa-sambandha* is not accepted, then the relation of *viśeṣaṇatva* being dependent on another relation will lead to infinite regress; hence, the opponent too has to concede that the *viśeṣaṇatva,* i.e. *abhāva* does not depend on any other relation.

The Prābhākara admits identity relation between *abhāva* and the substratum. If the relation of identity is established on valid grounds, then it will find acceptance. If, on the other hand, the identity relation is not an established fact, then it cannot be accepted. However, to say that the knowledge of *abhāva* is not produced by the sense-organ does not establish non-cognition as the *pramāṇa* for the knowledge of *abhāva.*

Lastly, it is futile to argue that non-cognition must be admitted as a separate proof, because it is necessary for producing the knowledge of *abhāva.* The necessity of non-cognition may consist in its being a *kāraṇa,* but not as a *karaṇa.* Only the sense-organ can be the *karaṇa* of the knowledge of *abhāva.* The sense-organ is admitted by all as the *karaṇa* of the perception of positive entities. So, it must be the *karaṇa* of the perception of negative entities also. Non-cognition is only an auxiliary cause.

If this is not accepted, then it may be admitted that like the knowledge of negative entity the knowledge of positive entity also is produced by the non-cognition. The knowledge of jar may be produced by the non-cognition of absence of jar, as the knowledge of jar may arise only when there is non-cognition of the absence of jar. Thus, non-cognition of *abhāva* may be the *pramāṇa* for the knowledge of positive entity and non-cognition of *bhāva* for the knowledge of negative entity. In that case, the sense-organ would not be a *pramāṇa,* and consequently, the *pratyakṣa-pramāṇa* itself will be abolished.

To avoid this position, it must be admitted that the sense-organ which is an established *pramāṇa* for the knowledge of positive entity is the *pramāṇa* for the knowledge of negative entity also. Non-cognition is not the *pramāṇa* for the knowledge of *abhāva*.

प्रत्यक्षादिभिरेभिरेवमधरो दूरे विरोधोदयः
प्रायो यन्मुखवीक्षणैकविधुरैरात्माऽपि नासाद्यते।
तं सर्वानुविधेयमेकमसमस्वच्छन्दलीलोत्सवं
देवानामपि देवमुद्भवदतिश्रद्धाः प्रपद्यामहे।।२३।।

Without dependence on God even perception, etc. cannot establish their own nature, (only when God is established as the subject, then only) there is possibility of the rise of any contradiction; with the greatest faith aroused we take resort to that Lord of the lords, who cannot be invalidated by any means of knowledge, who is One without a second, who delights in the sportive exercise (of the creation of the world, etc.), unparallel, independent.

STABAKA - IV

GOD—THE SUBSTRATUM OF VALID KNOWLEDGE

This chapter deals with the Mīmāṁsaka objection that even if God does exist, He cannot be the substratum of valid knowledge. One characteristic mark of valid knowledge, according to the opponent, is *agṛhītārthagrāhakatva,* that is, apprehending an object not apprehended before. The knowledge of God is eternal; He has no object unapprehended before. Thus, as the criterion of valid knowledge does not apply in the case of God's knowledge, God cannot be the substratum of valid knowledge. But Udayana refutes this and maintains God as the substratum of valid knowledge.

अव्याप्तेरधिकव्याप्तेरलक्षणमपूर्वदृक्।
यथार्थानुभवो मानमनपेक्षतयेष्यते।।१।।

Apprehending an object unapprehended before is not the characteristic mark of valid knowledge, because it involves the fallacy of less pervasiveness and of over-pervasiveness; valid knowledge is true experience which is independent (does not depend on any prior experience).

1.i. The Naiyāyika argues that if the validity of knowledge consists merely in apprehending an object *(arthagrāhakatvameva);* then memory also will turn out to be a valid knowledge, since in memory too some object is apprehended. But memory, according to the *siddhāntin*, does not fall within the category of valid knowledge. Therefore, according to the opponent it must be admitted that validity of knowledge consists in apprehending an object not apprehended before.

Udayana maintains that apprehending an object unapprehended before is not a characteristic mark of valid knowledge, as it fails to apply in such an affirmative instance as continuous stream of cognition and wrongly applies to such a negative instance as the erroneous judgement that 'this nacre before me is silver'. Thus, the definition of valid knowledge, put forth by the opponent, involves the fallacy of both *avyāpti* (less-pervasiveness) as well as *ativyāpti* (over-pervasiveness). Udayana defines valid knowledge as *yathārthānubhava* (true experience). The ancient philosophers did not extend the term *pramā* to memory, because memory is necessarily dependent, as its object is the same as that of the original impression which produced it. Therefore, the authoritativeness of memory must stand or fall with that of its productive factor. Hence, he adds the epithet *anapekṣatayā* (independent), that is, valid knowledge or *pramā* is *yathārthānubhava* which is independent and does not depend on any prior experience.

Since the definition of valid knowledge is *yathārthānubhava* and not *agṛhītārthagrāhakatva,* God does certainly possess valid knowledge, for He has *yathārthānubhava.* It may be asked: 'What does the opponent mean when he says validity of knowledge consists in apprehending an object not apprehended before? Does he mean that there does not arise experience *(anubhūti)* of an object which is already apprehended, or that even if it arises, it is not valid knowledge?' It cannot be the first, for there arises experience of an object which is already apprehended whenever causal factors are present. The second alternative also is untenable, because it cannot be proved that such experience is invalid.

It cannot be said that since in apprehending an already apprehended object the subsequent knowledge has the same object as that of the previous one, it is not valid. For, it can equally be said that the previous knowledge is not valid, because it has the same object as that of the subsequent one. To say that the previous knowledge is valid, because it does not depend on the subsequent one for its generation, would not solve the problem. The subsequent knowledge can also be valid in the same way, that is, on the ground of its not being dependent on the previous one.

The opponent may argue that as an axe does not cut a piece of wood which has already been cut into pieces, so also eyes, etc.

do not have efficacy towards apprehending an object which has already been apprehended. This simile is inappropriate. In the former case the axe does not cut the same piece of wood for a second time, because the effect, viz. cutting, does not arise for the second time in respect of the same piece of wood. But in the latter case, the knowledge of an object which has already been apprehended arises for the second time, and as the effect is present, the eyes, etc. also have efficacy towards it as instrumental causes.

This knowledge cannot be ranked with memory which is not valid knowledge on the ground that it has the same object as its content which has already been apprehended. The question arises as to how this knowledge can be viewed as invalid. If the criterion of validity is *yathārthānubhava,* then surely this knowledge is valid, because it is an experience and a true one.

If it is invalid, because there is absence of apprehending an object previously unapprehended, then it establishes what is already established. For, the *siddhāntin* also admits that there is apprehension of an object which is already apprehended. But he does not subscribe to the opponent's position that the criterion of validity lies in apprehending an object previously unapprehended.

1.ii. Moreover, the reason advanced by the opponent to prove the invalidity of knowledge, viz. absence of apprehending an object previously unapprehended, is the same as the *sādhya*, viz. *pramātvābhāva* or invalidity, as according to him invalidity means absence of apprehending an object unapprehended before. Therefore, both *hetu* as well as *sādhya* being the same, the former is as doubtful as the latter and cannot prove the latter.

If it is said that by invalidity is meant absence of use of the term *pramā* or validity, it is not so, for the term *pramā* is used for such knowledge which has as its content an object which is already apprehended.

It is not proper to maintain that apprehending an object not apprehended before is the ground for the use of the term *pramā.* For, in that case, as it has already been mentioned, erroneous knowledge too will be rendered as valid.

1.iii. It may be said that the content of erroneous knowledge is not real. The ground for the use of the term *pramā,* however, consists in apprehending a real object which is not apprehended before. This position is also untenable, for this will not apply to a continuous stream of cognition *(dhārāvāhijñāna).* The latter, according to the Mīmāṁsaka, is a valid knowledge.

In a continuous stream of cognition the successive cognitions apprehend the same object. Hence, all the cognitions excepting the first one cease to be valid. The Mīmāṁsaka may say that all these cognitions are valid, because the object is apprehended at different moments of time and the successive moments at which the object is apprehended are previously unapprehended. According to the Bhāṭṭa Mīmāṁsaka, newness marks every one of these continuous cognitions. The existence of a flower in a subsequent moment cannot be apprehended by its cognition in the preceding moment. If time-moments are symbolized by T_1,T_2,T_3 etc., and the perceived object by O, then the object of the first-moment cognition is OT_1, that of the second-moment cognition is OT_2, and so on. As each of the cognitions reveals a new thing, i.e. a thing which is not previously apprehended, all are valid.

This contention stands rejected, because the difference between two successive moments of time, which is calculated from the movement of the sun *(sūrya-parispanda)* is too subtle to apprehend. What the perception reveals at the subsequent moments does not appear to be different from what it reveals at the first moment. When we look at a flower continuously for a few moments, we do not find it to be different at successive moments from what we find at the first moment. The cognitions other than that at the first moment do not reveal anything which is not apprehended beforehand.

The opponent may say that though moments are not perceptible, the object qualified by the moments is perceptible. And by this, cognitions can make out the difference between objects qualified by different moments. But the problem is, if qualification is produced in the object by the moment itself, then past and future moments being absent at the time when the object is apprehended, qualification produced by the past and future moments in the object would not be apprehended at all.

If it is said that it is not the moment itself but the knowledge of the moment which produces the qualification, then also the aforesaid difficulty arises. That is, comprehension of the moment being impossible, knowledge of the moment will not arise, and consequently, qualification also will not be produced in the object. Moreover, it cannot be proved that moment produces qualification in the object.

The Mīmāṁsaka says that in a continuous stream of cognition the object of the successive cognitions is qualified by a quality called *jñātatā* (knownness) which is new in every cognition and is never apprehended beforehand. As the object qualified by such *jñātatā* is apprehended in every successive cognition, the criterion of the validity of knowledge applies in continuous stream of cognition.

With the Mīmāṁsaka, after the generation of the knowledge, viz. 'this is a jar', there is produced *jñātatā* in the jar. The knowledge 'this is a jar' gives rise to *jñātatā* in the jar. The object of knowledge at the first moment, in the continuous stream of cognition of a jar, is 'jar' which is not apprehended before. The next moment the object of knowledge is 'jar' qualified by *jñātatā* produced in the 'jar' by the first moment knowledge, which is also not apprehended before. Similarly, the object of other successive cognitions is also the 'jar' qualified by *jñātatā* produced by earlier and still earlier cognitions. Since *jñātatā* that is produced in every successive cognition is something new and is not apprehended before, continuous successive cognitions are all valid.

This position is not acceptable to the *siddhāntin*. The doctrine of *jñātatā* does not find favour with him. If this is upheld when a series of memory arises successively, the object of the successive series being qualified by *jñātatā* produced by the earlier series, the successive series will turn out to be valid.

1.iv. The knowledge, arising from apprehending an object which was apprehended before but forgotten, must be admitted valid. Otherwise, the knowledge of the Vedas will be devoid of all validity. To the Mīmāṁsaka the world is eternal; and if it were so, the meaning of the Vedic injunction *yajeta svargakāmaḥ* must have been known before, and it cannot be said that the knowledge of this injunction has its content previously unapprehended. Nor can it be maintained that the object of verbal knowledge is qualified by the moment which is not apprehended before, because in verbal knowledge barring the word-sense or the sentence-sense nothing else is apprehended. The validity of the verbal knowledge of the Vedic injunction cannot be sustained on the ground that the injunction is not comprehended in this life. For, even in this life one may comprehend the meaning of the injunction after repeated hearing though one might have known the meaning of the very same injunction before, yet eventually has forgotten.

Therefore, it is not judicious to hold that the criterion of the validity of knowledge lies in apprehending an object previously unapprehended.

The opponent may ask, if apprehending an object which is already apprehended is termed as valid knowledge, then it must be pointed out in what way memory is distinguished from valid knowledge, as memory also consists in apprehending an already apprehended object. It may be, in reply, observed that memory is not a valid knowledge, because it is not an *anubhava* and only *anubhava* is a valid knowledge. It is not that all knowledge that has as its object an object already apprehended is valid. Gautama, the author of the *Nyāya-sūtra*, defines valid knowledge as *anubhava* which is also *yathārtha* (true). Valid knowledge must be true experience, and also it must be free from error.

The opponent may argue that memory is true experience in the same way as perceptual experience is. He may offer the following inferential argument in support of his view — 'memory is a valid knowledge, because it is a true experience, like perceptual experience'.

This argument is refuted on the ground that *anubhava* which is the cause of memory is not admitted by the scholars of *pramāṇa-śāstra* as a means of valid knowledge. And whichever is not admitted as valid knowledge cannot suddenly acquire validity on any ground whatsoever.

To argue that *anubhava* is included in the four means of valid knowledge, viz. *pratyakṣa*, *anumāna*, *upamāna* and *śabda*, and that is why it is not mentioned as a separate means of valid knowledge, is unsound. Memory is an indirect knowledge, because it arises only when a previous experience is recollected. But perceptual knowledge is direct, as it arises then and there whenever there is sense-object contact. It is, therefore, obvious that as both are different in kind, the former cannot be included in the latter. Again, the *karaṇa* of memory, viz. *anubhava*, and that of *anumiti*, etc., viz. *anumāna*, etc. produce the respective knowledge in a different manner. *Anubhava* being present *(sattā-mātrena)* produces memory, whereas it is the knowledge of *anumāna*, *upamāna* and *śabda* that produces *anumiti*, etc. respectively. Therefore, *anubhava* cannot be included in *anumāna*, etc. also. Thus, as *anubhava*, the cause of memory, is not admitted as the fifth means of valid knowledge, there is no valid ground for accepting memory as valid knowledge.

That memory cannot be given the status of valid knowledge is supported by the following argument also. The object of memory is that of the object of *anubhava*, and *anubhava* being the cause of memory the validity or invalidity of memory depends on the validity or invalidity of *anubhava*. Even if memory were valid, it is the previous *anubhava*, the cause of memory, which is valid, because it does not depend on memory for its generation, and not memory which depends on previous *anubhava* for its generation. Validity of memory cannot be ascertained independent of the validity of *anubhava*, as the validity of the latter renders validity to the former, and the invalidity of it marks memory invalid.

1.v. The opponent may object that the validity of *sādhya* in inference depends on the valid perception of *liṅga-parāmarśa*, and so inferential knowledge is not valid knowledge.

This objection is not tenable, because in inferential knowledge the *sādhya* which is inferred from the perception of *liṅga* is not the same as the *liṅga*, whereas in memory the cause of memory as well as the object of memory is the same, *anubhava* being both the cause as well as the object. Owing to this difference in respect of the object, memory and inferential knowledge cannot be treated in the same level.

1.vi. The opponent now argues against the validity of verbal testimony. The validity of verbal testimony depends on the speaker's valid knowledge of the verbal statement. Since the object of the knowledge of verbal testimony and that of the speaker's knowledge of verbal statement is the same and the validity of the speaker's knowledge determines the validity of the knowledge of verbal testimony, verbal testimony is not a valid knowledge.

This objection stands refuted, for verbal testimony produces knowledge in the listener, and the speaker and the listener being different the validity of the speaker's knowledge of verbal statement does not determine the validity of the listener's knowledge of verbal testimony. In the case of memory, the validity of *anubhava* determines the validity of memory in respect of the same person, viz. knower, whereas in the case of verbal testimony the procedure is different. In the latter case, validity of the verbal statement determines the knowledge of the verbal statement in respect of the speaker and not in respect of the listener who is different from the speaker.

Continuous stream of cognition also does not lack validity, although all the subsequent perceptions have the same object as that of original one. The same sense-contact which is the cause of original one is the cause of the subsequent series also. It is not that the earlier series of cognition is the cause of the subsequent series. Unlike memory, there is no cause-effect relation between continuous stream of cognitions, and in the absence of any such causal relation there is no reason to maintain that subsequent cognition is not valid, because it has the same object as that of the earlier one. Memory stands invalidated on the ground that it has the same object as that of its cause, viz. *anubhava*.

The opponent may gainsay the fact that the object of memory is the same as that of *anubhava*. Memory arises in the following form — 'I have *anubhava* of the jar'. Thus, in memory there is the knowledge of *anubhava* along with the object of *anubhava*, but in *anubhava* there is no knowledge of *anubhava*. Therefore, as in memory one apprehends something more than the mere object of *anubhava*, memory must be a valid knowledge.

The fact of *anubhava's* being the object of memory cannot be denied, for otherwise the previous existence of *anubhava* itself cannot be proved. And if *anubhava* remains unproved, the causal relation between *anubhava* and memory also cannot be established.

1.vii. An attempt may be made to establish memory by inference and presumption. The arguments are as follows:

'Memory must have a cause, because it is an effect, and since memory cannot be explained without presuming *anubhava* as its cause, *anubhava* must exist prior to memory.'

These arguments are untenable, because by these only this much is established that memory must have a cause, but this does not prove that it is *anubhava* which is the cause of memory.

It may be objected that if *anubhava* is not admitted as the cause of memory, memory may arise in connection with an object which is not apprehended before. This objection is not sound, because until it is proved that *anubhava* is the object of memory, it cannot be proved that memory does not arise in regard to an object which is not apprehended before.

Anubhava may be established on the ground that the form *(ākāra)* of a prior *anubhava* is conferred on memory, and unless there exists *anubhava* it cannot be explained how the form of *anubhava* is conferred on memory.

The question arises as to what is this form of *anubhava* which is conferred on memory. According to Sautrāntika Bauddha, when the cognition of some object, say blue, arises, the form of blue is conferred on the cognition. As the form of blue is present in the cognition of blue, blue must exist, for unless an external object exists, the form of that object cannot be conferred on its knowledge.

Thus, existence of an external object is inferred as the cause of the form of knowledge, without which the form of knowledge cannot be explained. When it is said that the form of blue is conferred on the cognition of blue, it means that the content of knowledge is blue. Hence, the existence of blue is established by the cognition of blue on which is conferred the form of blue.

The point at issue is that the form of *anubhava,* according to the opponent, is conferred on memory. From this proposition either of the following alternatives may be drawn. First, *anubhava* may be inferred from the fact of the conference of the form of *anubhava* on memory; secondly, memory has as its object *anubhava*, the form of which is conferred on memory.

The first alternative is untenable; it may be inferred that memory exists prior to *anubhava*, because it takes the form of *anubhava.* There is no *vyāpti* relation to the effect that the form of knowledge must exist prior to knowledge. Surely, the object in the erroneous knowledge of silver in shell does not exist prior to knowledge. In the second alternative, if *anubhava* becomes the object of memory, the latter would turn out to be a valid knowledge. If *anubhava* is denied to be the object of memory, *anubhava* remains unestablished, as there is no means to prove its existence. Consequently, no causal relation can be established between *anubhava* and memory.

The *siddhāntin* points out that there is no proof to attest the fact that *anubhava* is the object of memory, and so it is not proper to acknowledge memory as valid knowledge on the ground that it has *anubhava* as its object which is not apprehended before. When memory arises in the form 'that jar', 'that' does not refer to any *anubhava*, but it refers to the object related to particular time and place of the previous *anubhava. Anubhava* and the causal relation between *anubhava* and memory are established by *pratyabhijñā* (recognition).

The first *anubhava* of the jar which is apprehended by *anuvyavasāya* (reflective cognition) leaves an impression of *anubhava* in the soul. When that jar is remembered, earlier impression of *anubhava* is also revived, and there arises the following *pratyabhijñā* in the soul — 'I remember the same jar which I apprehended before'. This *pratyabhijñā* entails the prior *anubhava* as well as the present remembrance of the jar. Thus, both *anubhava* and the causal relation between *anubhava* and memory are established by *pratyabhijñā.*

1.viii. The Mīmāṁsaka objects to the validity of God's knowledge on the ground that His knowledge is conditioned by *gṛhītārthagrāhitva*. *Gṛhītārthagrāhitva*, according to him, means apprehending the same object which is already apprehended by the same agent. But this charge is unfounded, for God possesses only one knowledge; therefore, the question of His having prior knowledge of the same object does not arise at all.

It cannot be objected that since the object of God's knowledge remains the same at every succeeding moment, His knowledge is not valid. *Gṛhītagrāhitva*, according to the *siddhāntin*, consists in the generation of the knowledge of an object in dependence on the knowledge of the same object, the substratum of the knowledge being the same. God's knowledge at successive moments does not arise in dependence on His earlier knowledge, so His knowledge is free from *gṛhītagrāhitva*, i.e. He does not apprehend what is already apprehended. And even if God's knowledge is supposed to be devoid of validity owing to apprehending the same object at every succeeding moment, then the continuous stream of cognition too must be termed as *apramā*.

It may be remarked that *agṛhītārthagrāhitva* which is the definition of valid knowledge according to the Mīmāṁsaka is also present in God's knowledge in the sense that His knowledge consists in apprehending that which is not not-apprehended. The expression is *nahyagṛhīto yo'rthastadgrāhitvam*.

The opponent says that *gṛhītagrāhitva* means apprehending an object which is the substratum of *jñātatā* or wherein is produced *jñātatā*. As the knowledge of God is eternal, *jñātatā* is already produced in the object of His knowledge, and so He apprehends what is the substratum of *jñātatā*. Thus, as His knowledge consists of *gṛhītagrāhitva*, it is not valid.

With the Mīmāṁsaka cognition must produce a particular quality, i.e. *jñātatā*, in the object, for otherwise there would not be in cognition the definite distinguishing of the object, that is, that this is a cloth and not a jar. Even in repeated cognition we apprehend a thing previously unapprehended, since each separate act of cognition on our part produces anew this particular quality in the object, and this, therefore, is ever cognized anew.

If *jñātatā* is not admitted in the object, then a cloth can be called a jar and a jar a cloth. It is *jñātatā* that residing in the object distinguishes one object from the other. *Jñātatā* is produced in the object, by the cognition of that object. In the absence of cognition *jñātatā* is not produced in the object, because the production of *jñātatā* in the object depends on the cognition of that object which is its material cause. Without the knowledge of the material cause no effect is produced. To maintain that an effect can be produced without the knowledge of the material cause will nullify the attempt to prove the existence of God by the argument that the creation of the world must imply a previous knowledge of the material cause, viz. the atoms, etc., out of which it was made, since all effects, such as jar, etc., imply such knowledge.

Therefore, the production of all effects must imply the knowledge of the material cause. The cognition that produces *jñātatā* is expressed in the form 'this is a jar' and the substratum of this knowledge is the cognizer, since the cognizer cognizes 'I know the jar'. Thus, it is decided that *jñātatā* is produced by the cognition of the object, which resides in the cognizer.

स्वभावनियमाभावादुपकारोऽपि दुर्घटः।
सुघटत्वेऽपि सत्यर्थेऽसति का गतिरन्यथा।।२।।

It is the nature of the cognition to distinguish itself from other objects; if the nature is not accepted, then the acceptance of cognizedness also will not help in this matter; even if it is granted that cognizedness is produced in an existing object, it cannot be produced in a non-existing object.

2. The *siddhāntin* does not accept the doctrine of *jñātatā*. His reply to this objection is as follows: If a jar is not distinguished as a jar unless *jñātatā* is produced in the jar, then before the

production of *jñātatā*, the jar being not distinguished as jar, it is not understandable how cognition can produce *jñātatā* in the jar. If it is said that it is the nature of the cognition to produce *jñātatā* in the jar, then the same nature can distinguish the jar from other objects also, there is no need for postulating *jñātatā* for that purpose. Again, even if it is granted that *jñātatā* is produced in an existing object, it cannot be produced in a non-existing object, e.g. a jar now destroyed, since its material cause, i.e. the jar, is non-existent. It follows that in this case, at any rate, there is nothing definite to distinguish the object.

Nor can it be argued that though the jar is destroyed, jarness exists and *jñātatā* is produced in jarness. For, in that case the object of cognition would be jarness and not the jar. That is, the cognition would be in respect of jar-in-general, and not a particular jar.

The Mīmāṁsaka admits identity-cum-difference relation between *dharma* and *dharmī*, and so the identity relation that exists in the *jāti* exists in the *vyakti* also. Thus, the relation of identity may be admitted between jarness and jar, both being *jāti* and *vyakti*. Consequently, *jñātatā* produced in jarness will be produced in the jar also. But then *rasa* and the jar having identity-cum-difference relation there arises the contingency of the perceptual knowledge of the jar being the perceptual knowledge of *rasa*. And as the jar and *rasa* are identical, the perception of *rasa* would become the perception of jar and the perception of jar as the perception of *rasa*. That being the case, it may be asked why *rasa* is not perceived as *rasa*. If it is said that *rasa* is not perceived as *rasa* because *jñātatā* is not produced in it by perceptual cognition, then similar would be the case with past or future cognition also. In respect of the cognition of an object that existed in the past or will exist in future, cognition produces *jñātatā* in general and not in any particular object; so no knowledge of a particular object arises. Therefore, the knowledge of a particular object, viz. a jar, as ditinguished from the knowledge of a cloth, must be established in some other way, since the knowledge of past and future object cannot be explained on the assumption of *jñātatā*.

अनैकान्त्यादसिद्धेर्वा न च लिङ्गमिह क्रिया।
तद्वैशिष्ट्यप्रकाशत्वान्नाध्यक्षानुभवोऽधिके।।३।।

Action cannot be the probans for proving cognizedness, since it is inconclusive or unfounded; perception does not prove in cognition anything other than the attribute, the substantive and their relation.

3. The Mīmāṁsaka puts forth one more argument to prove *jñātatā*. He argues that an action or *kriyā* must produce some attribute in its object or *karma*, and as cognition is also a *kriyā*, it also must produce an attribute in the form of *jñātatā* in its object, viz. jar, etc. The argument in its inferential form is thus: 'Wherever there is a *kriyā*, it produces an attribute in its object; cognition is also a *kriyā*, therefore, it must produce an attribute in its object, which is *jñātatā*'

The *siddhāntin* observes that *jñātatā* cannot be established on the basis of this inferential argument, because the probans in the inference, viz. *kriyātva*, is fallacious.

Kriyā may mean three things — (i) *dhātvartha* or root-meaning, (ii) *karaṇa-vyāpāra* or intermediary function of the instrumental cause, and (iii) *spanda* or motion.

If by *kriyā* is meant *dhātvartha*, then as in such instance as 'he unites his arrow to the sky' no effect is produced by the action of uniting in its object, viz. sky which is eternal, the *hetu* becomes *anaikāntika* or inconclusive. Again, in the case of a series of sounds, the first sound may give rise to the second sound, but the last sound does not give rise to anything. Thus, as the last sound which is a *kriyā* does not produce an effect, the Mīmāṁsaka position that anything that is a *kriyā* must give rise to an attribute in its object stands invalidated.

If by *kriyā* is meant *karaṇa-vyāpāra*, then also the probans is inconclusive *(anaikāntika)*, as no effect is produced in the object, e.g. a jar, by the contact of the organ of perception therewith. The sense-object-contact which is a *karaṇa-vyāpāra*, produces knowledge in the soul, but does not produce anything in the object. Thus, as the probans, viz. *kriyā* in the form of *karaṇa-*

vyāpāra, exists in sense-contact, yet it does not prove the probandum, i.e. it does not produce an attribute in the object, the *hetu* is *anaikāntika.*

Again, if *kriyā* means *spanda,* then as *spandatva* is not present in cognition, the probans, viz. *kriyā* as *spanda,* remains unestablished in the subject, viz. cognition. Thus, the probans involves the fallacy of *svarūpāsiddha.* Therefore, the inference advanced by the Mīmāṁsaka to establish *jñātatā* is proved unfounded.

> अर्थेनैव विशेषो हि निराकारतया धियाम्।
> क्रिययैव विशेषो हि व्यवहारेषु कर्मणाम्।।४।।
>
> *It is the object of cognition which distinguishes one cognition from the other, as cognitions themselves have no definite forms to distinguish them from each other; in usages about objects (such as* jñātoghaṭaḥ), *it is the verb which distinguishes objects.*

4.i. The opponent, further, attempts to prove *jñātatā* on perceptual evidence. He points out that such verbal usages as *jñāto'yamarthaḥ, sākṣātkṛto'yamarthaḥ,* etc. suggest that *jñātatā* is known by perception.

The *siddhāntin* does not lend support to this view. With Udayana the cognitions themselves have no definite forms to distinguish them from each other, and it is the object of cognition which distinguishes one cognition from the other. He does not concur with the Sautrāntika Baúddha that object confers its form on its cognition. Whenever we have a definitely distinguished cognition, we have a definitely distinguished attribute (*viśeṣaṇa*), the thing distinguished and the connection that exists between the two. In the case of *ghaṭa-jñāna,* the object *ghaṭa* is the *viśeṣaṇa* of *jñāna* which is *viśeṣya,* and apart from *ghaṭa,* its *jñāna* and their relation or *sambandha,* no such attribute as *jñātatā* is perceived. Again, it is the *kriyā* or the verb which distinguishes *karma-kāraka* or objects and not any attribute such as *jñātatā* residing in the object. In such usage as *jñāṭoghaṭaḥ,* the object *ghaṭa* is distinguished by the act of cognition. In usages like *ghaṭakriyā, paṭakriyā,* etc., the objects *ghaṭa* and *pata* are the distinguishing factors, whereas in *kṛto ghaṭaḥ, kariṣyate ghaṭaḥ,* etc. the *kriyā* is the distinguishing factor. In the former *karma* is *viśeṣaṇa* and *kriyā* is *viśeṣya*, in the latter *kriyā* is *viśeṣaṇa* and *karma* is *viśeṣya*. Thus, the cognition reveals only *viśeṣya*, *viśeṣaṇa* and their relation, and nothing more like *jñātatā*. Otherwise, if *jñātatā* is perceived in the cognition of a jar, in such cases as 'the jar is desired', 'the jar is produced', we have to presume such novel attributes as 'desiredness', 'producedness', etc.

The opponent says that *jñāna-vyavahāra* depends on knowledge, so *jñāna-vyavahāra* cannot be explained unless knowledge is established. Knowledge cannot be perceived, since it is super-sensuous. Hence, knowledge must be inferred from the effect it produces, and that effect is *jñātatā*. This argument may be put in the following form: 'Knowledge must be inferred from its effect, because of the existence of *jñāna-vyavahāra,* even though knowledge is not apprehended in any other way'. This is one more inferential argument advanced by the Mīmāṁsaka to prove *jñātatā*, which is refuted by the *siddhāntin* in the following manner. According to the opponent knowledge of *jñātatā* arises first, then is inferred knowledge from *jñātatā*, which is followed by *jñāna-vyavahāra,* and lastly there is apprehension of *jñāna-vyavahāra. Jñātatā* is inferred from the *jñāna-vyavahāra.* The whole position stands thus: 'Knowledge is inferred from the knowledge of *jñātatā,* and the knowledge of *jñātatā* is established on the ground that otherwise *jñāna-vyavahāra* cannot be explained'. This process ends up in the fallacy of mutual interdependence; therefore, *jñātatā* remains unestablished.

The Mīmāṁsaka contends that knowledge is super-sensuous *(atīndriya)* and brings forth various arguments to prove the same. The *siddhāntin* examines them and on ultimate analysis rejects them all.

4.ii. One of the arguments is that knowledge is supersensuous, because it is not perceived by the sense-organs *(indriyeṇa anupalabhyamānatvāt).* It may be enquired whether this non-perception is the *hetu* for proving super-sensuousness, or signifies that knowledge being supersensuous can be perceived by non-cognition *(anupalabdhi),* as super-sensuousness implies absence of sense-perception.

The first alternative is untenable, because both *sādhya*, viz. super-sensuousness *(atīndriyatva)* and *hetu,* viz. non-perception *(anupalabhyamānatva)* being the same, the latter cannot prove the former. In the second alternative, only a legitimate *(yogya)* non-cognition can perceive absence *(abhāva),* that is, non-cognition can perceive the absence of a thing only when its counter-correlate, viz. the thing, is fit to be perceived. So, super-sensuousness of knowledge can be perceived by non-cognition only when knowledge is fit to be perceived. And if knowledge is fit to be perceived, it can never be super-sensuous.

Knowledge cannot be proved as super-sensuous on the ground that the substratum of *jñātatā*, viz. knowledge, is not perceived. For, knowledge itself is not established owing to the absence of any means that proves knowledge as super-sensuous. The argument that unless knowledge is established *jñāna-vyavahāra* cannot be explained, as knowledge is the cause of *jñāna-vyavahāra,* is not sound. Knowledge cannot be proved unless there is perceptual evidence for it. So far as the cause of *jñāna-vyavahāra* is concerned, the causes of knowledge, viz. soul-mind contact, etc., can serve that purpose, there is no need of presuming knowledge for it. In the generation of knowledge, first there is soul-mind contact, then mind-sense-organ contact, then sense-object contact, then arises knowledge and after that *jñāna-vyavahāra*. The *siddhāntin* states that *jñāna-vyavahāra*, the last one in the series, may be due to the causes of knowledge and one need not postulate knowledge as the cause of *jñāna-vyavahāra*.

The opponent objects that soul-mind contact cannot be the cause of *jñāna-vyavahāra,* because during deep sleep *jñāna-vyavahāra* is absent in spite of the presence of soul-mind contact. It may be argued that mere soul-mind contact is not the cause of *jñāna-vyavahāra*. The entire causal aggregate which is the cause of knowledge can directly become the cause of *jñāna-vyavahāra*;

therefore, knowledge need not be posited as the cause of *jñāna-vyavahāra.* Nor can knowledge be established on the ground that memory depends on prior knowledge, and unless knowledge is admitted, memory cannot be explained. For, the *siddhāntin* does not admit memory as valid knowledge.

The opponent, further, tries to prove the super-sensuous character of knowledge on the ground of its being momentary. The argument is this: 'Knowledge is super-sensuous, because it is momentary.' This argument, however, suffers from the fallacy of either *vyabhicāra* or *asiddha.*

If by momentariness is meant destruction of a thing immediately after its production, there arises *vyabhicāra* (discrepancy) in respect of desire, volition, pleasure, pain, cognition of duality, etc., since they are not destroyed immediately after their origination.

If, on the other hand, momentariness means one moment existence, it is *asiddha* (unestablished) in respect of knowledge, because knowledge must exist at least for two moments, one moment for its origination and another for the origination of the subsequent knowledge which destroys the first one.

4.iii. The Bauddha advocates that perceptual knowledge deals with *vartamāna-viṣaya* and it is momentary. The *siddhāntin* examines this contention and proves that neither the object of knowledge nor the knowledge is momentary.

To begin with, when perception arises in the form 'this is a jar' — the jar is an existent object *(vartamāna-viṣaya)* and is momentary. Enquiry may be made whether the moment at which the perception of jar is originated is apprehended at the second moment also, or some other moment is apprehended, or none is apprehended.

It is not the first, because the moment at which the knowledge originates is destroyed at the second moment; so it becomes past, and according to the opponent perceptual knowledge deals only with *vartamāna-viṣaya.* Past or non-existent object cannot be taken as *vartamāna-viṣaya,* because what is valid at the first moment lacks validity at the second moment.

The second alternative is untenable, because it is not reasonable to maintain that something which is not apprehended at the first moment can be apprehended at the second moment. If anything has the capacity to be apprehended, it can be apprehended at the first moment as well. If, however, it is apprehended at the first moment, it can no longer be maintained that perception deals only with *vartamāna-viṣaya,* because what is apprehended is only the future moment and not existent moment.

The third alternative is rejected on the ground that if no moment is apprehended, there being absence of object, knowledge will not arise at all.

Further, the opponent opines that knowledge exists for one moment and it apprehends the object at that very moment. Let us examine the purport of the expression 'knowledge apprehends the object' *(jñānaṁ gṛṇhāti).* The expression does not mean either that knowledge is the substratum of the knowledge of object or that knowledge possesses *jñātatānukūla-vyāpāra* pertaining to the object. On the contrary, it means that knowledge is objectified *(viṣayī karoti).* In fact, says the *siddhāntin,* knowledge exists for the second moment also, and what is perceived at the first moment is perceived at the second moment also.

Nor can it be objected that knowledge of something non-existent as existent is erroneous, because the jar that is perceived at the first moment is perceived at the second moment also, and nothing contradictory is perceived.

It is not proper to maintain that a thing which is non-existent cannot be the object of perception, that is, the first moment which is absent at the second moment cannot be perceived at the second moment, For, when the first moment was apprehended, it was then existent, and a thing which is already perceived can be perceived later on also, even though it is destroyed. Knowledge may continue, even though the moment at which the object was perceived is destroyed. Knowledge and object of knowledge are not identical, even in the absence of one the other may continue. Just as object exists for sometime prior to the generation of knowledge, so also after the object is destroyed knowledge continues for one moment.

From this discussion it is clear that the Bauddha contention that perceptual knowledge is momentary cannot be sustained; as a result knowledge cannot be proved to be super-sensuous on the ground of its being momentary.

4.iv. The *siddhāntin* now shows how perception of momentary knowledge is not possible. Knowledge is not self-cognized, because it is an object like other objects, viz. jar, etc. *(na svaprakāśaṁ, vastutvāt itaravastuvat).* A jar is not perceived by itself, but it is the object of knowledge; similarly, knowledge too must be the object of some other knowledge. But knowledge cannot be apprehended by another knowledge, because simultaneous origination of two cognitions being impossible, one cannot apprehend the other at the time of its origination. It cannot be apprehended by the subsequent one that follows the original one, because the original knowledge does not exist at the time when the subsequent one arises, and the time when it exists the subsequent one being absent the latter cannot apprehend the former. Otherwise, if the subsequent knowledge apprehends the non-existent original one, the contention that perceptual knowledge deals only with existent object cannot be sustained.

Similarly, were *jñātatā* momentary, it would defy perception. *Jñātatā* is not self-perceived, because it is an object *(na svaprakāśaṁ vastutvāt)*, so it must be apprehended by the knowledge. *Jñātatā* cannot be apprehended by the knowledge which gives rise to *jñātatā*, because it being produced at a time later is absent at the time when knowledge originates. Nor can it be held that knowledge apprehends *jñātatā* after it is produced, because that which is not apprehended earlier cannot be apprehended later also. *Jñātatā* cannot also be apprehended by the knowledge which is produced at the time of the generation of *jñātatā* for the following reason. Perceptual knowledge requires prior existence of the sense-contact which gives rise to perception. Prior sense-contact which produces that knowledge at the time of the generation of *jñātatā* does not exist at the time when *jñātatā* is generated. *Jñātatā* cannot be apprehended by the knowledge which is produced after the generation of *jñātatā*, since at that time *jñātatā* ceases to exist.

Let us examine how *jñātatā* is momentary. In the perception of an existent jar, viz. 'this is a jar', the knowledge having as its object an existent jar qualified by an existent moment, is to be inferred from *jñātatā*. It is not possible to apprehend *jñātatā* which is momentary. The position may be stated thus: 'Is this *jñātatā* at the second moment qualified by the same moment at which it is generated, or qualified by some other moment, or not qualified by any moment at all?' It is not the first, because the moment at which it is generated does not exist at the second moment. It is not the second also, for *jñātatā* lacks the power to move, so it cannot be qualified by any moment other than the moment at which it is generated. Moreover, it is erroneous if the knowledge existing at a particular moment has *jñātatā* as its object existing at some other moment. Just as the knowledge of silver possessing *jñātatā* belonging to shell is erroneous, so also is erroneous the knowledge that exists at a particular moment having an object, viz. *jñātatā*, existent at some other moment. Nor can the knowledge existent at a particular moment have the next moment as its object, because the next moment is yet to exist.

The third alternative is untenable, because in that case *jñātatā* will be devoid of its nature, as it is neither qualified by the moment at which it is originated, nor by any other moment. Consequently, *jñātatā* will not arise at all.

It may be said that *jñātatā* qualified by the moment at which it is generated can be present at the next moment as being implied *(upalakṣitā)* by that moment, even though the first moment ceases to exist at that time. Just as colour, etc. exist for one moment even after the destruction of their substrata, so also *jñātatā* may continue after its destruction. The continuation of *jñātatā*, qualified by the moment of its origin after the destruction of that moment, may be explained by an example. A person who sees a tree with a crow perching on it on a particular occasion identifies the tree on a later occasion as the tree on which a crow was perching, although on the later occasion the tree is free from any relation to crow. As the tree is identified by the remembrance of the fact of a crow's perching on it on a previous occasion, similarly *jñātatā* is identified as qualified by the moment of its origin by the fact that it was previously apprehended so qualified. But the difficulty in this view is that in that case *jñātatā* will not be an existent object. If a previously existing entity can be acclaimed as an existent object, then the knowledge qualified by the moment of its origin can also continue after the destruction of that moment. According to the Mīmāṁsaka by the presented character of an object of perceptual knowledge is meant only existent *jñātatā* which is produced in the object. If the continuation of *jñātatā* even in the absence of that object is treated as an existent object, then an object which no longer exists and one which is yet to exist may as well be taken as existent objects.

Therefore, in order to retain the presented character of *jñātatā*, it must be admitted that *jñātatā* ceases to exist as soon as the moment at which it is originated comes to an end. From this discussion it is established that *jñātatā* is momentary, and for that very reason it cannot be perceived.

This problem can be avoided, according to the *siddhāntin*, if *jñātatā* exists for a longer time. In the perceptual knowledge of a jar, *jñātatā* produced in the jar continues for more than one moment and as the knowledge is inferred from *jñātatā*, knowledge also similarly exists for more than one moment. Since both *jñātatā* and knowledge exist for more than one moment, neither of them is momentary. And the contention that perceptual knowledge deals with *vartamāna-viṣaya* can also be sustained.

4.v. The opponent adduces one more reason to prove the supersensuous character of knowledge. If knowledge is apprehended by the sense-organ, says the opponent, then the knowledge of a jar will be due to the contact, viz. *manaḥ-saṁyukta-samavāya*. As such contact is always present, the perception of the knowledge of the jar will give rise to the perception of the perception of that knowledge, and so on. Consequently, there arises the contingency of a series of mental perceptions, and knowledge of other external objects will not arise. The outcome of this argument is that if knowledge is apprehended by the sense-organ, knowledge of different external objects will not arise at all. But the knowledge of various external objects does arise; therefore, it must be admitted that knowledge is super-sensuous. The argument may be stated thus: 'Knowledge is super-sensuous, because there arises knowledge of different external objects' *(bāhyaviṣayasañcāritvāt).*

It may be maintained that the perception of a thing arises only when there is the desire to perceive that thing; thus, sometimes there arises the perception of external object, at other times perception of knowledge according to the respective desire. This position is not acceptable to the opponent. The cause of desire is doubt, doubt arises from the knowledge of external object. The knowledge of external object may be generated either by mere sense-contact without depending on desire or may depend on desire for its generation. If it is the former, owing to the sense-mind contact there will arise an infinite number of mental perceptions and knowledge of other external object will not be generated. If the latter, that is, if desire gives rise to knowledge, then also there is a difficulty. The cause of desire is doubt, the cause of doubt is the knowledge of external object, the knowledge of external object again depends on desire — thus arises a vicious circle. As a result, knowledge of external object will not arise at all.

In reply to this objection the *siddhāntin* observes that the same difficulty arises in respect of the knowledge of *jñātatā* also. If perception of *jñātatā* which is produced in the jar by the knowledge of the jar is admitted, then there will arise the perception of that *jñātatā* which is produced in the *jñātatā* by the knowledge of *jñātatā*, and so on *ad infinitum.*

To say that perception of *jñātatā* arises only when there is desire to perceive it and in the absence of that desire perception of other object arises, is untenable. For, desire depends on the prior knowledge of *jñātatā* which in turn depends on desire, consequently the knowledge of *jñātatā* does not arise at all.

If the knowledge of *jñātatā* arises without depending on desire, then owing to sense-mind contact there arises an infinite series of the knowledge of *jñātatā*. If, on the other hand, it depends on desire, the knowledge of *jñātatā* itself turns out to be an impossibility. Thus, the same difficulty which has been pointed out in respect of the knowledge being sense-perceived besets the perception of *jñātatā* too.

It may be said that after the cessation of the sense-contact with the external object, e.g. a jar, knowledge of *jñātatā* ceases to arise owing to the absence of sense-contact and then there may arise knowledge of other external object that comes in contact with the sense.

This position is untenable, because in this way knowledge of *jñātatā* residing in the object may not arise, and there may arise a series of knowledge of *jñātatā* inherent in the soul, which is generated by such cognitions, as 'I am happy', etc., because the contact, viz. *manaḥ-saṁyukta-samavāya,* the cause of such cognitions is always present.

If it is said that it is the nature of the knowledge of *jñātatā* that some knowledge of *jñātatā* arises even in the absence of desire to apprehend it but not others, then the same would be the case with knowledge also. It is the nature of knowledge to apprehend some object, even though there is absence of desire to apprehend it and not to apprehend some other object. Thus, when knowledge ceases to apprehend itself, it apprehends external object. Objection may be raised that nature lacks authenticity to be the determinant of anything, and only a cause can be a determining factor. Therefore, desire to apprehend must be the cause of the knowledge of *jñātatā*. This objection has already been replied to as it has been mentioned that desire, in the absence of the knowledge of external object, does not arise.

The opponent may say that though a particular knowledge of *jñātatā* produced in a jar might not have generated in the past, yet such kind of knowledge of *jñātatā* might have generated in the past and desire in general may arise from the remembrance of such knowledge, and owing to that desire knowledge of *jñātatā* which is produced by the knowledge in the object due to visual sense-contact can very well generate. In the absence of that desire knowledge of other external object takes place.

To this the *siddhāntin* rejoins that perception of knowledge and perception of external object too take place in the same way. Though a particular knowledge might not have been generated in the past, yet owing to desire produced by the remembrance of such kind of knowledge which had been generated in the past the perception of knowledge takes place and in the absence of that desire perception of other external object is generated.

Therefore, as the probans, viz. perception of different external objects, does not prove the super-sensuous character of knowledge, i.e. the probandum, it involves the fallacy of *vyāpyatvāsiddha*. The inference, thus, stands invalidated and thereby rejected.

4.vi. The next argument formulated by the opponent to prove that knowledge is super-sensuous is this: 'Knowledge is super-sensuous, because it is *nirvikalpaka* (indeterminate) and not *savikalpaka* (determinate)'. A *nirvikalpaka* knowledge is that in which there is no *prakāratā,* that is, that knowledge is devoid of the property of a particular cognition which distinguishes it as the cognition of a particular object from other cognition. The *savikalpaka* knowledge is the knowledge of the qualified object. *Nirvikalpaka* knowledge is a knowledge of bare object without the qualities. As *savikalpaka* is the knowledge of qualified object, according to the Nyāya *nirvikalpaka* precedes *savikalpaka.*

The opponent objects that knowledge cannot be *savikalpaka,* because it does not exist for three moments. Knowledge is originated at the first moment, at the second moment arises the *nirvikalpaka* knowledge, and at the third moment knowledge in its *savikalpaka* form is generated. The object of *savikalpaka* knowledge must be present at the time when the knowledge is generated, because perceptual knowledge consists in perceiving an existent object. But knowledge does not continue for three moments, as knowledge which is produced at the first moment is destroyed as soon as *nirvikalpaka* knowledge is produced. The special quality of an all-pervading substance is destroyed by the next special quality produced immediately after it. Thus, knowledge which is a special quality of all-pervading soul being produced at the first moment is destroyed by the *nirvikalpaka* knowledge which is generated at the second moment. And a *nirvikalpaka* knowledge without being *savikalpaka* is not knowledge proper, because *nirvikalpaka* form according to the Nyāya is not actually perceived but is inferred from the *savikalpaka* one. Thus, as knowledge does not continue for three moments, it cannot be *savikalpaka;* it is *nirvikalpaka* which cannot be perceived. Therefore, knowledge, says the opponent, is supersensuous.

It cannot be maintained that as in the knowledge of *samavāya* and *abhāva, savikalpaka* knowledge arises without *nirvikalpaka* one, so also *savikalpaka* knowledge may be generated at the second moment after the origin of knowledge without the *nirvikalpaka* one. *Samavāya* and *abhāva* are apprehended as qualified by their counter-correlates which are their *viśeṣaṇas.* Apart from these *viśeṣaṇas,* the knowledge of *samavāya* and *abhāva* do not have any other *prakāra* or property, viz. *abhāvatva,* etc. Counter-correlates, i.e. *viśeṣaṇas,* of *samavāya* and *abhāva* are apprehended beforehand, and so knowledge of *viśeṣaṇa* being already there *savikalpaka* knowledge of *samavāya* and *abhāva* arises even in the absence of *nirvikalpaka* knowledge. Similar is the case with inferential knowledge also. For example, in the inferential knowledge 'there is fire in the hill', 'fire', the *viśeṣaṇa,* is apprehended during the knowledge of *parāmarśa,* and so the *savikalpaka* knowledge, viz. the knowledge that 'there is fire in the hill', arises immediately after perceiving 'smoke' in the hill, though the knowledge of *viśeṣaṇa,* viz. fire, does not arise at that time.

The opponent objects that *jñānatva, pratyakṣatva, anumititva,* etc. which are the *viśeṣaṇas* in the respective knowledge cannot be apprehended beforehand, and in the absence of the knowledge of *viśeṣaṇa,* the knowledge of *viśeṣya* being impossible, it cannot be maintained that *savikalpaka* knowledge arises without the *nirvikalpaka* one.

To this the *siddhāntin* replies that all *savikalpaka* knowledge are not preceded by *nirvikalpaka* knowledge, the former may arise without being preceded by the latter. Thus, first the knowledge of a jar is generated, then arises the *nirvikalpaka* knowledge, viz. the knowledge of *jñānatva,* then again the knowledge of the jar arises, and this time the knowledge of the jar is *savikalpaka* which is generated without being preceded by the *nirvikalpaka* knowledge of the jar. This time contact of the sense-organ with the jar is competent to give rise to *savikalpaka* knowledge, because the knowledge of *jñānatva,* which is a *nirvikalpaka* and which is a necessary condition for the generation of *savikalpaka* knowledge, has already been generated at the time of the production of the second knowledge.

Though *jñānatva* is apprehended in connection with the previous knowledge of the jar, and not as the *viṣaya* of *savikalpaka* knowledge which arises later, yet there is nothing wrong in apprehending it earlier, as *jñānatva* is the same.

If it is maintained that *savikalpaka* knowledge arises only when *viśeṣaṇa* is known as residing in the particular object, then inferential knowledge in its *savikalpaka* form will never arise. For, in the inferential knowledge 'there is fire in the hill', fireness *(vahnitva),* the *viśeṣaṇa* of fire, is not apprehended before as residing in the fire in the hill. So, in the absence of the knowledge of *viśeṣaṇa, liṅga-parāmarśa* which gives rise to inferential knowledge, being devoid of accessory condition, viz. the knowledge of *viśeṣaṇa,* will not give rise to the *savikalpaka* inferential knowledge.

Therefore, it must be admitted that the knowledge of *jñānatva* which is the *viśeṣaṇa* of a previous knowledge can be the accessory condition to *savikalpaka* knowledge when the same kind of knowledge arises again.

Thus, the opponent's argument that knowledge is super-sensuous, because it is *nirvikalpaka* and not *savikalpaka,* stands rejected, for the probans is unestablished *(asiddha).* The probans for the denial of knowledge as *savikalpaka* lies in the fact that knowledge does not exist at the third moment when it arises in its *savikalpaka* form. But the *siddhāntin* has proved that *savikalpaka* knowledge arises even at the second moment, provided there is previous knowledge of *viśeṣaṇa* which is *nirvikalpaka* knowledge. Thus is the probans *asiddha.*

4.vii. The opponent has a further say in this matter. He argues that knowledge cannot be perceived by the sense-organ, because it is knowledge like the knowledge of super-sensuous object. The purport of this argument is that mental perception of knowledge is possible only when the object of knowledge comes in contact with the mind through the direct external sense-object contact. But knowledge which has super-sensuous object, viz. atom, etc., cannot be perceived by the mind, because atom, etc. cannot come in contact with the mind through direct external sense-contact. Atoms being super-sensuous cannot be in contact with the external sense-organ which is in conjunction with the mind. Hence, direct perceptual knowledge of atoms is not possible. Consequently, atoms cannot be the objects of *savikalpaka* knowledge which is a direct perceptual knowledge.

The opponent intends to say that just as mental perception of the knowledge of super-sensuous object is not possible due to the absence of contact of the object with the mind through external sense-contact, so also mental perception of the knowledge of other objects is not possible. Therefore, knowledge is super-sensuous.

The *siddhāntin*, in reply, points out that external sense-object contact is the cause of external perception. In internal mental perception external object is perceived through *jñāna-lakṣaṇa-pratyāsatti.*

Jñāna-lakṣaṇa-pratyāsatti is a kind of perceptual knowledge in which one perception gives rise to another, as when one perceives a piece of sandal-wood at a distance, one at once knows that it is fragrant. Here fragrance could not be perceived by the nose, as the sandal-piece was at a distance; it is, therefore, apprehended by a kind of extraordinary perception, called *jñāna-lakṣaṇa-pratyāsatti.* That is, though fragrance is not perceived by the olfactory sense-organ at the time when visual sense-organ comes in contact with the sandal-wood, yet it was perceived before. This time the knowledge of fragrance arises as the *viśeṣaṇa* of the knowledge of sandal-wood, because as soon as sandal-wood is perceived the mind comes in contact with the previous knowledge of fragrance, which, therefore, becomes the *viśeṣaṇa* of the former.

The mental perception of knowledge of atoms is also possible through *jñāna-lakṣaṇa-pratyāsatti.* The knowledge of atoms arises through inference, and so it can be the object of mental perception through the contact of the mind with the previous inferential knowledge. Similarly, when the knowledge, viz. 'this is a jar', is generated earlier, the knowledge of the jar can be perceived by *jñāna-lakṣaṇa-pratyāsatti,* that is, through the contact of the mind with the previous knowledge. If *jñāna-lakṣaṇa-pratyāsatti* is not admitted, then memory of the things that are not subject to sense-perception also will not arise, since memory arises only of those things that have already been apprehended, that is, memory depends on previous knowledge.

It is not that memory arises from impression, impression arises only when *anubhava* ceases to exist. *Anubhava* that exists even one moment earlier can be again apprehended by the mind without depending on its impression.

Since mind perceives atoms which have already been known through inference, it cannot be concluded that this knowledge is memory. For, the previous inferential knowledge of atoms which is apprehended in mental perception through *jñāna-lakṣaṇa-pratyāsatti* is not apprehended before; therefore, it is not memory. Nor can it be said that as the object of knowledge, viz. atoms, is already apprehended; so far as the object-part is concerned it is memory. Just as in *pratyabhijñā* (recognition), e.g. 'this is that jar', the object qualified by the previous place and time is apprehended as the *viśeṣaṇa* of the knowledge, so also are apprehended atoms as the *viśeṣaṇa* of the mental perception. It cannot be said that in *pratyabhijñā* the knowledge of object is acquired through memory. The same knowledge cannot be both perceptual as well as memory. *Anubhava* consists in knowledge other than *smṛti* (memory). Hence, *anubhava* and *smṛti* being completely different from each other, *pratyabhijñā* which is an *anubhava* cannot be *smṛti*.

In *pratyabhijñā* the object recollected actually exists before our eyes, and the novelty of the knowledge consists only in the identity of the object now perceived with some one previously seen, as when on seeing Devadatta one recollects that it is the same Devadatta whom one saw before. Here the actual perception of Devadatta is as much the cause of knowledge as the impression left in the mind by a former perception. In memory the impression is the immediate cause. Memory consists in an idea which recurs to the mind without the operation of an external object on the sensory nerve and is, thus, opposed to perception.

Impression is a common causal factor both in *pratyabhijñā* and *smṛti*. But there is a difference between them. In *pratyabhijñā*, e g. 'this is that jar', there is sense-object contact, i.e. contact of the visual organ with the jar, apart from the impression left by the previous perception of the jar. In *smṛti* there is no such sense-object contact; thus, *smṛti* is produced by the impression left in the mind by a former perception of an object. Since *pratyabhijñā* consists of a knowledge which is not apprehended before, it is *anubhava* and not *smṛti*. But even this fact, the opponent may say, does not prove *pratyabhijñā* to be perceptual in character, since in this sense inferential knowledge is also *anubhava*. The *siddhāntin* replies that *pratyabhijñā* is perceptual in character, because besides impression, it is generated by the mind-external sense contact, without depending on the knowledge of *liṅga*, etc. Inferential knowledge cannot be categorized with *pratyabhijñā*, because it depends on the knowledge of *liṅga*, etc.

The question may arise as to the exact nature of the contact in respect of *pratyabhijñā*. *Pratyabhijñā* is qualified knowledge. In respect of knowledge in general the contact is *manaḥ-saṁyukta-samavāya*, i.e. the inherence of the knowledge in the soul which is in conjunction with the mind. In respect of the object which is *viśeṣaṇa* of *pratyabhijñā* knowledge, the contact is *saṁyukta-samaveta-viśeṣaṇatva*, i.e. the visual organ is in conjunction with the object of knowledge *(saṁyukta)* which is in inherence relation with the soul *(samaveta)* and of which the object is *viśeṣaṇa (viśeṣaṇatva)*.

It may be mentioned that in mental perception of external objects mind does not work independently of the external sense-organ. For, if the mind works independently, then a blind man can perceive colour and a deaf person can hear sound, as external sense-organ is not needed for the generation of such knowledge. This difficulty does not arise, because mind perceives external objects in dependence on the knowledge produced by the external sense-organs and not independently.

From this discussion it may be concluded that as the knowledge of super-sensuous object can be perceived by the sense-organ, the inference, viz. 'knowledge cannot be perceived by the sense-organ, because it is knowledge, like the knowledge of super-sensuous object', stands invalidated, because the example is unestablished.

4.viii. So far the *siddhāntin* has shown how knowledge cannot be proved to *be atīndriya* by any of the probans advanced by the opponent. Now he puts forth the *pramāṇa* that proves knowledge as *aindriyaka* (sensuous). *Pratyakṣa-pramāṇa* (perception) is the means of perceiving knowledge. All have such mental perception as 'I perceive this', 'I do not perceive this' — whereby the former suggests that knowledge is present in the soul and the latter as absent in the soul. From the perception of presence and absence of knowledge in the soul, it can be safely concluded that knowledge is perceived by the sense. Thus, states Gautama in his *Nyāya-sūtra: jñānavikalpānāṁ ca bhāvābhāvasaṁvedanādadhyātmam* (5.1.31).

The opponent attempted to prove *jñātatā* on the ground that knowledge is super-sensuous and so *jñātatā* must be admitted as an effect produced by the knowledge in the object from which knowledge is inferred. The *siddhāntin* has proved that knowledge is not super-sensuous; therefore, the opponent's attempt to establish *jñātatā* on that ground is futile. Consequently, the opponent's contention that God's knowledge is not valid, because it consists of *gṛhītagrāhitva*, i.e. it consists in apprehending an object wherein is produced *jñātatā*, stands disproved.

> **मितिः सम्यक् परिच्छित्तिस्तद्वत्ता च प्रमातृता।**
> **तदयोगव्यवच्छेदः प्रामाण्यं गौतमे मते ॥५॥**
>
> *Valid knowledge is true experience and right knowing is the possession thereof; authoritativeness, according to Gautama, consists of being devoid of all absence of knowledge.*

5.i. The opponent, however, is not ready to give in; he seeks to disprove the validity of God's knowledge in some other way. Valid knowledge being a *kriyā* is always produced by *pramāṇa* or means of valid knowledge. God's knowledge being eternal is not produced by any *pramāṇa;* hence, His knowledge is not *pramā.* Nor is God's knowledge *pramākaraṇa* (instrument of valid knowledge), because it does not give rise to valid knowledge. As the knowledge of God is neither *pramā* nor *pramākaraṇa,* God cannot be established as the *pramātā* (substratum) of valid knowledge.

The *siddhāntin*, in reply, points out that *pramā* is *yathārtha anubhava*, and *pramātṛtva* (right knowing) is the possession thereof; authoritativeness *(prāmāṇya),* according to Gautama, means being separated from all absence of knowledge, i.e. being possessed of knowledge always.

Since *pramā* is defined as *yathārthānubhava,* it is not inconsistent with God's knowledge, even though His knowledge is eternal and not produced. Right knowing, i.e. being a right knower *(pramātā),* means being connected with *pramā* by intimate relation, i.e. the relation of inherence which connects a substance and its qualities. This can be established in the case of God also. God is *pramātā,* i.e. agent of *pramā,* but His agency consists in His being *āśraya* or substratum of *pramā.* Agency in respect of God must not be taken in its usual meaning, i.e. in the sense of *pramāṇa-kartā,* as His knowledge is eternal, but in that of *āśraya.*

In the same manner God is an authority *(pramāṇa)* as being ever connected with *pramā,* i.e. as being ever separated from all absence thereof. *Pramātva* in respect of God cannot be defined as *anityatve sati yathārthānubhavatva.* The epithet *anityatva* (non-eternality) does not apply to God's knowledge. If God's knowledge is proved as *pramā*, He is being always connected with knowledge, the epithet *anityatva* is futile. Again, if eternal knowledge of God is unestablished, it is futile to distinguish His knowledge by *anityatva.* Nor can God's knowledge be considered as *apramā*, because it is not produced knowledge. For, if God Himself is unestablished, there being no substratum the question of His possessing valid knowledge does not arise. And if His existence is established, the same reason that establishes His existence will establish His possession of valid knowledge too.

Pramātṛtva does not mean inherence of *pramā* in a substratum which is *kartṛ-kāraka.* The epithet *kartṛ-kārakatva* is not applicable to God. If the existence of God is established, His *pramātṛtva* too is established, as *pramātṛtva* cannot be separated from God. If, on the other hand, God's existence itself is unestablished, the question does not arise whether He is *kartṛ-kāraka* or not. *Pramātṛtva,* according to the *siddhāntin*, means *pramāsamavāyānuyogitva,* i.e. the fact of *pramā's* being present in the substratum *(anuyogi)* through the relation of inherence.

5.ii. It may be objected that in the *Nyāya-sūtra* God has been declared as *pramāṇa,* but not as *pramātṛ*. Therefore, His *pramātṛtva* is not *prāmāṇya.* The *sūtra* runs thus: *mantrāyurvedaprāmāṇyavacca tatprāmāṇyamāptaprāmāṇyāt* (2.1.68). That is, 'the trustworthiness of the word of the Vedas (i.e. an instrument of right knowledge) is based upon the trustworthiness of the reliable (veracious) expositor, just like the trustworthiness of Incantations (against poison, etc.) and Medical Scriptures'. The reliable expositor, according to the Nyāya, is God.

In answer to this objection the *siddhāntin* says that as *pramātṛtva* means *pramā-samavāya* and *pramāṇa* means relation of *pramā* with the substratum always, both the words, *pramātṛtva* as well as *pramāṇa,* can be used in respect of God.

The opponent further objects that God is neither endowed

with *pramāṇatva* nor with *pramātṛtva.* The word *pramāṇa* means instrument of valid knowledge *(pramīyate anena iti pramāṇam)* — hence, *pramāṇa* signifies *karaṇa-kāraka.* The word *pramātṛ* means one who is the knower of valid knowledge *(pramiṇoti iti)* — hence, it signifies *kartṛ-kāraka.* As God is neither *karaṇa-kāraka,* nor *kartṛ-kāraka,* He cannot be either *pramāṇa* or *pramātṛ.*

The *siddhāntin* replies that the word *pramātṛ* signifies *kartṛ-kāraka* according to its derivative meaning. But the ground for the use of the word *pramātṛ* is *pramā-samavāya-anuyogitva,* i.e. the fact of the presence of *pramā* in the substratum by the relation of inherence. In this sense God is no doubt a *pramātṛ.* God is *pramāṇa* in the sense that He is ever connected with His knowledge which is eternal.

The opponent may say that if God is taken as a *pramāṇa* in the sense mentioned above, then there must be five means of valid knowledge, four instrumental causes (*karaṇa*) and the fifth one is *pramā'yoga-vyavacchinna.* Consequently, the Naiyāyika's old division of four means of valid knowledge will not hold good. This *pramāṇa* is not included in perception, inference and verbal testimony, because its *karaṇa* is not *indriya, liṅga* and *śabda.* It goes without saying that this *pramāṇa* is not *upamāna.*

This objection is untenable, because God's knowledge being direct valid knowledge *(sākṣātkāripramā)* is perceptual in character, hence, it comes under *pratyakṣa-pramāṇa.* Perceptual knowledge has been defined in the *Nyāya-sūtra* as that which is produced out of sense-object contact, but this definition applies only to the knowledge of *jīva,* it does not comprise God's knowledge.

5.iii. The knowledge of God is not valid, because it embraces not only truth but error too — this is the last argument set forth by the opponent to disprove the validity of God's knowledge. God's knowledge is erroneous in the sense that He cognizes the erroneous knowledge which we possess. Knowledge is knowledge of object, so God's knowledge of error is also knowledge of erroneous object. If God's knowledge does not comprise error, He would lack omniscience and one cannot rely on the instructions laid down by a non-omniscient God.

This charge is unsound. Error falls short of valid knowledge, but the knowledge that 'this is erroneous knowledge'is not erroneous, it is valid knowledge. Otherwise, there will be no criterion to single out a particular knowledge as erroneous. It is valid knowledge on the basis of which the existence of an object is established. But error is not valid knowledge. The nature of valid knowledge is not violated so long as the knowledge is not associated with a contradictory object, i.e. so long as one does not apprehend silver in what is not silver but nacre. Error consists in the cognizing of attribute in a locus where it is absent. For example, silver is absent in the nacre, silverness exists in silver not in nacre. Cognition of silverness in nacre which has the absence of silverness owing to its not being the locus of silverness is called error. But God's knowledge does not consist in cognizing an attribute in a locus where it is absent. It is a fact that in error there is a definite object as nacre, and also that it is viewed under the notion of silver, and God knows this fact. The validity of God's knowledge cannot be impaired by His apprehending this fact. Thus, the definition of valid knowledge, viz. *yathārthānubhava*, tallies with God's knowledge about the erroneous knowledge possessed by human beings. It is, therefore, established that God possesses valid knowledge, and He is *pramātā* as well as *pramāṇa*.

साक्षात्कारिणि नित्ययोगिनि परद्वारानपेक्षस्थितौ
भूतार्थानुभवे निविष्टनिखिलप्रस्ताविवस्तुक्रमः।
लेशादृष्टिनिमित्तदुष्टिविगमप्रभ्रष्टशङ्कातुषः
शङ्कोन्मेषकलङ्किभिः किमपरैस्तन्मे प्रमाणं शिवः।।६।।

He whose direct eternally related true knowledge which is independent of other means of knowledge consists of the proper succession of all the various objects and which is free from the least doubt owing to the absence of any incidental defects arising out of the slightest ignorance — He, the Śiva, is my authority; why should I care for other authority tainted as it be with rising doubts ?

STABAKA - V

PROOFS FOR THE EXISTENCE OF GOD

An analysis of the first four *stabakas* suggests the need for arguments to prove the existence of God. The first *stabaka* aims at proving that the attainment of the fruits of the sacrifice, etc. is possible through *adṛṣṭa* ensuing from the sacrifice, etc. as directed by God. The second one explains that the validity of the Vedas that enjoin sacrificial rites, etc. depends on God. The third one shows that none of the *pramāṇas* can prove the non-existence of God. The fourth one attempts to prove that God who is the author of the Vedas possesses valid knowledge. Thus, it is clear that the subject-matter of the first four *stabakas* centres around the existence of God. In the fifth *stabaka,* the last section, Udayana proves the existence of God on the basis of inference, and adds that the Upanisadic texts also confirm what has been arrived at on the basis of reasoning.

कार्यायोजनधृत्यादेः पदात् प्रत्ययतः श्रुतेः।
वाक्यात् संख्याविशेषाच्च साध्यो विश्वविदव्ययः।।१।।

An enternal omniscient Being is established on the grounds of products, concretive activity, sustaining effort, etc. (destruction), introduction of empirical usages, infallibility, Śruti, sentences thereof, and particular number.

1. Udayana gives forth his arguments to prove the existence of God hereunder:

i) Earth and such other products *(kārya)* that constitute the created world must be created by a conscious agent who possesses full and definite knowledge of all the details relating to the required causal apparatus. Such an agent who creates the whole world cannot be *jīva* (individual soul), but the Supreme Soul *(Īśvara).*

ii) At the beginning of creation *(sṛṣṭi),* the volitional effort *(yatna)* that leads to the concretive activity *(āyojana)* and out of which contact between two atoms is produced must inhere in a conscious being. This conscious being cannot be *jīva* and must be *Īśvara.*

iii) The fact that various planets are sustained in their position and do not collide with each other must be due to the sustaining effort *(dhṛti)* of some conscious being who is *Īśvara.* And the entire universe beginning with *brahmāṇḍa* and ending with diads undergoes destruction, and destruction must be due to the volitional effort of a conscious being. That conscious being must be *Īśvara.*

iv) The intelligent being who has originally introduced *(pada)* into the world empirical usages, like verbal convention, crafts and arts, like weaving, pot-making, etc., must be God and not the individual soul.

v) The infallibility of the Vedas depends on the validity of knowledge derived from them. And the knowledge derived from the Vedaș is always valid because of the eternal purity of the source from which they have originated, and that source is the omniscient God.

vi) The Vedas are composed by an omniscient being in view of the fact that they are the Vedas, and that which is not composed by an omniscient being is not the Veda as in the case of empirical sentence.

vii) The Vedic texts consisting of sentences must be composed by some intelligent author, and that author is the omniscient God.

viii) The cause of the size of diads *(dvyaṇuka)* is the number 'two' *(dvitvasaṅkhyā)* belonging to two atoms; and two and the higher numbers are all products resulting from the enumerative cognition *(apekṣābuddhi)* of the person who counts. At the beginning of creation, it is the omniscient God to whom such enumerative cognition can be attributed and to none else.

न बाधोऽस्योपजीव्यत्वात् प्रतिबन्धो न दुर्बलैः।
सिद्ध्यसिद्ध्योर्विरोधो नो नासिद्धिरनिबन्धना।।२।।

The probans (in the inferential argument) does not involve the fallacy of incongruity, because (the very attempt to prove the probans fallacious) depends on it; it is not counter-balanced, because the alleged counter-probans are too weak; whether (the creation of the earth, etc. by a non-corporeal agency) is established or not, in either case the probans is not contradictory; the probans is not unfounded, for there is no reason for its being unfounded.

2.i. We shall now explain the above eight grounds of inference successively. The first ground of inference is *kāryatvāt.* The inferential argument is that the earth, etc. *(pakṣa)* must have a creator *(sādhya),* because they are effects *(kāryatvāt),* and that creator must be God. It may be objected that this ground of inference is fallacious in this that it seeks to establish a particular factor in the *pakṣa* wherein that factor does not exist. Absence of *sādhya* in the *pakṣa* is known as *bādha,* and the ground of inference which seeks to establish such a *sādhya* is known as *bādhita.* Here, absence of *sādhya* in the *pakṣa* is decided on the basis of perception. Perception shows that a creator (say, potter) is invariably possessed of a body. From this it is deduced that one who is a creator must possess a body In the present case God who is sought to be established as the creator does not possess body, and so He cannot be the creator. Therefore, God as the creator which is the *sādhya* is not present in the *pakṣa,* and so the ground of inference, viz. *kāryatvāt,* as it seeks to establish a *bādhita-sādhya*, is a *bādhita-hetu.* The result of this argument is that the above inferential argument possessing such a *bādhita-hetu* cannot give rise to a valid inferential cognition.

The objection raised in the foregoing paragraph is refuted by Udayana. It is not correct to say that the *sādhya* in the above inferential argument is absent in the *pakṣa.* It is because that in order to negate the presence of God as a creator in the earth, etc.

which constitute the *pakṣa*, the knowledge of God is essential. For, negation of a thing always presupposes the prior knowledge of that thing. When viewed in this light, if God's existence is negated as related to the earth, etc., then what is necessary is that there must be the prior knowledge of the existence of God as related to the earth, etc. Such a knowledge of God we could arrive at only on the basis of the inferential argument which consists of *kāryatva* as the ground of inference. Thus, since *kāryatva* is the one on which the existence of God depends, it is not proper to say that this ground of inference is *bādhita*.

It may be said that the *hetu-kāryatva* in the above inferential argument has another equal *hetu* on the opposite side. This means that another equal *hetu* seeks to establish that the world is not at all created by any one. The former *hetu-kāryatvāt*, thus, has another *hetu* which seeks to prove the contrary. The *hetu-kāryatvāt* is technically known as *satpratipakṣita*. Udayana argues that no such equal *hetu* can be presented to prove the contrary, viz. absence of a creator with reference to the earth, etc.

It may be objected that the ground of inference, viz. *kāryatvāt*, is *viruddha* in the sense that it is invariably related to the absence of *sādhya*. In the above inferential argument *kāryatva* seeks to establish creation by one who does not possess a body *(aśarīra - kartṛtva)*. The opponent says that *kāryatva* is always invariably related to *saśarīra-kartṛtva* which is *aśarīra-kartṛtvābhāva*. Since in the present case *aśarīra-kartṛtva* is the *sādhya* and since it is said by the opponent that the *hetu-kāryatva* is associated with the absence of *aśarīra-kartṛtva*, the *hetu* is said to be *viruddha*.

2.ii. Udayana replies that if *śarīra-kartṛtva* is established by a *pramāṇa* in the *pakṣa* like the present *sādhya—aśarīra-kartṛkatva*, then the two, viz. *śarīra-kartṛkatva* and *aśarīra-kartṛkatva*, are both present in the *pakṣa*, and in that case we can conclude that certain objects like pot, etc. have *śarīra-kartṛkatva* and certain other objects are *aśarīra-kartṛka*. Thus, there is no reason to believe that the *hetu-kāryatva* is *virduddha*. If *śarīra-kartṛkatva* is not established in the *pakṣa* by a *pramāṇa*, then also there is no reason to believe that the *hetu—kāryatvāt,* is invariably related to *śarīra-kartṛkatva* or *aśarīra-kartṛkatvābhāva,* i.e. *sādhyābhāva.* Thus, the *hetu—kāryatvāt,* is not *viruddha.*

The ground of inference does not suffer from the fallacy of *asiddha* too. This fallacy is of three kinds, *āśrayāsiddha, svarūpāsiddha* and *vyāpyatvāsiddha.* Of these three the first one consists in the total non-existence of the subject of inference itself. The second one consists in the absence of the ground of inference in the subject of inference. The third one consists in the ground of inference being associated with an *upādhi.* The ground of inference must always be the *vyāpya,* and the thing that is to be established, i.e. *sādhya,* must be the *vyāpaka.* In the well-known inferential argument 'the mountain has fire, because it has smoke', there is invariable concomitance between the smoke and the fire in the form 'wherever there is smoke, there is fire'. Here 'smoke' which is the ground of inference is *vyāpya,* and 'fire' which is the thing that is to be established is *vyāpaka.* But in the inferential argument 'the mountain has smoke, because it has fire' — the invariable concomitance must be of the form 'wherever there is fire, there is smoke'. But it is not so. We find a glowing iron-ball, but we do not find smoke therein. So, there cannot be invariable concomitance of the above form, and this is known as *vyāpyatvāsiddhi.*

Udayana argues that none of the three is present in the inferential argument which is under consideration, as the subject of inference is an established fact and the ground of inference exists in the subject of inference. This is as it should be, because the earth, etc. are composed of parts, and since whatever is composed of parts is an effect, the earth, etc. also are effects. And finally, there is invariable concomitance between the state of being an effect and the state of having a creator.

2.iii. The *siddhāntin* analyses the opponent's proposition that 'one who is possessed of a body alone can be a creator'. This may mean one of the following alternatives. (1) God is corporeal, since He is the creator of the universe. (2) God is not the creator, since He does not possess physical apparatus. (3) The earth, etc.

are produced by some corporeal creator, since they are effects. (4) The earth, etc. are not products at all, since they are free from corporeal agency. (5) The earth, etc. are not produced by any agent, since they are not produced by a person possessed of body. (6) The proposition is advanced either in order to contradict the inference of the *siddhāntin* or to cite a rival inference with a view to counter-balancing it.

All these alternatives, according to the *siddhāntin*, are riddled with fallacies. The first and the second counter-inferences are vitiated by a number of fallacies, viz. *āśrayāsiddhi, bādha, apasiddhānta* and *pratijñā-virodha.* Though the opponent does not believe in the existence of God, yet God is presented as the subject of inference in both the arguments. Hence the *āśrayāsiddhi* (non-existence of the subject). If, on the other hand, the existence of God who is the subject of inference is established, the very *pramāṇa* that establishes God's existence would establish His non-corporeal agency also. This invalidates the opponent's position; hence, there results the fallacy of *bādha.* In any way the opponent is constrained to admit the existence of God which leads to the fallacy of *apasiddhānta* (deviation from accepted tenet). Moreover, the word 'God' denotes non-corporeal agency. Hence, any attempt to prove God as corporeal is to commit the fallacy of *pratijñāvirodha.*

The third alternative that 'the earth, etc. are produced by some corporeal creator, since they are effects' is also beset with discrepancy. In this counter-argument the opponent has drawn a universal concomitance between the state of being an effect *(kāryatva)* and precedence of corporeal creatorship. Udayana points out that these two are in no way related through concomitance in presence and in absence. One contrary instance *(vipakṣa)* may be cited that the earth which is a product, as the opponent also agrees, is not preceded by any corporeal creator.

The fourth objection, viz. 'the earth, etc. are not products at all, since they are devoid of corporeal agency', is unsound. For, such a position stands contradicted by perceptual experience. Hence, the probans turns out to be inconclusive *(anaikāntika).*

The fifth argument involves the fallacy of futile attribution *(asamarthaviśeṣaṇatva)*. 'Not being produced' can very well prove that the earth, etc. are not produced by any agent. The adjunct 'by a body' in the reason is superfluous. Hence, the inference suffers from the fallacy of *vyāpyatvāsiddhi* (futile adjunct).

If, however, the opponent discards the adjunct in the reason, then he has to meet with the fallacy of *svarūpāsiddhi*. For, 'not being produced' which would then stand as the probans in the inference does not necessarily subsist in the subject. Neither the opponent nor the *siddhāntin* would admit that the earth, etc. are not produced at all. In connection with the sixth alternative it may be observed that these counter-inferences cannot be advanced by the opponent either to contradict the inference of the *siddhāntin* or to counter-balance it. For, the counter-inferences of the opponent suffer from the absence of subsistence of the probans in the subject *(pakṣadharmatā)*, the knowledge of which determines the validity of inference.

The first counter-inference of the opponent is 'God is corporeal, since He is the creator'. The opponent draws an invariable concomitance between agency and embodiedness. *Pakṣadharmatā* consists in the subsistence of the probans, as pervaded by the probandum, in the subject. Here, the probans, agency, is absent in the earth, etc. which are the subject of the inference. In the negative form the invariable concomitance stands thus: 'Whoever is not corporeal is not the creator'. But non-corporeality, the probans, also does not exist in the subject, the earth, etc. Hence, the contrary invariable concomitance also suffers from the absence of *pakṣadharmatā*. Such inferences, therefore, cannot counter-balance the argument of the *siddhāntin*.

The third and the fourth arguments advanced by the opponent have already been proved to be fallacious.

The opponent may argue that the fifth inference, viz. 'the earth, etc. are not produced by an intelligent agent, since they are not produced by a corporeal being', cannot be charged with the absence of *pakṣadharmatā*. It has already been noticed that this particular inference is riddled with the defect of futile attribution in which case positive concomitance *(anvaya-vyāpti)* turns out to be an impossibility. The positive concomitance in such cases is 'whatever is not produced by a body, is not produced by an intelligent agent'. Such concomitance in its negative form would be 'whatever is produced by a sentient being is produced by an embodied being'. The probans of this negative concomitance, i.e. production by a sentient being, is free from futile attribution. When an inference is based upon the observation of agreement in absence alone, it becomes an exclusively negative inference *(kevala-vyatirekin)*. Negative inference holds good in such cases where positive instance *(sapakṣa)* is not available. But the opponent's inference cannot be an exclusively negative one, because positive instance is available there. A positive instance is that in which the probandum is known to exist. 'Ether' serves as the positive instance in the opponent's inference, since 'ether' is not produced at all, so there is no question of its being produced by a sentient being. It forms the positive instance as the probans 'not being produced by a corporeal being' exists in it. Thus, the opponent's inference cannot be an exclusively negative one. Nor can it be a positive one, since in the probans 'not being produced by a corporeal being', the word 'corporeal being' is superfluous. Therefore, the inference, be it a positive or a negative, is destitute of validity.

To avoid the charge of futile attribution the opponent may change the probans and argue that 'the earth, etc. are not produced by an intelligent being, since they are not pervaded by the precedence of sentient agency'. Though this positive inference is free from the fallacy of futile attribution, it remains to be determined what exactly would pervade the precedence by an intelligent agent. The *vyāpaka* (determinant) of this particular probans has to be found out, otherwise the probans would be defective involving the fallacy of *svarūpāsiddhi.* For, the absence of being a product does not exist in the earth, etc. The precedence of body cannot be the *vyāpaka* here. The absence of the *vyāpaka* leads to the absence of the *vyāpya* (determinate). But in this inference the absence of the precedence of a physical body cannot lead one to infer the absence of the precedence by an intelligent agent. The earth, etc. are certainly preceded by an intelligent agent, but not by a physical body. Thus, as there is no *vyāpaka* of the probans in the opponent's inference, latter cannot counter-balance the *siddhāntin*'s inference.

2.iv. The *pūrvapakṣin* further argues that the *siddhāntin's* inference that 'the earth, etc. are preceded by an intelligent creator, because they are effects', involves the fallacy of *virodha* (contradiction)

In the inference the jar, etc. are similar instances where the probandum, 'the nature of being preceded by an intelligent creator', and the probans, 'the nature of being an effect', co-exist with each other. Precedence of an intelligent agency is admitted in the case of each and every product; hence, it is proved in the case of jar, etc. also. The probans *kāryatva* here proves the existence of an embodied agent of limited intelligence. But difficulty arises in the case of earth, etc. where the very same probans which attempts to prove the existence of an intelligent creator of the universe also tries to prove that the same intelligent creator is an omniscient and omnipotent being. The inference is vitiated by the fallacy of *virodha* so far as the particular feature of the probandum is concerned owing to the fact that the self-same probans *kāryatva* proves the existence of an embodied agent of limited intelligence in one case and an omniscient and non-corporeal agent in the other.

Udayana meets this charge by saying that if an omniscient non-corporeal agent is established in the case of the earth, etc. and a non-eternal and corporeal agent in the case of jar, etc., then both being *pramāṇasiddha,* there is no *virodha.* If, however, one is unestablished, then also there is no *virodha.* For, in that case there being no need of the rejection of the probans and the probandum on either side, *virodha* does not arise. Further, if an omniscient non-corporeal agent is established in the case of the earth, etc., the probans and the probandum being co-existent with each other in the substratum, there is no *virodha.* If, on the other hand, such agent is unestablished, the probandum itself being unestablished there is no question of the probandum's being non-concomitant with the probans *(hetau sādhyāsāmānādhikaraṇya)* or the probans being non-concomitant with the probandum *(sādhye hetvasāmānādhikaraṇya).* As a result there does not arise the contingency of the probans and the probandum being discarded; therefore, the fallacy of *virodha* too does not arise.

Moreover, the Nyāya does not admit that *viśeṣavirodha* is a logical fallacy. The character of the probandum is only mutilated by *viśeṣa-virodha,* the inference is not falsified by it. According to Jayantabhaṭṭa if *viśeṣa-virodha* were taken to be a vitiating factor, the possibility of argumentation itself would be completely annihilated.

2.v. The *pūrvapakṣin* further observes that a particular type of event testifies to the existence of a particular type of cause. The jar, etc. are particular events that guarantee the existence of a particular corporeal efficient cause. But the universe in its entirety never justifies the existence of an embodied intelligent being behind it. The probans *kāryatva* establishes an agent only when that agent is an embodied being. Hence, as the probans is vitiated by the conditioning factor, viz. possession of body, the *siddhāntin's* inference that tries to establish an universal concomitant relation between *kāryatva* and *sakartṛkatva* suffers from the fallacy of *vyāpyatvāsiddha.*

This charge of the opponent arises due to the lack of understanding of the logical basis of *vyāpti* relation (universal concomitance). The opponent accepts the universal concomitance between a particular product and a particular corporeal agency. But he does not accept the possibility of eternal co-existence between product as such and the nature of being preceded by an intelligent agent. This is untenable. For, in that case, universal concomitance between smoke in general and fire in general cannot be established, in so far as scented smoke is the cause of scented fire and dark smoke is the cause of some other kind of fire, and so forth. In fact, a probans in general, for example, smoke in general, is the cause of the probandum in general, namely, fire in general, and a particular probans, for example, particular smoke, is the cause of a particular probandum, namely, particular fire. If universal concomitance between two particular events is accepted and that between two general events is denied, then the so-called particular inference also cannot be established. For, the particular inference also may turn out to be of a general type in respect of the inference which is based on the universal concomitance of the particular effect and the nature of being preceded by a particular corporeal efficeint cause. The universal concomitance, according to Udayana, must be accepted as valid, even though co-existence among individual instances subsists in such cases. In spite of the eternal relation subsisting between the jar as product and corporeal agency behind it, there is no bar in inferring universal concomitance between the product as universal and the nature of being preceded by an intelligent agent.

It may be observed that fire is seen to exist even when only smoke, and not scented smoke, exists. In the same way there may be the precedence of an efficient cause, wherever there is an effect, without there being present any *upādhi*, viz. possession of body.

As there is universal concomitance between unconditioned smoke and fire, similarly universal concomitance can also be established between precedence of an efficient cause and an unconditioned effect. Therefore, the opponent's position that corporeality is concomitant with efficient causality stands rejected. The probans *kāryatva* is not vitiated by any *upādhi*.

The opponent may refuse to accept this position on the ground that production of an effect by non-corporeal agent is not attested by perceptual experience. To this the *siddhāntin* rejoins that if this were the case, then it will be difficult to explain how diad is produced out of two atoms. For, nobody has evidently perceived the production of diad by a corporeal agent. Further, the fact that eternal colour is the material cause of the colour inherent in the gross bodies also does not have perceptual evidence. Again, to say that only a particular effect is produced by a particular corporeal efficient cause and that there is no universal concomitance between effect-as-such and efficient cause-as-such would make it impossible to establish an invariable concomitant relation between triad and the gross body produced out of it. Therefore, inference of a general type must be admitted; otherwise, there arises the contingency of the production of an effect without a cause.

2.vi. The opponent points out further difficulty in the probans *kāryatva*. The *siddhāntin's* inference, according to the opponent, is riddled with the fallacy of doubtful negative *(sandigdha-vyatirekin)*. An inference to be valid needs to have concomitance in presence of the probans and the probandum in the homologue *(sapakṣasatta)* and their agreement in absence in the heterologue *(vipakṣāsatta)*. In the case under consideration concomitance in presence, viz. what is a product is conditioned by some intelligent agent, is available. But concomitance in absence, viz. whatever is not conditioned by some intelligent agent is not a product, cannot be proved. Hence, it is doubtful.

The *siddhāntin* may argue that concomitance in absence is available, as ether may serve as the heterologue. Ether is not conditioned by anything, and it is also not a product. The opponent rejoins that what is proved by the above-mentioned concomitance in absence is that 'whatever is not conditioned by a cause is not a product'. But the same cannot establish that 'whatever is not produced by an intelligent agent is not a product'.

The *siddhāntin*, in reply, says that absence of agency *(sakartṛkatvābhāva)* also leads to the absence of productivity *(kāryatvābhāva)*. Agency is a kind of cause, and the absence of any kind of cause, whether material, instrumental, non-inherent or efficient, guarantees that the object is not a product. The effect does not necessarily come into existence, when there is the absence of an efficient cause, because in this case there is the absence of one type of cause. The opponent should admit this; otherwise, if the effect is supposed to be produced even in the absence of the cause, causality would be thrown overboard.

Non-production of an effect should be determined by the absence of any particular cause. Otherwise, if non-production were due to the absence of cause in general, it cannot be decided whether the absence of effect is caused by the absence of material cause, or of instrumental, or of non-inherent, or of efficient cause.

The opponent enquires whether the agent is a cause and says that unless the agent is proved to be a cause all these aforesaid arguments advanced by the *siddhāntin* will not hold good. To this objection it is replied that 'to have an agent' *(sakartṛka)* means *kartṛjanya,* i.e. agent is the cause of the thing produced. A weaver's agency in respect of the production of a piece of cloth, or an intelligent being's agency in respect of the earth, etc., consists in their being the cause of cloth or of the earth, etc.

2.vii. The opponent advances three more objections against the causal argument put forth by Udayana for the existence of God. The Nyāya holds that the entire universe is created out of atoms, and the atoms are supervised by some intelligent being who is God. The next objection is directed against this conception of God. The opponent argues that the atoms cannot be supervised by God, either directly or indirectly. What is directly supervised by a sentient being is his own body, because direct supervision is possible only in the case of one's own body. Indirect supervision, on the other hand, is conducted through the mediation of body, e.g. a potter's supervision in the case of a lump of clay is indirect, as the former supervises the latter with the help of his body. Since God does not possess a body, the opponent says, He can neither supervise the atoms directly nor indirectly. The argument in its inferential form may be expressed in the following way: 'Atoms are not supervised by any intelligent being (God), because direct supervision being impossible, (because of the absence of body) indirect supervision is also not possible'.

In this argument the probans, viz. direct supervision being impossible, indirect supervision is also not possible, is free from any fallacy; hence, it is a valid one. Atoms are not directly supervised by an intelligent being, because they are other than the body. This argument in its negative form stands thus: 'Whatever is directly supervised by an intelligent being is his own body'. Here our body is the negative instance, because our body is directly supervised by the soul.

Similarly, atoms are not supervised indirectly, because there is absence of mediation of the body. Our body is the positive instance here. Our body is not supervised indirectly, because there is absence of the mediation of body. On the contrary, it is supervised by the soul directly.

This argument in its negative form may run thus: 'Whatever is supervised indirectly is through the mediation of body, for example, stick in respect of a jar'. Stick is supervised by a potter indirectly with the help of his body. Therefore, the opponent says that the probans in the argument 'atoms are not supervised by any intelligent being, because direct supervision being impossible, indirect supervision is also not possible' being a valid one, the argument is also valid.

The opponent intends to say that if atoms are to be supervised by an intelligent being, viz. God, He must have a body. Since He does not have a body, He cannot be the supervisor of atoms, either direcly or indirectly. The opponent further argues that the earth, etc. also cannot be supervised by God. The earth, etc. are not in need of any sentient supervision, because they are caused by one who is not an embodied being. This argument he advanced as a counter-argument *(satpratipakṣa)* against the *siddhāntin's* argument that 'the earth, etc. must have a creator, because they are effects'.

The Nyāya advocates that God, omniscient and omnipotent as He is, is the supervisor of everything. All human productions, even a piece of cloth, stand in need of divine supervision. God's knowledge, desire and volitional activity are boundless, His supervising capacity is also unlimited.

Against this contention the opponent sets forth his next objection. He contends that if God supervises the thread, yarn, etc. Himself, then a weaver's supervision of them stands useless. If in spite of divine supervision the voluntary activity on the part of the weaver is held to be essential, the necessity of one more supervisor who will be other than God and man would be needed. And this process, when continues indefinitely, would lead to the fallacy of infinite regress. Experience, however, shows that it is the weaver who supervises the thread, yarn, etc. for production of cloth. Hence, it is quite in the fitness of things that the weaver is the supervisor of the thread, etc. and not God. God who is postulated as omniscient and supervisor of everything, thus, stands rejected.

It may be argued that though the production of cloth stands in need of divine supervision, yet weaver is also necessary. The weaver's utility consists in its being a reproductive factor *(kāraka).* This contention leads to a further difficulty. The Nyāya holds that whatever is an effect is preceded by a sentient being, e.g. a piece of cloth. Now, if the weaver is needed in respect of the production of cloth only as a *kāraka,* i.e. reproductive factor, then the probandum, viz. precedence by a sentient being in the above-mentioned example, viz. a piece of cloth, stands invalidated. For, reproductive factor in general is insentient, and therefore, the weaver would turn out to be insentient.

Nor can it be argued that God is a reproductive factor in respect of human production. For, in that case the probans 'effecthood' *(kāryatva)* would be concomitant with the probandum 'precedence by a reproductive factor' and not with 'precedence by a sentient being'. Further, this establishes what is already established, for the opponent too admits that the earth, etc. are produced out of non-sentient atoms, etc. Therefore, the existence of a sentient being, viz., God, remains unestablished.

2.viii. The last objection against the theistic inference is as follows. The opponent says that *the siddhāntin's* argument that 'whatever is a product is conditioned by some intelligent creator' is vitiated by a condition, viz. precedence of non-eternal volitional activity *(anityaprayatnapūrvakatva).* Whatever is conditioned by some intelligent being is preceded by non-eternal volitional activity. The condition, thus, is concomitant with the probandum, viz. precedence by an intelligent being. But it is not concomitant with the probans, viz. being a product. Whatever is a product need not necessarily be preceded by non-eternal volitional activity. For the earth, etc. that are products are not preceded by such temporal volitional activity. The opponent contends that since the ascription of universal concomitance between a product and precedence by an intelligent agent depends on another concomitance, viz. whatever is preceded by intelligent being is preceded by non-eternal volitional activity, the concomitance turns out to be invalid.

It cannot be argued, the opponent says, that as possession of body is not the condition of precedence by an intelligent creator, so also non-eternal volition is not the condition of the latter.

Body is the cause of limited, non-eternal intelligence; hence, the absence of the former leads to the absence of the latter. Body is concomitant with non-eternal intelligence only, and not with eternal, unlimited intelligence. That is to say, it is not the determinant of God's intelligence that is eternal. Therefore, it is maintained that possession of body is not the condition of intelligence as such *(sāmānyato-buddhimat-pūrvakatva).* But so far as non-eternal volition is concerned, it is, unlike possession of body, not the cause of intelligence. Hence, it is not the case that with the absence of non-eternal volition, only non-eternal intelligence is absent. On the contrary, absence of the former leads to the absence of intelligence-as-such, both eternal as well as non-eternal. Hence, non-eternal volition, being concomitant with intelligence-as-such, is the condition of the latter.

All these objections have been examined and finally rejected by the *siddhāntin.* In connection with the first objection the opponent has stated that what is directly supervised by an intelligent being is his own body. The *siddhāntin* posits that atoms are directly supervised by an intelligent being, viz. God. Then naturally question would arise: 'Is the opponent ready to admit atoms as the body of God?' Before answering this question the *siddhāntin* enquires what, according to the opponent, is the nature of a body *(śarīratva).* The *Nyāya-sūtra* states that 'body is the substratum of action, senses and enjoyment'. If by body is meant *ceṣṭāśrayatva* (substratum of action) and if *ceṣṭāśrayatva* were the same as *sākṣātprayatnavad-adhiṣṭheyatva* (direct supervision by sentient volitional activity), then the *siddhāntin* would consider the question of atoms becoming the body of God.

If, however, body is defined as the substratum of sense-organ *(indriyāśrayatva),* it may be enquired what is the purpose that a sense-organ subserves. Sense-organ gives rise to cognition, which in turn generates creative effort *(prayatna).* Thus, limited volition *(avacchinna-prayatna)* is determined by sense-organ. But

God's knowledge and volition are eternal, hence, they are not in need of any sense-object contact. His knowledge is unlimited, it relates to all objects. A body in the sense of the substratum of sense-organ is not invariably concomitant with eternal knowledge and volition. Therefore, physical apparatus in relation to divine knowledge cannot be defined as the substratum of sense-organ.

Body as the substratum of enjoyment *(arthāśrayatva)* also is not needed by God. Volition is one of the causes of enjoyment. Enjoyment is due to merit and demerit, which are conditioned by such defects as attachment *(rāga)*, aversion *(dveṣa)*, etc. And these defects are caused by error *(bhrama)*. Error is due to non-apprehension of the specific character of the object. Every object has got two aspects, object as something and object as determined by a specific feature, for example, rope as determined by ropehood. In erroneous perception rope is not cognized as determined by ropehood which is the specific feature of rope. Thus, in erroneous knowledge the specific feature of the object, which differentiates it from other objects, is not cognized. To be precise, erroneous knowledge *(mithyājñāna)* gives rise to defects *(doṣa)*, from defects comes activity *(pravṛtti)* and activity leads to enjoyment *(bhoga)*. Thus, erroneous knowledge is the root-cause of enjoyment. God, however, is free from all erroneous knowledge, because being omniscient He is cognizant of every object with all its peculiarities. As God does not possess erroneous knowledge, He is free from all defects and also merit and demerit which are conditioned by defects. In the absence of merit, etc. He does not have enjoyment also. Therefore, He is not in need of a body which is the substratum of enjoyment.

The opponent's first objection intends proving that atoms are not supervised by any intelligent being, either directly or indirectly. The arguments he advanced are as follows:

i) Atoms are not supervised directly by any sentient being, because they are other than body.

ii) Atoms are not supervised indirectly by any sentient being, because there is absence of mediation of body.

In the previous passage body has been defined in three ways, and if we apply these definitions in these arguments mentioned above, each of them would take three different forms. Following the three definitions the first argument will be in the following forms.

(a) Atoms are not supervised directly by any sentient being, because they are not supervised directly by sentient volitional activity (the definition of body, *ceṣṭāśrayatva,* being the same as *prayatnavadadhiṣṭheyatva*).

(b) Atoms are not supervised directly by any sentient being, because they are other than the substratum of sense-organ *(indriyāśrayetaratvāt).*

(c) Atoms are not supervised directly by any sentient being, because they are other than the substratum of enjoyment *(bhogāyatanetaratvāt).*

All these alternatives, when analysed, are found out to be fallacious. The probans of the first inference being the same as the probandum involves the fallacy of *sādhyasama* (being equipollent with the probandum). The second and the third inferences suffer from the fallacy of *vyāpyatvāsiddha,* because both the probans are not concomitant with the respective probandum.

Both the probans are advanced to prove that atoms are not supervised directly by a sentient being. But in fact, atoms are supervised by God directly, because they are insentient (as all insentient things stand in need of intelligent supervision). In respect of God both the probans are applicable, i.e. atoms are neither the substratum of sense-organ nor that of enjoyment in the case of God, but the probandum does not exist in God. According to the Nyāya, God supervises the atoms directly, because atoms are insentient. Hence, the two probans advanced by the opponent are not concomitant with the probandum.

Following the three definitions of body the second inference takes the following forms.

(d) Atoms are not supervised indirectly by any sentient being, because there is absence of mediation of direct supervision by a sentient volitional activity.

(e) Atoms are not supervised indirectly by any sentient being, because there is absence of mediation of the substratum of sense-organ.

(f) Atoms are not supervised indirectly by any sentient being, because there is absence of mediation of the substratum of enjoyment.

Like the first inference the second inference is not sound, and like the former it suffers from the same defects. In the fourth inference the probans being as doubtful as the probandum involves the fallacy of *sādhyasama*.

The probans of the last two inferences involve the fallacy of *vyāpyatvāsiddha*, as there is no invariable concomitance between the probans and the probandum. Even without the mediation of the substratum of sense-organ or that of enjoyment, atoms may be set into motion by air, etc. Evidently air does not have a physical apparatus. Again, as atoms are insentient, they may be guided by a sentient being indirectly. Therefore, it is decided that atoms may be guided by a sentient being indirectly even without the mediation of a body.

It has been stated that the opponent advances a counter-argument, viz. 'the earth, etc. are not in need of any sentient supervision, because they are produced by one who is not an embodied being', against the *siddhāntin's* argument that 'the earth, etc. must have a creator, because they are effects'.

This counter-argument also can be expressed in three ways following the three definitions of body.

(g) The earth, etc. are not in need of any sentient supervision, because they are not directly conditioned by sentient volitional activity *(sākṣātprayatnādhiṣṭhite-tarajanyatvāt)*.

(h) The earth, etc. are not in need of any sentient supervision, because they are not produced by the substratum of sense-organ.

(i) The earth, etc. are not in need of sentient supervision, because they are not produced by the substratum of enjoyment.

Here also, as in the previous arguments, the first inference involves the fallacy of *sādhyasama*, because the probans and the probandum are the same. In a valid inference the probans is known to exist in the subject, but the existence of the predicate or the probandum in the subject should be dubious. But in this argument the probans is as dubious as the probandum. Hence, the argument is riddled with the fallacy of *sādhyasama.*

The probans of the last two inferences are also fallacious, because they are not concomitant with the *sādhya* or the probandum. The earth, etc. stand in need of God's supervision, because they are insentient. As God possesses eternal knowledge, desire and volition, He is in no need of having a substratum of sense-organ or that of enjoyment in order to supervise the earth, etc.

Further, in the inference 'the earth, etc. are not in need of any sentient supervision, because they are produced by one who is not an embodied being', the probandum is not established in the subject *(pakṣa).* The positive instance in this argument is 'ether'. 'Ether', being an eternal verity, forms the locus of the probandum, i.e. absence of sentient supervision. But the probans, viz. being produced by someone other than an embodied being, does not exist in ether. 'Ether' being eternal is not produced at all. Thus, ascertainment of concomitance in presence is thwarted, turning the probans as uncommon inconclusive *(asādhāraṇa).* Again, negative instance in this argument is jar, etc. A jar forms the locus of the absence of the probans, i.e. absence of production by someone other than an embodied being. But the absence of the probandum, i.e. absence of the absence of sentient supervision, does not exist in jar. Hence, the probandum in the argument is unreal *(sādhyāprasiddhi).* Hence, the counter-argument of the opponent being fallacious stands rejected. So far the examination of the first objection.

2.ix. The second objection aims at disproving God on the ground that if God were the supervisor of everything, including even a piece of cloth, then there is no utility of a weaver in respect of the production of a cloth. Since experience shows that it is the weaver who produces cloth, God, therefore, has been rejected as the supervisor of cloth.

The question raised by the opponent, whether the weaver is the supervisor or a productive factor for the production of cloth, is irrelevant. The weaver is undoubtedly a productive factor in respect of the production of cloth, because he is the agent in respect of the production of cloth. But his being a productive factor consists in his having knowledge of the materials out of which the cloth is produced, desire to produce and volitional effort to put forth. And when the weaver produces cloth in this way, i.e. having the knowledge of materials, desire to produce and volitional effort, he also supervises the cloth.

The opponent has raised an objection that if God supervises even yarn, thread, etc., there is no utility for a weaver's supervision. The *siddhāntin* analyses this objection and says that it may prove one of the following alternatives:

i) futility of human efforts in the production of an effect;
ii) denial of the existence of God; or
iii) the fallacy of infinite regress resulted from the postulation of voluntary activity on the part of the weaver over and above divine supervision.

The first alternative is untenable. The weaver and the piece of cloth are related by concomitance in presence and absence *(anvaya-vyatireka).* Presence of a piece of cloth presupposes supervision by a weaver, and absence of weaver proves non-production of cloth. As the necessity of the weaver in weaving out the cloth is an observed fact, the importance of the practical efficiency of the weaver cannot but be taken into account.

The second charge is also unsound. God has been postulated on the ground that eternal atoms cannot combine to create the molecules and universe out of them, unless they are regulated by some intelligent being. Individual selves possessing limited knowledge cannot regulate, and hence, the postulation of a supreme regulating principle, viz. God, is necessary. As God, according to the Nyāya, is omniscient and omnipotent, it is possible for Him to supervise the process of creation of a piece of cloth also. Had the existence of God been proved merely on the ground that He is the regulator of the yarn, etc., then the criticism of the rival school would have been to the point. For, in that case the weaver would have been the efficient cause through *anvaya-vyatireka* relation, and the postulation of God as a further supervising agent would have been redundant. But as there are other reasons to prove God's existence, the existence of God cannot be invalidated.

The third alternative is also unfounded. This alternative purports to say that if in spite of divine supervision the weaver is essential for the production of cloth, then there would be the necessity of postulating one more supervisor who will be other than God and man, and so on *ad infinitum.* This charge is, however, futile. An effect is produced by an agent, and God being omniscient and omnipotent is the cause of everything. In the production of an effect, apart from these two, no third person is seen to be needed. It may be asked what is the reason behind the postulation of two supervisors? The word 'reason' may mean either cause *(hetu)* or purpose *(phala).* According to the first meaning, the cause that determines supervision on the part of God and human agency has to be pointed out. God's supervision is eternal, hence is uncaused. The weaver's supervision is, however, determined by his desire for fruit, etc. So far as purpose of supervision is concerned, purpose is enjoyment attained after accomplishing the work, God's supervision is not determined by the attainment of enjoyment. The weaver's supervision is, however, determined by the attainment of enjoyment. In fact, cloth, etc., are produced by an intelligent weaver, and the volitional urge of the weaver for the production of cloth springs from his desire for enjoyment. Cessation of activity results from enjoyment. Even if God is held to act as the supervising agent of the production of cloth, etc. the volitional activity of the weaver, etc. does not stand futile. This point the Nyāya-Vaiśeṣika seeks to clarify by citing an example. According to the Nyāya-Vaiśeṣika the magnitude *(parimāṇa)* of a product comes from three different sources — plurality of number of constituent causes *(kāraṇabahutva),* gross magnitude of the constituent causes *(kāraṇamahatva),* looseness of conjunction of constituent causes *(pracaya-viśeṣa).* The magnitude of a jar is caused by the magnitude of two pot-sherds.

The magnitude of triad is resulted from the three diads, and the magnitude of a bundle of cotton is caused by the loose contact of its constituent elements. Sometimes these three conditions produce magnitude jointly, sometimes they do it independently. But when magnitude is produced by one, others do not become destitute of practical efficiency. In the same way, even if God were the supervisor of production of a piece of cloth, the weaver in no way becomes futile. Cloth may be produced by their joint efforts. If it is said that there is difference between the magnitude produced independently by one cause and that produced jointly by three causes, this is the same in the case of divine and human agency also. There is difference between the earth produced by one cause, viz. God, and cloth produced by God as well as the weaver.

Īśvarahood does not consist in nullifying human agency. Human agency is needed for the production of effect, because only in that way human beings can enjoy the fruits of their actions. God's supervision consists in making human beings enjoy their fruits of actions. The volitional urge of a weaver or a potter for the creation of cloth or jar springs from his desire for enjoyment. If the weaver or the potter is not considered as the efficient cause, enjoyment would not be possible. God is the general cause and the weaver, etc. are the specific causes in the production of effect, and both are necessary.

2.x. The *siddhāntin* now examines the last objection of the opponent that the theistic inference, 'whatever is a product is conditioned by some intelligent agent', is conditioned by non-eternal volitional activity. The opponent has attempted to prove that intelligent agency is conditioned by non-eternal volition. Non-eternal volition can be, says the *siddhāntin*, a condition of intelligence *(buddhi)*, only when absence of the former leads to the absence of the latter. But absence of non-eternal volition does not lead to the absence of intelligence, i.e. non-eternal volition is not the *vyāpaka* of intelligence. For, even in the absence of volition on the part of an agent who is not engaged in producing an effect, the agent may possess intelligence. Nor can it be said that conditioned intelligence *(hetubhūtā-buddhi)* has non-eternal volition as *upādhi*, because intelligence of an agent not engaged in production is also conditioned by impression *(saṁskāra)*, memory *(smaraṇa)*, etc. It cannot be maintained that knowledge of productive factors *(kārakaviṣayagatabuddhi)* is conditioned by volition. For, even in the absence of volitional activity in respect of jar, etc. the knowledge of productive factors of a jar, viz. wheel, etc., exists. The knowledge of productive factors that are determined by cause *(hetubhūtākārakabuddhi)* is also not conditioned by *prayatna* (volition), since the knowledge of wheel, etc. the productive factors of a jar, is determined by *saṁskāra*, etc. Nor can it be said that the knowledge of productive factors conditioned by the knowledge of result produced by productive factors *(kārakaphalahetubhūtā-kārakabuddhi)* is conditioned by volition. The knowledge of thorn-prick, conditioned by the knowledge of suffering due to coming in contact with thorn, may exist even without accompanied by non-eternal volition. The knowledge determined by desire also does not have volition as a condition, because the former exists even in the absence of the latter. If it is said that the knowledge determined by the desire of what should be done the next moment *(avyavahitottarakṣaṇe kṛtijananīyā cikīrṣā tadhetuṁ buddhi)* is conditioned by non-eternal volition, then only conditioned knowledge will be concomitant with non-eternal volition. Knowledge as such is not concomitant with the latter; hence, absence of precedence by non-eternal volition would not lead to the absence of precedence by knowledge or intelligence in general. Thus, precedence of non-eternal volition is not the conditioning factor of the precedence of intelligent agent.

The opponent infers knowledge *(buddhi)* from volition on the ground that when volition exists, knowledge exists, and when knowledge is absent, volition is absent. This is analogous to the position of the existence of body when knowledge exists, and non-existence of knowledge when body does not exist. This argument is not approved by the *siddhāntin*. He infers knowledge, desire and volition together from the same probans, viz. *kāryatva*.

The inference of knowledge from volition is not the same as the inference of body from knowledge. Body is inferred from knowledge, because body is the cause of knowledge. But the inference of knowledge from volition is not regulated by causal relation. Volition does not depend on knowledge for its origination; it depends on knowledge in the sense that the object of knowledge is the object of volition also. When volition is present, knowledge is present, and in the absence of knowledge, volition being devoid of its content is absent. The absence of non-eternal knowledge *(kāryabuddhi)* leads to the absence of non-eternal volition *(kāryaprayatna)*. But the content of eternal volition which God possesses is not borrowed from non-eternal knowledge. Eternal volition has its eternal content. Hence, eternal knowledge cannot be inferred from non-eternal volition. If eternal volition of God has to depend on non-eternal knowledge for its content, before the creation of body, knowledge being absent, volition would also be absent. And in the absence of volition, earth, etc. would not be created at all.

To conclude, the *siddhāntin* says that as the adjunct non-eternal volition is not concomitant with precedence of intelligent agency, the word 'non-eternal' is a futile adjunct. For, the absence of non-eternal volition does not lead to the absence of eternal knowledge.

तर्काभासतयाऽन्येषां तर्काशुद्धिरदूषणम्।
अनुकूलस्तु तर्कोऽत्र कार्यलोपो विभूषणम्।।३।।

The pūrvapakṣin's *arguments being defective are fallacious, hence, they cannot invalidate the* siddhāntin's *argument; whereas the* siddhāntin's *inference is strengthened by the supporting argument of absence of effect (i.e. if there is no creator, there can be no effect).*

3.i. All the arguments advanced by the opponent have been proved to be fallacious. The arguments, viz. 'if the earth, etc. have a creator, they must be produced by a body', 'if the earth, etc. are preceded by an intelligent agent, they must be conditioned by non-eternal volition', etc., involve the fallacy of *bādhita*, *vyāpyatvāsiddha*, etc. Hence, they cannot invalidate the *siddhāntin's* argument, viz. 'the earth, etc. must have a creator, because they are effects'. On the contrary, the *siddhāntin's* inference has been strengthened by the supporting argument *(anukūlatarka)* that 'if there is no creator, there can be no effect'. Further, since the absence of any of the causes leads to the absence of effect and since the sentient agent is also one of the causes for production of effect, a sentient agent is established as a cause in respect of effect. As in the absence of other insentient causal factors, effect is not produced, so also in the absence of sentient agent effect is not produced. These are the supporting arguments for the *siddhāntin's* inference.

3.ii. The Bauddha maintains that causality determined by *anvaya-vyatireka* relation subsists in respect of those facts that come within the purview of perception. In *anvaya* relation cause is present, effect also is present, and in *vyatireka* effect is absent, cause is also absent. Causal relation can be ascertained only when both these relations are amenable to perception. Thus, causality subsists between smoke and fire, air and shaking, because causal relation between smoke and fire, etc. is attested by experience. Otherwise, if causal relation is ascertained even between invisible entities, there arises the contingency of gastric fire *(udaryāgni)* being the cause of smoke, and mild wind that is devoid of touch and velocity being the cause of shaking of branches. An earthly fire and air possessing manifested touch and velocity are seen to be the cause of smoke and shaking respectively, the former two should be regarded as the causes of the latter two. As the causal relation between smoke and gastric fire or that between shaking and mild wind is not based on perception, causality cannot be established in respect of smoke and gastric fire, etc.

Following the same line of argument the opponent points out that as an embodied sentient being is perceived to be the cause of jar, etc., so also the cause of the earth, etc. must be an embodied sentient being.

This objection of the Bauddha is refuted in the following way. It is true, says the *siddhāntin*, that causality in respect of perceptible entities is determined by *anvaya-vyatireka* relation. But perceptibility is not the determining factor of causality.

Causality is determined by generality inherent in cause which is perceived and by that inherent in effect which is perceived. Thus, both perceived as well as unperceived entity determined by generality, can be the cause. According to the Bauddha, touch is the cause of touch, colour is the cause of colour, and so forth. He who advocates momentariness says that touch belonging to the prior series of atoms that are imperceptible gives rise to the subsequent touch belonging to the subsequent series of atoms that are perceived. Hence, as the cause touch, belonging to the prior series of atoms, is not perceived, the Bauddha has to leave aside his theory that causality should be determined only in respect of perceived facts. In the case of touch, the *siddhāntin* says, it is the generality inherent in touch, i.e. touchness *(sparśatva)*, that determines causality. Exactly similar consideration applies to colour, etc. The Nyāya admits the relation of inherence *(samavāya)* which is imperceptible, between substance and attribute *(dravya* and *guṇa)*, part and whole *(avayava* and *avayavi)*, etc. And imperceptible *ākāśa* and air are established as the substrata inhering in which sound and touch are produced. Therefore, causality has nothing to do with perceptibility, both perceived as well as unperceived entities can be the cause. Otherwise, if perceptibility were the criterion for causality, fire in the frying-pan cannot be inferred from seeing smoke, because that fire being colourless is not perceived.

The opponent's objection that if an unperceived entity were the cause, then smoke can be produced from gastric fire or shaking can result from mild air, is untenable. For, only earthy fire is the cause of smoke, and in the absence of the former, the latter cannot be produced. Similarly, air possessing manifested touch and velocity is the cause of shaking, therefore, shaking cannot result from mild wind which is devoid of manifested touch and velocity.

In the case of the earth, etc., the earth, etc. are obviously effects, but no corporeal being is seen to be the cause of them. Hence, the *siddhāntin* says that it is sentience in general or sentiencehood which is the cause of effect in general or effecthood.

The fact that causality is determined by the generality inherent in cause and effect, is the supporting argument for the *siddhāntin's* inference which draws a universal concomitance between *kāryatva* and *sakartṛkatva.* Apart from this, the argument that there is invariable concomitant relation between *kāryatva,* the probans, and *sakartṛkatva,* the probandum, since the probans is not vitiated by any condition *(upādhi),* is also another supporting argument.

It may be argued that if a non-corporeal agent be established on the ground of *kāryatva,* hare's horn too can be established on the ground of *paśutva* (animality). The purport of this argument is that though non-eternal horn in hare is not perceived to exist, unperceived eternal horn may exist.

In reply to this objection, it may be observed that had there been a universal concomitance between animality and possession of horn, the cogency of this argument could have been accepted. In fact, there does not exist any such relation between animality and possession of horn. Evidently, the hare possesses animality, but it does not have horn. Hence, hare's horn cannot be established on the ground of animality.

The *siddhāntin* says that as the relation of *vyāpti* has been established between *kāryatva* and *sakartṛkatva,* no such objection can counter the *siddhāntin's* argument.

3.iii. The opponent further asks, does God create the world independent of *adṛṣṭa* or does He create depending on *adṛṣṭa?* If it is the former, then there arises the contingency of production of everything always, since there is nothing to limit the production. If the latter, *adṛṣṭa* alone can create the world, there is no need of postulating God.

This objection, according to Udayana, is too futile to negate the existence of omniscient and omnipotent God. In fact, every effect is dependent on various auxiliary conditions for its emergence. The opponent's objection, as it were, amounts to prove that the production of a jar out of cause other than a stick would lead to the perpetual production of it, and the production of the same with the help of a stick would make the cause superfluous. This will do away with the causal principle itself, since a permanent cause does never produce an effect without being assisted by accessory conditions.

Therefore, the inference, viz. 'the earth, etc. must have a creator, because they are effects', is valid and is not riddled with any fallacy. Moreover, an argument that is supported by the Śruti is stronger than the one which is not supported by the Śruti. As Manu says, 'He, and none else, knows religion, who investigates the Veda and the religious teachings of the sages by means of such reasoning as is not contrary to the Veda and the Śāstras' (*Manu,* 12.106).

The *siddhāntin's* argument that proves God's existence is stronger, because it is supported by the Śruti, while the opponent's argument that attempts to disprove God's existence, being contrary to the Śruti is weaker. The following is the Śruti text that Udayana has cited as his supporting argument.

viśvataścakṣuruta viśvatomukho viśvatobāhuruta viśvataspāt
sambāhubhyāṁ dhamati sampatatrairdyāvābhūmī janayan deva ekaḥ.
(Śvetāśvatara Up., 3.3)

This Upaniṣadic text is interpreted by Udayana thus: 'The one God creates the heaven and the earth, is the omniscient one, the source of all, the one by whom all else becomes active, omnipresent, and the one who unites the beings with *dharma* and *adharma,* and one who directs the atoms'.

Udayana cites evidence from the Smṛti also, 'I am the origin of all, from me all proceeds' (*Gītā,* 10,8).

स्वातन्त्र्ये जडताहानिर्नादृष्टं दृष्टघातकम्।
हेत्वभावे फलाभावो विशेषस्तु विशेषवान्।।४।।

If it (the atom) acts independently (without the guidance of any external intelligent principle), it ceases to be insentient, adṛṣṭa *(as a cause) does not abolish (the causal efficiency of) other perceived causes; in the absence of cause (supervision by an intelligent being) there will be absence of effect (production of insentient objects); a particular effect is produced from a particular cause (and effect-as-such is produced from cause-as-such).*

4.i. The first groud for God's existence advanced by Udayana is *kāryatva.* The second ground is *āyojana.* The argument from *āyojana* runs thus : 'The motion in the atoms, leading to the formation of gross bodies, must be contingent on some intelligent being, since motion in insentient entity is conditioned by some sentient principle'. This argument is based on the atomic theory recorded in the early Vaiśeṣika texts. According to this theory earth, water, fire and air—these four material substances are ultimately composed of indivisible and infra-sensible particles, called atoms *(paramāṇu).* These four types of ultimate units of matter are distinct in character. The atomic theory as accepted by the Vaiśeṣika maintains that two atoms combine to produce a binary compound *(dvyaṇuka),* and three such binary compounds make a triad *(truṭi or trasareṇu).* This triad, according to the Vaiśeṣika, is the remotest visible molecules.

The Nyāya-Vaiśeṣika philosophy explains in the following manner how cosmic creation is initiated through the conjunction of atoms. Prior to cosmic creation, i.e during the period of cosmic rest, the atoms, even though they are dynamic entities, do not possess causal efficiency. It is just on the eve of creation that motion leading to the emergence of binary compound is produced in the individual atoms. But as atoms are insentient, they stand in need of intelligent guidance, so that they may be set into motion. Udayana points out that this intelligent principle through whose guidance atoms are made active is God. Motion is produced in the atoms due to their contact with the Soul, i.e God, possessing volition. It may be argued that atoms come into contact by themselves without the guidance of any external intelligent principle. This involves self-contradiction. For, if atoms can combine themselves, they would turn out to be conscious entities, a position that would not be accepted either by the opponent or by the *siddhāntin.*

4.ii. The Nyāya believes that when God takes to creation, He is assisted by the *adṛṣṭa* of individual beings. Individual *adṛṣṭa* serves as the necessary condition for all products. According to the Nyāya, the primary actions in the atoms leading to the creative conjunction also depend on *adṛṣṭa.* The opponent contends that effects can be produced by *adṛṣṭa* alone, there is no need to postulate God as the supervisor of this *adṛṣṭa*.

In reply to this objection, Udayana points out that the Nyāya accepts *adṛṣṭa* as the necessary condition of all products. But at the same time he does not deny the existence and causal efficiency of other perceived causes. If unseen *adṛṣṭa* alone could produce objects and no other cause be needed, there would have been no necessity of the receptacle of enjoyment, i.e. body, the external objects to be enjoyed and the instruments for enjoyment, i.e. the sense-organs. But since locus, the objects and the instruments are the perceived conditions for enjoyment, it is improper to deny their practical efficiency. Moreover, the fact that insentient products are supervised by a sentient being is attested by perceptual experience. Therefore, it is not proper to repudiate the causal efficiency of such an agent. Supervision by an intelligent being is one of the causes for the production of insentient objects. If the former is denied, the latter too has to be denied *(hetvabhāve phalābhāvo),* since an effect is not produced in the absence of even one of the causes.

The opponent argues that it is only a particular activity that is supervised by a sentient being. In other words, in particular cases there may be concomitance between a particular type of activity and a particular type of volition. But there is no concomitance between activity-as-such and volition-as-such. A particular effect is produced from a particular cause *(viśeṣastu viśeṣavān),* but there is no causal relation between effect-as-such and cause-as-such.

In reply to this objection, the *siddhāntin* enquires whether the opponent intends saying that activity *(ceṣṭā)* is the vitiating condition of the inference that 'motion in the atoms is contingent on a sentient being, since motion in the insentient entity is conditioned by a sentient being'. If so, the opponent has to explain what does he mean by activity. Activity may be defined as that which is produced by the non-inherent cause in contact with the soul having volition *(prayatnavadātmasaṁyogāsamavāyikāraṇikā kriyā).* Or it may be defined as that which is produced by volition which is incidental cause *(prayatnarūpanimittakāraṇajanyatvaṁ).*

Activity in both these sense cannot be a vitiating condition of the above-mentioned inference. A vitiating condition is concomitant with the probandum *(sādhya)*, but it is not concomitant with the probans. The probans, therefore, is absent in the subject *(pakṣa)*. The probandum in the inference concerned is supervision by an intelligent being, and supervision means the production of action in the atoms due to their contact with the Soul, viz. God, having volition. Supervision in this sense is the same as the activity defined before. Hence, both probandum and the condition being the same, the latter cannot be the condition of the former.

Activity may be defined as an act which is conducive to the acquisition of what is desirable and the prevention of what is undesirable. This definition is, however, untenable. Consumption of poison for committing suicide is also an act, but this is not conducive to the acquisition of what is desirable and the prevention of what is not desirable.

It may be said that in committing suicide there is attainment of what is desirable, namely, killing of oneself which is desired in the present case. Hence, the previous definition of activity, i.e. an act which is conducive to the acquisition of the desirable and the prevention of the undesirable , is a correct one.

To this issue the *siddhāntin* joins that if activity is defined in that way, it is applicable in the case of the activity of the atoms also. The question arises: 'Is this attainment of the desirable and prevention of the undesirable, in respect of the agent or in respect of others?' In either way, as the action of atoms leads to the creation of the universe, both God as well as individual souls attain what is desirable. In respect of God it is the creation of the universe which is desirable, and in respect of the individual souls it is the enjoyment they derive from the created universe that is desirable. Therefore, as activity defined in the sense mentioned above is the same as the activity of the atoms, the former cannot be a vitiating condition of the inference advanced by the *siddhāntin* to prove God's existence.

This definition of activity is, however, fallacious. A patient may mistakenly take some medicine which instead of subsiding his disease may aggravate it. In this case though activity is there, attaining the desirable and avoiding the undesirable are absent. Hence, the definition is less pervasive *(avyāpta)*.

Nor can activity be defined as that which is produced inhering in the body *(śarīrasamavāyikriyā)*. For, an action may be produced even in a corpse owing to wind. Production of activity in a corpse would not be accepted by the opponent, because with the latter a particular activity is supervised by a sentient being, which is not possible in a corpse. Therefore, the definition of activity as that which is produced inhering in the body is more pervasive *(ativyāpta)*.

Activity cannot also be defined as that which is produced inhering in a living body *(jīvato yaccharīram tatsamaveta-kriyātvaṁ)*. Movement of eyelashes which is an action may be produced owing to air having touch, but this is not supervised by a sentient being. Hence, the definition of activity as that which is produced inhering in a living body also suffers from the fallacy of *ativyāpti*.

The opponent may argue that an action is produced provided there is absence of the mediation of entity having touch. But even then his contention that a particular activity is supervised by a sentient being would not hold good. According to the opponent, burning of fire is produced due to contact with the soul having *adṛṣṭa (adṛṣṭavadātmasaṁyogajanya)*, and it is not supervised by any sentient being. As this is the case, the opponent cannot maintain that a particular activity is supervised by a sentient being.

Further, activity cannot be defined as that which is produced inhering in a body Body has been defined in the *Nyāya-sūtra* as the substratum of action, sense-organ and enjoyment. Again, if activity is also defined in terms of body, there arises the fallacy of inter-dependence *(anyonyāśraya)*. Therefore, all these definitions of activity being fallacious cannot be the vitiating conditions of the *siddhāntin's* inference.

4.iii. The opponent may argue that there is no need of defining activity, it is activity-as-such *(ceṣṭātvajātiviśeṣaḥ)* which gives rise to the inference that an activity is preceded by a sentient volition.

The *siddhāntin* does not admit this, according to him activity-as-such is concomitant with the precedence by sentient volition-as-such, and the activity-as-such *(kriyātva)* refers to all actions that are preceded by sentient beings.

The opponent may further argue that a particular activity is concomitant with embodied intelligent enjoyer, because it is out of desire for enjoyment that an intelligent agent engages in activity. As God is neither an embodied being nor an enjoyer, He cannot supervise activity in atoms. Udayana, in reply, observes that activity of human body is concomitant with the volition of an embodied enjoyer, but activity-as-such is concomitant with volition-as-such. Volitional activity *(prayatnakriyā)* may be with body or without a body. In the former case it is preceded by the volition of an enjoyer, but not in the latter. Thus, it cannot be said that the activity in atoms is not supervised by a sentient being, because atoms are not physical organism. For, when an activity is supervised by the volition of an enjoyer, a body is needed, but divine volition is not the volition of an enjoyer, and so He is not in need of a physical apparatus in order to supervise the activity in atoms.

Udayàna gives the following Śruti and Smṛti texts in support of this proof for the existence of God.

yadā sa devo jāgarti tadedaṁ ceṣṭate jagat,
yadā svapiti śāntātmā tadā sarvaṁ nimīlati.

This text is interpreted by Udayana thus: 'All actions are dependent on God's volition. God's volition is the auxiliary cause for action; and, inaction is due to the absence of that auxiliary cause'. Again,

ajño janturanīśo' yamātmanaḥ sukhaduḥkhayoḥ,
īśvaraprerito gacchet svargaṁ vā śvabhrameva vā.

This is interpreted by Udayana to mean that 'individual souls are guided by God, because they are ignorant and devoid of eternal volition *(nirapekṣa-prayatna)*'. Since *hetu* in this inference, viz. ignorance etc., applies to atoms also, atoms are also guided by God. *Svarga* and *śvabhra* refer to *iṣṭa* and *aniṣṭa* respectively.

Further,

mayā'dhyakṣeṇa prakṛtiḥ sūyate sacarācaram;
(*Gītā,* 9.10)
tapāmyahamahaṁ varṣaṁ nigṛhāṇmyutsṛjāmi ca.
(*Gītā,* 9.19)

Udayana interprets this thus: 'Actions of all things, sentient as well as insentient, are guided by God; and like action, inaction of insentient things also depends on God'.

4.iv. The third proof for the existence of God as set forth by Udayana is *dhṛti* (support). Our experience in everyday life shows that the tendency of a material substance having weight is to drop off. Only a volitional action can prevent that substance from falling downward. It is the conscious self who is the supporting agent. This support may be direct or indirect. When a flying bird does not drop down, it is the individual self of the bird that prevents its body from falling down. Here the self directly supports the body from falling downward. The support is indirect when a piece of grass carried by a bird does not drop down. In this case the self indirectly supports grass through the mediation of the bird's body.

Similar is the case with the planets, etc. The fact that luminaries and planets evolve in their respective orbits without colliding with one another cannot be explained, unless they are sustained by the volition of some agent. And that sustaining agent of the entire universe is postulated to be God. This argument based on the ground *dhṛteḥ* takes the following inferential form.

'The whole universe, including the earth and other planets, is sustained, either directly or indirectly, by the volition of God, because having possessed of weight it does not drop down, like the body of a flying bird and like the substance in conjunction with the bird.'

Where the scripture says that Indra, Yama, and others, sustain their respective world, there according to Udayana, it should be understood that God being immanent in them sustains the world. This identity of God with other things is referred to by such Śruti text as *ātmaivedaṁ sarvam, sarvaṁ khalvidaṁ brahma,* etc. Otherwise, God and the world being different, there cannot be identity relation between them. God enters into the person of Indra, and others in the same way as a magician out of his power projects himself as horse, bear, tiger, etc.

It may be argued that *adṛṣṭa* can sustain the universe, there is no need of postulating God's volition for the sustentation of the universe. This contention is untenable, because volition is ascertained as the sustaining cause of the universe on the basis of *anvaya-vyatireka* relation. It is established by *anvaya-vyatireka* relation that wherever there is sustentation, there is volition as a sustaining cause, and wherever there is absence of volition, there is absence of sustentation.

Moreover, volition must be considered as the cause towards the sustentation of the universe even though *adṛṣṭa* is present, because it has already been stated that the presence of *adṛṣṭa* does not preclude other causes. It cannot be said that *adṛṣṭa* is *upādhi* or condition in this inference, since *adṛṣṭa* is also the cause of the sustentation of the universe. For, like *adṛṣṭa* volition is also concomitant with the sustentation of the universe, and the presence of one cause does not become a vitiating condition towards other cause. If the presence of one cause turns out other cause as *upādhi*, nowhere can causality be ascertained. Universal concomitance cannot be established between smoke and fire, because though smoke is the probans towards the inference of fire, there may exist wet-fuel which is the condition of smoke.

It may be said that volition is concomitant with the sustentation of body and not with sustentation-as-such. It is not so, because the vital air and sense-organ which support the body cannot be sustained without volition. Nor can it be maintained that volition is concomitant with the sustentation of those things that are enjoyed and not with sustentation-as-such. A particular product is the effect of a particular cause and product-as-such is the effect of cause-as-such. Sustentation of things that are enjoyed is preceded by a particular volition accompanied by desire and not by volition-as-such. Volition-as-such is concomitant with sustentation-as-such.

The Śruti and Smṛti texts cited by Udayana in support of this argument are as follows:

etasya vā akṣarasya praśāsane gārgi, dyāvā-pṛthivyau vidhṛte tiṣṭhataḥ (Bṛhadāraṇyaka Up., 3.8.9).

Udayana interprets this thus: 'Oh Gārgi! the heaven and the earth are sustained by the volition of the supreme being'.

Further,

uttamaḥ puruṣastvanyaḥ paramātmetyudāhṛtaḥ,
yo lokatrayamāviśya bibhartyavyaya īśvaraḥ.

(*Gītā,* 15.17)

This is interpreted by Udayana in the following way: 'God who is not worldly and is omniscient, who is worshipped by all, who possesses eternal qualities and eternal will, sustains all the worlds which are in conjunction with his knowlege, desire and volition'.

The scriptural text which states that God in the form of tortoise, etc. sustains the earth is also explained by this.

4.v. Another ground for God's existence is denoted by the word *ādi* in the *kārikā; ādi* stands for destruction *(saṁharaṇāt).* God is not only the creator, but also is the destroyer of the universe. The entire universe beginning with *brahmāṇḍa* and ending with diads undergoes destruction, and destruction must be due to the volitional effort of a conscious being. That conscious being must be God.

In support of this argument Udayana cites the following Śruti and Smṛti texts:

eṣa sarvāṇi bhūtāni samabhivyāpya mūrtibhiḥ,
janmavṛddhiksayairnityaṁ saṁbhrāmayati cakravat.

'God pervades the entire universe and makes it revolve as it were by causing creation, sustentation, and destruction of them.'

Further,

sarvabhūtāni kaunteya prakṛtiṁ yānti māmikām,
kalpakṣaye punastāni kalpādau visṛjāmyaham.

(*Gītā,* 9.7)

That is, 'All beings, O son of Kunti, pass into My nature at the end of a unit of cosmic duration and at the beginning of the same unit, I emanate them'. 'All beings pass into My nature' means all products pass into atoms. This also explains, according to Udayana, the scriptural text that speaks about Rudra (destruction).

In the beginning of this chapter it has been shown how *kāryatvāt* the first ground for the existence of God is free from any *upādhi* (condition). The other three grounds for proving the existence of God, viz. *āyojana*, *dhṛtiḥ* and *saṁharaṇa*, are also *kārya;* hence, it can be said that they are also free from any *upādhi.* The first argument states that the entire universe is a product, the second shows how the first creation is initiated, the third points out how the universe as a product is sustained and how the entire universe is destroyed. Creation, sustentation, destruction all these are effects, God being the cause. Thus, there is cause-effect relation between God and creation, sustentation, etc. The probans *āyojanāt, dhṛteḥ* and *saṁharaṇāt* are free from *upādhi*, because they are supported by causal relation.

कार्यत्वान्निरुपाधित्वमेवं धृतिविनाशयोः।
विच्छेदेन पदस्यापि प्रत्ययादेश्च पूर्ववत्।।५।।

Thus, sustentation and destruction are free from any condition, because they are effects; empirical usage (pada) *is also not vitiated by any condition, because it is an effect owing to its destruction at the end of a cycle; for the same reason authoritativeness, etc. are also free from any condition.*

5.i. The fourth ground for God's existence is *padāt,* that is, God is the agent for introducing empirical usage. The Nyāya advocates that the universe has a origin, and hence, the usages and conventions do not exist from eternity. At the end of a cycle empirical usages come to an end; therefore, there must be one person responsible for introducing these usages in the beginning of a cycle. And this person is God. Since empirical usages are destroyed at the end of a cycle, they are effects, and so the ground *padāt* is also free from any *upādhi* for the reason mentioned earlier. Empirical usages may be of three kinds, verbal *(vācika)*, physical *(kāyika)* and mental *(mānasa)*. It may be said that verbal usages consist of words, and so the word *pada* applies to them. But physical and mental usages have nothing to do with words, so the word *pada* does not apply to them. In reply to this objection it may be said that the word *pada* is used here in the sense of *vyavahāra* (usage), and since *kāyika-* and *mānasa-vyavahāras* are also usages, the word *pada* applies to both. Verbal convention of human beings is *vācika-vyavahāra,* the art of writing is *kāyika-vyavahāra,* and the skill in weaving cloth, etc. is *mānasa-vyavahāra.*

There is no dispute about the fact that all empirical usages are effects, and therefore, they must have an originator. But from this, the opponent may argue, it cannot be concluded that this originator is God and not human beings. It may be that empirical usages were introduced by human beings.

In reply to this objection it may be observed that if human beings introduce empirical usages, then introduction of empirical usages at the beginning of a cycle cannot be explained, since at that time no human beings exist. Therefore, it must be God who introduces all the empirical usages at the beginning of a cycle.

It cannot be maintained that all the empirical usages are eternal in the sense that they depend on prior and still prior usages. For, in that case also difficulty arises with initial usage. Initial usages take place after dissolution, naturally they cannot depend on prior usages, because all the prior ones are destroyed during the period of dissolution.

It may be said that though empirical usages are subject to destruction, yet they may be eternal in the sense that when one type of usage is destroyed, another type is introduced, and this process continues eternally. And human beings can introduce these usages, there is no need of postulating God for this purpose.

This contention is untenable on the ground that though this process may continue to certain extent with the help of human beings, yet initial usages remain unexplained. In the beginning of creation there being no human being, no example would be there to follow, and as a result no usages would be introduced. It is God who sets up example regarding all the empirical usages in the beginning of creation and makes people follow them.

It may be objected that God who is non-corporeal cannot introduce empirical usages, specially *kāyika-vyavahāra*, etc. To this Udayana rejoins that God does not possess any body for His own enjoyment, but He is competent to assume physical form from time to time and impart instructions to human beings, whenever occasion arises. By this He exhibits His supreme power.

The Śruti and Smṛti texts in support of the argument based on *padāt* are as follows:

pitā'hamasya jagato mātā dhātā pitāmahaḥ (*Gītā*, 9.17).

That is, 'I am the Father of this world, the Mother, the supporter and the Grandsire'.

Again,

yadi hyahaṁ na varteyaṁ jātu karmaṇyatandritaḥ,
mama vartmānuvartante manuṣyāḥ pārtha sarvaśaḥ
utsīdeyurime lokā na kuryāṁ karma cedaham

(*Gītā*, 3.23-24).

That is, 'If I do not remain active (in Principle) never relaxing, men in every walk of life, O Pārtha, would take to my way. These (various) worlds (value systems) would be in ruins should I refrain from activity'. Further, the *Śuklayajūḥ-saṁhitā* proclaims:

namaḥ kulālebhyaḥ karmārebhyaḥ (XVI.27).

That is, 'salutation to the potter, to the blacksmith'.

5.ii. *Pratyaya* (authoritativeness) is the fifth ground for God's existence. The Nyāya believes that the authority of the Vedas depends on God who has pronounced them. Unlike the Mīmāṁsakas the Naiyāyikas do not believe the Vedas as revealed literature, not created by any being human or divine. The words of the authoritative persons are distinct sources of valid knowledge. The Vedas comprise of words and are verbal on that account. Therefore, their authority would depend on the veracity and infallibility of the speaker who has pronounced them. The infallibility of a person consists in his having the valid knowledge of all objects. The argument on *pratyaya* in its inferential form is thus: 'The validity of knowledge of the Vedas depends on the infallibility of the speaker, because it is valid knowledge, like perceptual valid knowledge'. As the validity of perceptual knowledge depends on contact, etc., so also the validity of verbal testimony depends on the authoritativeness of the speaker. The authoritativeness of the speaker consists in his having the exact knowledge of the word-meaning.

People believe in the authoritativeness of the Vedas, and this belief would not arise unless the knowledge of the Vedas is authoritative. The validity of the Vedas, according to Udayana, is not intrinsic. In the second *stabaka* he has refuted the doctrine of the intrinsic validity of knowledge and proved that the validity of knowledge is always extrinsic. It is not proper to maintain that the belief in the authoritativeness of the Vedas arises due to error. The seers too have accepted the authority of the Vedas, and surely they would not harbour any such erroneous belief. Nor can it be maintained that the validity of the Vedas depends on the infallible knowledge of individual beings, because being devoid of omniscience they cannot have the valid knowledge of super-sensuous things, like *dharma*, *adharma*, etc. mentioned in the scripture.

Thus, the infallibility of the Vedas depends on the unfailing validity of the knowledge derived from them; that knowledge is always valid on account of the eternal purity of the source from which the Vedas are originated, and that source is the omniscient God.

5.iii. Udayana has advanced *śruteḥ* as the sixth ground for God's existence. The argument is as follows: The Vedas are composed by an omniscient being in view of the fact that they are the Vedas; and, that which is not composed by an omniscient being, is not the Veda, as in the case of empirical sentence. This is negative inference or *vyatireka-anumāna*. The probans in this inference is *vedatva*. So, *vedatva* should be defined first; in other words, it has to be stated what is meant by the Vedas.

The Vedas cannot be defined as consisting of mere sentences. For, in that case empirical sentences too would be considered as the Vedic sentences, since they are also sentences. But empirical sentences are not composed by an omniscient being; so if the Vedas were defined as consisting of mere sentences, the probans, viz. *vedatva,* would prove the absence of the probandum, viz. the absence of composition by an omniscient being. As a result, the inference would involve *viruddha-hetvābhāsa* in that a conclusion is drawn just contrary to that which should be justified by the probans.

It is also not proper to define the Vedas as comprising sentences relating to the super-sensuous objects, viz. *dharma, adharma,* etc. The Bauddha literature too consists of such sentences, but the Bauddha literature is not the creation of an omniscient being. Therefore, the definition of the Vedas as comprising sentences relating to *dharma, adharma,* etc. involves the same *viruddha-hetvābhāsa* mentioned previously.

The Vedas are defined by Udayana as consisting of sentences, the cause of which is not perceived and the validity of which is accepted by the seers. *Anupalabhyamāna-mūlāntaratva,* i.e. non-perception of the cause, implies that the Vedas are not composed by human beings *(jīvājanyatva).*

To maintain that the Vedas are erroneous or that they are composed just to deceive the mankind, is untenable on the ground that the seers also accept the validity of the Vedas. Nor can it be maintained that the Vedas are not created at all, but possess blind continuity *(andha-paramparā).* The possibility of blind continuity in the case of the Vedas is ruled out on the ground that at the time of dissolution all the Vedas are destroyed, and in the beginning of a new cycle there arises the contingency of their creation again. The argument based on Śruti has *anvayānumāna* (positive inference) too. The positive inference runs thus: 'The Vedas are the composition of an intelligent being, because they consist of sentences'. This is the seventh inferential argument for the existence of God; the ground or probans in the inference is *vākyatva*. The Mīmāṁsaka raises an objection against this inference. He says that the above reasoning is applicable only to those works whose authorship is known. In the case of the Vedas this reasoning would not hold good as the authorship of the Vedas is not known *(asmaryamānakartṛkatvāt)*.

This objection is refuted on the ground that in every new cycle God composes new Vedas, and as the Vedas differ from cycle to cycle, it is not possible to know their authorship in the same way as the aurhorship of other works is known. Therefore, the *hetu*, viz. *asmaryamānakartṛkatvāt*, advanced by the opponent is *asiddha*, since it is not applicable in respect of the Vedas.

In support of his argument Udayana cites the following scriptural evidences.

anantarañca vaktrebhyo vedāstasya viniḥsṛtāḥ,
pratimanvantarañcaiṣā śrutiranyā vidhīyate.

'The Vedas come out of His face; and, it is different in each and every cosmic age'.

Again,

vedāntakṛdvedavideva cāham (Gītā, 15.15).

'I am the one who create the Vedas, and the one who know the Vedas'.

The opponent may say that these scriptural evidences are not *vidhis*, since they are not expressed in *vidhi-pratyaya*. They are only *arthavādas*, i.e. they are only recommendatory passages, and so they cannot prove an omniscient being as the author of the Vedas.

To this Udayana rejoins that in that case, since *kartṛsmaraṇa* (knowing of the author), also does not come under *vidhi-vākya*, Kālidāsa cannot be established as the author of *Kumārasambhava*, etc. on the ground of *kartṛsmaraṇa*. That is, if the authorship of the Vedas cannot be established on the ground of Śruti-Smṛti evidences, on account of their not being *vidhi-vākyas*, then the authorship of *Kumārasambhava*, etc. also cannot be established on the ground of *kartṛsmaraṇa* which is also not *vidhyartha*.

5.iv. The opponent may advance a counter-argument in order to prove that the Vedas are not the composition of an intelligent being. The counter-argument is this: 'The Vedas are not composed by an intelligent being, because they are not known by any other means of knowledge *(pramāṇāntarāgocarārthatvāt)*'. The probans in the argument is non-cognition of the Vedas by any other means of knowledge (viz., perception, inference, etc.). It may be asked: 'Is it the author of the Vedas who cannot know the Vedas by any other means of knowledge, or is it the hearer?' It cannot be the first, because God composes the Vedas after having the direct perceptual knowledge of the Vedas. Hence, the probans in the opponent's argument is unestablished in respect of the author of the Vedas.

The second alternative is also untenable, because it cannot be proved that only that which is known to the hearer by other means of knowledge is universally related with *pauruṣeya-vākya.* Sometimes it happens that suddenly some one smiles remembering some happy event of the past. The cause of the smile cannot be known to the hearer by any of the means of knowledge and being asked by the hearer what exactly is the cause of the smile the person himself tells the cause. Here the probans, viz. non-cognition by any other means of valid knowledge in respect of the hearer *(śrotṛpramāṇāntarāgocarārthakatva)* exists along with *pauruṣeya-vākya.* Therefore, this cannot be advanced to prove the absence of *pauruṣeya-vākya* in respect of the Vedas, that is, this cannot be advanced to prove that the Vedas are not the composition of an intelligent being.

It may be objected that the Vedas are not composed by an intelligent being, because there is absence of the causal aggregate for the knowledge of the Vedas. Thus, God cannot possess the knowledge of the Vedas, because He is a sentient being like us; His knowledge does not consist of the Vedas, because He possesses knowledge like us; His sense-organs cannot perceive the Vedas, because He possesses sense-organs like us; He cannot have mental perception of the Vedas, because mental perception is possible only with regard to the qualities of the soul and nothing else.

These are the arguments advanced by the opponent to substantiate the view that the Vedas are not the composition of God, because there is absence of the causal aggregate for the knowledge of the Vedas.

All these arguments, however, are invalid, because the subject of all these arguments, viz. God, is unestablished. Moreover, the arguments are futile, because God's knowledge being eternal, He does not stand in need of the causal aggregate for having the knowledge of the Vedas. The causal aggregate is needed for non-eternal knowledge but not for eternal knowledge.

The opponent now puts forth the reason, 'absence of the causal aggregate' as a ground for non-perception of the atoms. The argument goes thus: 'The atoms are not perceived by any one, because there is absence of the causal aggregate for the perception of the atom'.

In reply to this objection it may be asked what the word 'anyone' here refers to. If it refers to God, the reason 'absence of the causal aggregate' would not hold good, for God possessing eternal knowledge does not need the causal aggregate for the perception of the atoms. If, on the other hand, the word 'anyone' refers to persons other than God, it establishes what is already established, for the *siddhāntin* too admits that human beings cannot have the perception of the atoms.

To prove that the Vedas are composed by an intelligent being, the reason set forth by the *siddhāntin* is the fact of the Veda's consisting of sentences. The opponent says that the reason or the probans is vitiated by the condition *pramāṇāntaragocarārthatva*, i.e. the fact of the Vedas being known by other means of knowledge. The Vedas are the composition of an intelligent being, because they are known by other means of valid knowledge, and not because they consist of sentences. In other words, it is *pramāṇāntaragocarārthatva* which is concomitant with *pauruṣeyatva* and not *vākyatva*.

This objection has been refuted by Udayana in the following manner. The Bauddha scripture, according to Udayana, has been composed just to deceive people, and so it is erroneous. And that which is erroneous cannot be known by the means of valid knowledge. Thus, as the Bauddha scripture is not known by the means of valid knowledge, it follows that it is not composed by an intelligent being, because the opponent maintains that only *pramāṇāntaragocarārthatva* is concomitant with *pauruṣeya-vākya* and not *vākyatva.* This position, however, cannot be accepted. Therefore, the objection that *pramāṇāntaragocarārthatva* is the conditioning factor of the *pauruṣeyatva* of the Vedas stands rejected.

The opponent may argue that the condition *pramāṇāntaragocarārthatva* is concomitant with *pauruṣeyatva* (the composition by an intelligent being) only when that composition is authentic. The Bauddha scripture is not authentic, so *pauruṣeyatva* of the Bauddha scripture is not conditioned by *pramāṇāntaragocarārthatva.*

The *siddhāntin* rejects the view that *pramāṇāntaragocarārthatva* is the condition *(upādhi)* of *pauruṣeyatva.* He maintains that *pramāṇāntaragocarārthatva* is not concomitant with *pauruṣeyatva,* the probandum, since the former is included in the latter. *Pauruṣeyatva* implies authorship of an independent intelligent being *(svatantra-puruṣapraṇītatvaṁ)* who has the knowledge of the subject-matter to be composed, the desire to compose and who exercises volitional effort for that purpose. Since *pauruṣeyatva* implies that the author should have the knowledge of the subject-matter to be composed, *pramāṇāntaragocarārthatva* is also implied by that. Having the knowledge of the subject-matter implies *pramāṇāntaragocarārthatva.* Therefore, as *pramāṇāntaragocarārthatva* is included in *pauruṣeyatva* or the *sādhya* and not concomitant with the latter, it cannot be a condition *(upādhi)* of the latter.

It may be said that *pramāṇāntaragocarārthatva* means *śabdetarapramāṇagocaratva,* and *śabdetarapramāṇagocaratva* is *upādhi* of *pauruṣeyatva.* That is, that work which is known by the means of knowledge other than the Veda *(śabdapramāṇa)* is the composition of an intelligent being. This view is refuted on the ground that the works of Manu and others are based on the Vedas, and *so śabdetarapramāṇagocaratva* is absent in such works. In that case the works of Manu and others would turn out to be *apauruṣeya,* because according to the aforesaid view only *śabdetara-pramāṇagocaratva* is concomitant with *pauruṣeyatva,* the former being the *upādhi* of the latter. As this position cannot be accepted, the view, therefore, stands rejected.

Moreover, *pramāṇāntaragocarārthatva* cannot be a condition of *Vedapauruṣeyatva.* God, the author of the Vedas, possesses eternal knowledge, and so His knowledge of the Vedas is also eternal and not acquired through other means of knowledge. Therefore, the probans *vākyatva* is valid and establishes God as the author of the Vedas.

5.v. The last of the eight grounds put forth by Udayana to prove God's existence is *saṅkhyāviśeṣāt.* That is, God is established on the ground that His enumerative cognition gives rise to the notion of duality which is the cause of the magnitude of diad.

The Nyāya-Vaiśeṣika philosophers hold that the earth, water, fire and air are ultimately composed of infrasensible particles called atoms *(paramāṇu).* These four types of ultimate units of matter form four types of composite bodies. Two atoms combine to form a binary compound or diad *(dvyaṇuka)* and three such binary compounds form a triad *(tryaṇuka).* Atoms are the ultimate constituent elements of gross bodies. The magnitude of a diad is atomic, and since the diad is a product, its magnitude and other qualities must be non-eternal. As we know that attributes of non-eternal objects can never be eternal, so the magnitude of a diad must have a genesis. Therefore, the Nyāya-Vaiśeṣika has to explain how does a diad derive its magnitude. The magnitude of a composite body results either from the magnitude of the constituents *(avayavaparimāṇa)* or from their number *(saṅkhyā).*

It may be said that just as the magnitude of the part is the cause of the magnitude of the whole, so also the magnitude of atoms and the magnitude of diads are the cause of the magnitude of diad and triad respectively. If it were so, it may be enquired whether the magnitude of atoms and the magnitude of diad give rise to the magnitude of diad and triad depending on the number of atoms or do they give rise independently without depending on the number of atoms. If it is the first, since the number of atoms is also the cause of the magnitude of diad and triad, the number which is advanced as the ground for God's existence is established.

The second alternative is untenable, because the magnitude of the atoms and the magnitude of the diad being eternal are devoid of causal efficiency in respect of magnitude. Moreover, a magnitude can only produce a magnitude of superior type of the same category. Thus, gross magnitude *(mahat-parimāṇa)* produces grosser magnitude. If it is maintained that the magnitude of the diad and that of the triad have their genesis in the magnitude of the atoms and that of the diad respectively, which are their constituent elements, then the magnitude of the diad would be minuter than the atomic magnitude and the magnitude of the triad would be minuter than the magnitude of the diad. But a minuter magnitude than that of the atom is inconceivable, because the magnitude of an atom has been admitted as the minutest possible magnitude. To obviate this difficulty the Nyāya-Vaiśeṣika philosophers maintain that the atomic magnitude is devoid of all causal efficiency.

If it is held that eternal magnitude produces non-eternal magnitude, then there arises the contingency of eternal magnitude producing manifested magnitude (*mahattva*), since non-eternal magnitude possesses manifested magnitude. As a result diad which is produced by two atoms of eternal magnitude too would possess manifested magnitude. But the Nyāya-Vaiśeṣika does not admit that diad possesses manifested magnitude. Thus, the view that eternal magnitude produces non-eternal magnitude is not accepted by the Nyāya-Vaiśeṣika system.

It may be said that since both atom as well as diad possess atomic magnitude, both do not give rise to magnitude. That which does not possess atomic magnitude gives rise to magnitude. The reason behind this view lies in the fact that if atomic magnitude were the cause of the magnitude in respect of diad and triad, then like triad which possesses *mahattva*, diad too would possess *mahattva*, because the atoms that are the component parts of diad are also of atomic magnitude. The diad, however, is not admitted as possessing *mahattva*. Therefore, it is said that *kāraṇabahutva* or the plurality of number of the constituent elements is the cause of *mahattva* in respect of triad and not the atomic magnitude.

It may be said that the difference between the magnitude of the diad and that of the triad is determined by the number of their constituent elements. And the magnitude of the diad which is different from the magnitude of the triad may be the cause of the *mahattva* of the triad. This contention is untenable, because sometimes magnitude of the diad, i.e. atomic magnitude, being the cause of *mahattva*, as in the case of the triad, and sometimes *mahattva* being the cause of *mahattva*, as in the case of the compound object, there arises discrepancy *(vyabhicāra)*. Moreover, it is illogical to say that the same effect, viz. *mahattva*, is produced by different causes, viz. atomic magnitude (*aṇutva*) and manifested magnitude (*mahattva*).

If it is argued that number of the constituent elements is not needed to account for the *mahattva*, *aṇutva* is the cause of *anutkṛṣṭamahattva* and *mahattva* (manifested magnitude) is the cause of *utkṛṣṭamahattva*, then the difficulty is that even three atoms or two diad can produce *mahaddravya* (compound object), since number of constituent elements is immaterial. Therefore, to avoid this difficulty, it is admitted that *aṇutva* does not give rise to *mahattva*. And it is the number of atoms that gives rise to the *mahattva* of diad and triad. If the number of component parts is not taken into account, then both atom and diad will be of the same kind, since both possess atomic magnitude and there will be no distinction between atom and diad. Moreover, the accepted rule that the combination of two atoms makes a diad and the combination of three diads forms a triad would be useless. In fact, it is not reasonable to say that a product constituting of two atoms is the same as the product constituting of three diads. Therefore, it must be admitted that there is a difference in the magnitude of the products constituted by two atoms and three diads respectively. And in that case the cause of the difference also should be mentioned. The cause of the magnitude of a product produced by two atoms is the number two, and the number three is the cause of the magnitude of the product produced by three diads.

The Nyāya-Vaiśeṣika admits *pracaya* (loose conjunction of component parts) as one of the causes of producing magnitude. It may be said that the magnitude of diad and the magnitude of triad are also due to loose conjunction of atoms which are their component parts. This view is untenable, because loose conjunction of component parts is possible only in respect of the composite whole possessing parts. Atoms are held to be partless entities, and so loose conjunction of atoms is not possible.

According to the Nyāya-Vaiśeṣika, manifested magnitude *(mahattva)* of products from triad upwards is caused by *kāraṇamahattva* (magnitude of component parts) as well as *pracaya* (loose conjunction of component parts). But atomic magnitude is caused by *bahutva* or number of parts only. Thus, it is decided that the magnitude of diad and that of triad are due to the number, i.e. two and three of the constituent elements.

The Vaiśeṣika maintains that all integral numbers higher than one are derived from enumerative cognition *(apekṣābuddhi)*. When two objects are presented to our visual sense we have a complex cognition which we express as 'this is one' and 'that is one', and the third cognition is the accessory cause for the emergence of duality which inheres as a quality in the two objects on a collective basis. In the same way is derived the complex notion of three, etc. from the simple notion of unity. The notion of duality, etc. results from the complex cognition of individual being in common experience. But during cosmic rest individual selves remain unconscious, and so magnitude in the diad and the triad cannot be said to be due to the enumerative cognition of the individual soul. Therefore, it is posited that just prior to the cosmic creation there must be some intelligent being whose enumerative cognition creates the notion of two and three. This being must be God who is possessed of unimpeded cognition. If God does not create the notion of duality, then the magnitude of the diad would not be produced, because it is the notion of duality which produces magnitude in the diad. And unless the magnitude of the diad is produced, the triad which is formed by three diads would not be produced, because in the production of the triad the magnitude of diad is necessary. In the absence of the production of the triad, the universe itself would not be produced. Individual beings cannot produce the notion of duality, etc., because they do not exist before the production of the gross world. It is only after the production of the gross world that the atoms are inferred and the fact that the gross world is produced out of the combination of atoms in the form of diad, triad, etc. is also known. Thus, to say that the notion of duality is produced by the enumerative cognition of human beings would end in the fallacy of interdependence *(itaretarāśraya),* that is, human beings come to exist only after the production of the gross world and the notion of duality which gives rise to the magnitude of the diad is produced by the enumerative cognition of human beings; in other words, existence of human beings depends on the production of the gross world and the production of the world depends on human beings.

It is not proper to say that *adṛṣṭa* can produce the notion of duality, and so there is no need of *apekṣābuddhi.* For *adṛṣṭa,* without being assisted by other causes, cannot be the cause of the notion of duality.

Therefore, reflection shows that it is the omniscient God whose *apekṣābuddhi* produces the notion of duality which is the cause of the magnitude of diad. This is the last proof Udayana advances for the existence of God.

उद्देश एव तात्पर्यं व्याख्या विश्वदृशः सती।
ईश्वरादिपदं सार्थं लोकवृत्तानुसारतः।।६।।

(God's) intention (i.e. will) is the purport (of the Vedas); interpretation (of the Vedas) by an omniscient Being is valid; the words Īśvara, *etc. must have meanings as in secular usages.*

6.i. The *siddhāntin* now explains kāry*āyojana*.... etc. (*kārikā*-1) in a different manner. All the arguments, according to this explanation, are based on the different aspects of the Vedas. The ground *kāryatvāt* here means *tātparyāt.* It has been maintained that words are authoritative only in reference to their *tātparya* (purport). The Vedas in order to be authoritative must have *tātparya,* and one whose *tātparya* is declared in the Vedas is God. The Nyāya advocates that the Vedas primarily deal with something that is to be accomplished *(bhāvyārtha).* Teachings in the Vedas are twofold, either to enjoin us to do something *(pravṛtti)* or to refrain us from doing something *(nivṛtti).* The injunctions, 'one who is desirous of attaining heaven should perform sacrifice' *(yajeta svargakāmaḥ)* and 'one should not eat *kalañja' (na kalañjaṁ bhakṣayet),* directly refer to *pravṛtti* and *nivṛtti* respectively. And the *tātparya* of these injunctions is these *pravṛtti* and *nivṛtti.* The word *kārya* in the *kārikā* refers to this *pravṛttinivṛtti-tātparya.*

That part of the Vedas, e.g. the Upaniṣads, which is concerned with accomplished facts *(bhūtārtha)* like *'ātman'* and the *'brahman',* may not directly inspire us to do something or to refrain from doing something. Hence, it may be objected that the *tātparya* of accomplished facts *(bhūtārtha)* is not *pravṛtti* and *nivṛtti.*

The above objection may be answered thus. Though *bhūtārtha* does not directly refer to *pravṛtti* and *nivṛtti,* yet it refers to *pravṛtti* and *nivṛtti* being connected with those facts that refer to *pravṛtti* and *nivṛtti.* Thus, the *tātparya* of *bhūtārtha* is to be determined in reference to those facts that refer to *pravṛtti* and *nivṛtti.*

It cannot be maintained that *vidhi* (command) has the power to induce *pravṛtti* and *nivṛtti.* Even hearing the injunction, 'one who is desirous of attaining heaven should perform sacrifice', one may not take to sacrifice out of aversion. Again, hearing the injunction, 'one should not eat *kalañja*', one may be tempted to eat *kalañja* out of intense longing for it. In these cases, it is obvious that the command-power *(vidhi-śakti)* fails to induce *pravṛtti* and *nivṛtti* respectively. In those cases where injunctions do not induce *pravṛtti* and *nivṛtti,* it is *arthavāda* that strengthens the force of the Vedic injunction. People engage in such activities that are praised, and avoid such actions that are condemned. Thus, the *tātparya* of the *arthavāda* is *pravṛtti* and *nivṛtti.* The ground *kāryatvāt* in the *kārikā* entails the twofold action, viz. *pravṛtti* and *nivṛtti* which are the *tātparya* of the Vedas. And it is God whose *tātparya* is declared in the Vedas.

6.ii. The meaning of *tātparya* is different from word-meaning and sentence-meaning. The meaning of a word consists in its having the power of conveying its sense. And the meaning of a sentence consists in the mutual relation of significant words in conformity with syntactic expectancy *(ākāṅkṣā)*, semantic competency *(yogyatā)* and spatio temporal contiguity *(sannidhi)*. The Nyāya believes that language is the creation of God, and each word is capable of conveying a particular sense, because God has so willed it. Thus, God is established from the fact that His Will invests the words with their meanings. This theory of divine origin of the language is not accepted by many. Hence, the *siddhāntin* intends proving God's existence on the ground that His *abhiprāya* is declared in the Vedas.

The word *tātparya* may be explained as *tat eva paraṁ yasya tat tatparaṁ, tasya bhāvaḥ tātparyam.* That is *tat-paraṁ* which has *tat* (that) as its *paraṁ* (on which it is intent). The abstract noun from *tat-paraṁ* is *tātparyaṁ*. First, the meaning of *tat* (that) which is *paraṁ* in this case has to be ascertained. *Paraṁ* may mean *sādhya* (that which is to be established), or *pratipādya* (that which is conveyed), or *prayojana* (what is desired), or *uddeśya* (intention). Udayana says that it is *uddeśya* which is the meaning of the word *tātparya*. *Sādhya* cannot be the connotation of the word *tātparya*. The Vedas which are one of the four means of valid knowledge *(pramāṇa)* consist of sentences. Hence, any kind of action or *sādhya* can never be their *tātparya*. A *pramāṇa* can never be the cause of the object of the knowledge that it produces or the cause of any kind of action. Thus, it cannot be the cause of either *pravṛtti* or *nivṛtti*. It cannot be maintained that the Vedas as *pramāṇa* are, therefore, useless. For, conveying some particular meaning constitutes their usefulness. The Vedas convey their meanings, but they are not the cause of anything that is to be accomplished.

It may be argued that the *arthavādas* first convey their meaning either in the form of extolling or condemning something and in this way indirectly lead to *pravṛtti* or *nivṛtti*. This stands rejected, because in such sentence as 'the hamlet is situated on the Ganges' the word Ganges conveys its intended meaning, viz. its bank. But there is neither *pravṛtti* nor *nivṛtti* in regard to the bank. It is futile to say that as *pravṛtti* or *nivṛtti* is to be accomplished in respect of the bank, the bank is *para* here. For, when a past fact is stated 'there was a hamlet on the Ganges', there does not arise the contingency of *pravṛtti* or *nivṛtti* in respect of the bank.

6.iii. *Pratipādyatva* also cannot be the meaning of the word *tātparya*. If *pratipādyatva* were the meaning of *tātparya*, then to say that the Vedas are *karma-para* would amount to saying that the Vedas teach *karma* or that their meaning is *karman*. This view cannot be sustained. For, in the sentence 'God Vāyu is the swiftest of all' what is taught is not some kind of *karman*. In fact, words in a sentence convey their own individual meaning, while the sentence as a whole expresses their relation *(anvaya)*. Thus, the aforesaid sentence conveys only the meaning of individual words, viz. *vāyu,* and others, and their relation thereof. It may be said that though *karman,* viz. a particular sacrifice in favour of God Vāyu, is neither the meaning of any individual word constituting the sentence nor is understood by the power of the relation of different individual word-meanings, yet it is the *pratipādya* of the sentence, being invariably concomitant with God Vāyu. This argument cannot be sustained, because whatever is invariably concomitant with the connotation of a particular word cannot be regarded as the *para* or *pratipādya* of that word. Were it so, in the sentence, 'there is water in the Ganges', the word Ganges would convey the idea of its bank, since the two are invariably associated. It may be said that something associated with primary meaning is understood only when the primary meaning is incompatible. The primary meaning of the word Ganges here is not incompatible, hence, it would not convey the idea of its bank. This position is unsound. In the sentence 'My love, go if you so desire! May your path be auspicious! May I be born in the same place whither you are bound', the primary meaning is not incompatible. Yet the speaker intends to convey to the person addressed the sense — 'he should not go'. Moreover, *para* or that which is meant to be understood is not the *vyāpaka* of the primary meaning. For example, whent it is said 'the cradles are crying', what is meant is that babies lying in the cradle are crying. But neither cradles and babies, nor babies and crying are universally associated with each other.

6.iv. *Prayojana* is also not the connotation of the word *tātparya*. *Prayojana* may refer to the person addressed to *(pratipādya)* or to the speaker himself *(pratipādaka)*. But *śabda-pramāṇa* (the Vedas) does not depend upon the *prayojana* of the person addressed to, because even in the absence of the *prayojana* of the latter, the Vedic sentences possess *tātparya*. Further, if the Vedas depend on the *prayojana* of the person addressed to, the same sentence would possess different *tātparya* for different persons in accordance with their different *prayojanas*. As a result, there would be no distinction between *āpta* and *anāpta*.

It cannot be maintained that, that which is the *sādhya* of the meaning of the sentence *(vākyārtha)* is *prayojana* in this context, and the *sādhya* would constitute *tātparya*. *Sādhya* is of two kinds, viz. *kārya* (effect) and *jñāpya* (that which is revealed or made known). And each of these two can be of various kinds according to the person addressed to. This would lead to the defect called *vākya-bheda* which arises when various meanings are admitted in connection with the same sentence. This point may be clarified by an example. Smoke has many effects, blackening of the spot where it exists, dispelling of mosquitos, etc. Similarly, it also reveals many things, such as wet fuel, etc. Hence, when it is said 'smoke is rising in this place', it would not be possible to ascertain the *tātparya* of the sentence, because one would not be able to make out the *prayojana* of the different hearers. *Prayojana* of the speaker also cannot be the *tātparya*, because the opponent, the Mīmāṁsaka, does not admit any speaker in respect of the Vedas.

According to Udayana *uddeśya* is the meaning of the word *tātparya*. That which is intended by the speaker or the author is the *tātparya* (purport) of a word or a sentence. Actual practice proves this fact. When a laudatory sentence is uttered, what is meant is that the hearer should go in for the thing commended. Such a sentence has *upādāna* or *pravṛtti* as its *tātparya.* Similarly, the *tātparya* of a sentence that condemns something is *nivṛtti* or *hāna,* i.e. avoidance of that thing. In the same way is to be explained the Vedic sentences also, otherwise *arthavādas* would turn out to be meaningless. This *uddeśya* (intention) is the will *(icchā)* of the speaker or the author. *Uddeśya* is variously termed as *vyavasāya, adhikāra, abhiprāya, bhāva* or *āśaya.* Therefore, it follows that the Vedas must have an author whose intention or will is conveyed through the Vedas, and that author is God. In inferential form the argument runs thus: 'The Vedic laudatory sentences presuppose the will *(abhiprāya)* of the speaker, viz. things praised should be taken up, because they are laudatory sentences, like ordinary sentences, such as 'the ripe mangoes are full of sweet juice'. Similarly, the Vedic condemnatory sentences presuppose the will of the speaker, viz. things condemned should be given up, like ordinary sentences, such as 'the ripe jack-fruit is bad in taste'.

If *abhiprāya* of the speaker is not admitted in the Vedas, then not only all laudatory and condemnatory sentences which are known as *arthavādas,* would be meaningless, but *śrutārthāpatti* which is accepted by the Mīmāṁsaka also would be rejected. In *śrutārthāpatti* an unperceived fact is presumed in order to explain a fact already heard of. For example, the sentence *viśvajita yajeta* cannot be explained unless the word *svargakāma* is presumed. One who is desirous of attaining heaven, performs sacrifice. Hence, the aforesaid sentence is to be explained by presuming the word *svargakāma.* In the case of ordinary *(laukika) śrutārthāpatti* when it is said 'fat Devadatta does not eat during day', what is presumed is that 'Devadatta must be eating during night'. Otherwise, the fact of his being fat and not eating during day cannot be explained. The fact of Devadatta's eating during night is not heard, but is presumed. It is not uttered, because there is absence of the *vyāpāra* of the organ of speech which is the causal apparatus for the utterance of words or sentences. Moreover, had it been uttered, it would had been heard of. Nor can it be said that Devadatta's eating during night is something absurd, because an absurd thing like 'hare's horn' can never be presumed. Just as in the case 'fat

Devadatta does not eat during day' the fact 'he eats during night' is presumed in order to explain two inconsistent facts 'being fat' and 'not eating during day', so also the word *svargakāmaḥ* is presumed in the case *yajeta*, etc. in order to explain the meaning of the latter fact. When the sentence *viśvajita yajeta* is uttered, the word *svargakāmaḥ* is not uttered. Since without presuming the latter word, the former sentence cannot be explained; it is said that *svargakāmaḥ* is the intended meaning *(abhiprāya)* of the speaker.

Thus, if *abhiprāya* is not admitted in the Vedas, *śrutārthāpatti* too stands rejected. Since all the Vedas primarily teach *karman* in the form of *pravṛtti* or *nivṛtti*, *abhiprāya* is to be admitted in order to explain the Vedic words and their meanings. And a speaker is established as the substratum of this *abhiprāya,* who is God.

6.v. *Āyojanāt,* the second ground for God's existence, stands for interpretation *(vyākhyānaṁ).* The proper understanding of the meaning of the Vedic literature and the carrying out their injunctions require the correct interpretation of the Vedas. The interpreter cannot be one who knows only part of the Vedas, because a word judged without reference to what has preceded it and what has followed it, would yield quite a different meaning. Ideas conveyed by the Vedas lie beyond the ken of the senses and are very difficult to comprehend with difficult words intervening. Therefore, Udayana postulates the first interpreter of the Vedas as *sarvajña,* i.e. all knowing person who is cognizant of the entire Vedic literature. He also knows the extent of the Vedas, that is, He knows that 'this much forms the Vedas' *(etāvānevāyamāmnāya).* And such an omniscient interpreter is God. The fact of God's having the knowledge of the import of all the Vedas is called *dhṛti,* and *dhṛteḥ* is the third ground for God's existence.

The knowledge that 'only this much constitutes the Vedas' cannot be possessed by any other being excepting God. For, human beings with their limited knowledge can neither master the entire Vedas nor can teach them. If the Vedas were interpreted by non-omniscient beings, then the interpretations would have differed from place to place and time to time due to the limitation of the persons who interpreted them. As a result, performances of the Vedic rites too would have differed from person to person, there being no *jñānavyavasthā*. Since there is uniformity of the performances of the Vedic rites, it is inferred that the Vedas are interpreted by an omniscient being, viz. God. The inferential argument is thus: *vedāḥ kadācit sarvavedārthavidvyākhyātāḥ, anuṣṭhatṛmaticalane'pi niścalārthānuṣṭhānatvāt.*

The Vedas cannot be interpreted by those who perform the various rites prescribed by the Vedas, because they do not have the proper understanding of the Vedas. Moreover, no confidence can be reposed in any non-omniscient interpreter. Nor can it be maintained that the tradition of the performance of the Vedic rites is beginningless, as it depends on prior and still prior tradition. For, if the tradition is not based on the Vedas, it lacks sound basis. If, on the other hand, it is based on the performer's knowledge of the Vedas, the tradition will not be regulated. In fact, correct performance of the rites will be possible only when it is initiated by a person of unquestionable competence. And it is only God who is competent to initiate the tradition of the ritual performance.

6.vi. The argument from *padāt* has been interpreted in the following manner. Words like *Oṁ, Īśvara, Īśāna,* etc. used in the Vedas prove God's existence. The words *Īśvara*, etc. cannot but be meaningful, because they are used in the Vedas, Smṛti and Itihāsa along with other meaningful words, like jar, etc. The following *arthavāda* statement tells us what *Īśvara* is like:

> *uttamaḥ puruṣastvanyaḥ paramātmetyudāhṛtaḥ,*
> *yo lokatrayamāviśya bibhartyavyaya Īśvarah.*

Similarly, there is a passage on *Īśvarapraṇidhāna:*

> *sarvajñatā tṛptiranādibodhaḥ svatantratā nityamaluptaśaktiḥ,*
> *anantaśaktiśca vibhorvidhijñāḥ ṣaḍāhuraṅgāni maheśvarasya.*

As we believe in the transcendental things connoted by the words *svarga,* etc. used in the Vedas, similarly we should believe in the existence of God who is referred to by the words *Īśvara, Paramātman,* etc. in the Vedas.

The Mīmāṁsaka objects that the fact that the words *Īśvara,* etc. refer to God, is not known by any *pramāṇa*. Hence, the existence of God cannot be proved in that way. But the same objection may be raised against the Mīmāṁsaka too. He believes in the existence of *svarga* which is admitted as an abode of eternal blisss, unassociated with pain and miseries. But there is no *pramāṇa* to prove the existence of such a *svarga.*

It may be said that people perform sacrificial rites with a view to attaining heaven. If there were no heaven in existence, people would not have performed such rites. Therefore, the existence of heaven must be admitted; otherwise, the performance of sacrificial rites cannot be explained.

This view gives rise to a difficulty. The knowledge of heaven precedes the performance of the sacrificial rites and the existence of heaven is admitted as otherwise performance of the rites cannot be explained — this leads to the fallacy of interdependence *(itaretarāśraya).*

The above fallacy cannot be avoided by saying that the existence of heaven has been admitted by the earlier wise men. For, this leads to the plight of a row of blindmen *(andhaparamparā),* as the trustworthiness of the person who has admitted the existence of heaven for the first time can never be verified.

The Mīmāṁsaka maintains that when a word is used by the learned people to refer to an object, that word signifies that object, provided that word does not have any other *vṛtti,* such as *lakṣaṇā* and *gauṇī.* Thus, the word *svarga* is used to refer to bliss; hence, the former signifies the latter. To this Udayana rejoins that in the same manner the meaning of the words *īśvara,* etc. also is to be determined. *Īśvara* is used in the Vedas to refer to God, the creator of the world. Otherwise, if the words *īśvara,* etc. were meaningless, there is no reason why they should be used in the Vedas together with other meaningful words.

Again, the word 'I' *(asmad)* is used in the Vedas in the sentences like 'one should worship Me', etc. In mundane usage, the word 'I' does not refer to any insentient thing, because it is never used in reference to such thing. It does not refer to the soul in general also, because in that case it may be used with reference to other's soul. But the word 'I' never refers to other's soul. In fact, it is used by the independent speaker or the author of a sentence with reference to oneself. Just as the word 'I' in ordinary usage refers to an independent speaker of a sentence, so also the word 'I' used in the Vedas refers to their author. To argue that the word 'I' refers to the person who merely pronounces the word, is untenable, because he merely repeats the particular word. Otherwise, the sentence 'one should worship Me' would mean that the person repeating the sentence is to be worshipped. Again, when the statements 'I am the origin of everything in this universe, everything proceeds from Me' are repeated by the successive teachers and their pupils, it would mean that they are endowed with the power of creating the universe, etc. Consequently, all worships would turn out to be a big hoax, all activities would come to a stand still. Thus, it must be admitted that the word 'I' does not refer to one who merely repeats the word *(anuvaktā)*, but it refers to the independent speaker or the author *(svatantravaktā)*. And the existence of God who is the author of the Vedas is inferred from the connotation of the word 'I', the first person pronoun, used in the Vedas.

Further, there is usage of such words as *yaḥ, kaḥ, saḥ,* etc. in the Vedas. The words *yaḥ,* etc. refer to *cittavṛtti* pertaining to prior knowledge *(buddhyupakramaḥ)*, desire to know *(jijñāsā)* and recollection *(pratisandhāna)* respectively, and God is established as the substratum of that *cittavṛtti.* Similarly, there are usages of such words in the Vedas that express doubt *(saṁśaya)*, and God is established as the agent who expresses that doubt.

One pertinent question may be raised: 'How can God who is omniscient, have *jijñāsā* or *saṁśaya?'* In reply it may be said that though God Himself does not have either *jijñāsā* or *saṁśaya*, such words are used in the Vedas anticipating that the hearers may have that. Just as Śabarasvāmī in his *Mīmāṁsābhāṣya* has written *ko dharmaḥ kathaṁ lakṣaṇakaḥ,* etc. anticipating that the readers may raise such question, though he himself did not have any doubt as to what is *dharma*, etc., similarly words expressing *jijñāsā*, etc. are used in the Vedas anticipating the hearer's *jijñāsā* or *saṁśaya*. But God does not have any such *jijñāsā* or *saṁśaya*.

6.vii *Pratyayāt*, another ground for God's existence, refers to *vidhi-pratyaya* suffixed to verbal roots expressing command, etc. Suffixes, like *liṅ*, etc. convey the command of the Supreme Person, viz. God, and thus, the existence of God is inferred from the use of *vidhi-pratyaya* in the Vedas. In empirical usage when an injunctive sentence is uttered by a trustworthy author, the intention of the author is that the hearer should either do something or desist from doing something. And the hearer infers that doing the thing in question would enable him to achieve his object, or that by not desisting from the thing prohibited he would reap some undesirable reward. As in the case of empirical injunction, so also in the case of the Vedic injunctions the *vidhi-pratyaya* expresses the intention *(abhiprāya)* of the trustworthy person who is their author. The use of *vidhi-pratyaya* in the Vedas presupposes a trustworthy person as their author whose intention is conveyed by the *vidhi-pratyaya* and that author must be God. Thus, '*vidhi-pratyaya* according to Udayana, signifies intention (*abhiprāya*) of the trustworthy person *(āpta)* or the speaker *(vaktā)*.

प्रवृत्तिः कृतिरेवात्र सा चेच्छातो यतश्च सा।
तज्ज्ञानं विषयस्तस्य विधिस्तज्ज्ञापकोऽथवा।।७।।

Activity is really volition (kṛti *or* yatna), *and this (volition) springs from the will to act, and this (will to act) arises from knowledge, and the object of that knowledge is the meaning of* vidhi-pratyaya *or (as Udayana would hold the meaning of* vidhi-pratyaya *is the* abhiprāya *of God) which is the cause of the knowledge of that object.*

7. Hearing the suffixes, viz. *liṅ*, etc. people are led to activity *(pravṛtti)*. Hence, it is clear that *liṅ*, etc. convey something in favour of *pravṛtti*. The question arises as to what *pravṛtti* is. Udayana expresses the view that *pravṛtti* is really volition (*kṛti* or *yatna*), volition springs from the will to act, the will to act springs

from knowledge, and the object of that knowledge is the meaning of *vidhi-pratyaya*; or as Udayana would hold, the meaning of *vidhi-pratyaya* is *abhiprāya* of God, which is the cause of the knowledge of that object.

Pravṛtti cannot be defined as mere muscular action *(kāyaparispandamātra),* because were it so in the case of such *vidhi-pratyaya* as 'one should know the Ātman', there would be no *pravṛtti.* For, one is not in need of any muscular action in order to know one's own soul. *Pravṛtti* is also not a mere desire *(icchāmātra),* because if *pravṛtti* were a desire, one can attain the fruit just by having a desire to perform the sacrifice without putting any volitional effort to perform it. As a result there will arise inactivity *(karmā'nārambha-prasaṅga).* In fact, *pravṛtti* is neither muscular action nor desire, but is *prayatna* (volition). Desire can be there in respect of a thing that can be accomplished as well as a thing that cannot be accomplished. But *prayatna* is possible only in respect of a thing that can be accomplished. Thus, *pravṛtti* in the sense of *prayatna* is invariably accompanied by action, whereas *pravṛtti* in the sense of desire is not so. *Pravṛtti* in the former sense also exists in such cases as 'one should know the Ātman'.

After determining the nature of *pravṛtti,* the *siddhāntin* defines *vidhi* or *liṅ-artha. Vidhi* is not that which gives rise to *pravṛtti,* because desire *(icchā)* gives rise to *pravṛtti* existing prior to *pravṛtti.* Desire does not depend on the knowledge of *vidhi-vākya* in order to give rise to *pravṛtti,* therefore, desire cannot be *vidhi.*

Nor can it be said that *vidhi* gives rise to desire which is the cause of *pravṛtti.* The cause of desire is the knowledge of *liṅ-artha,* and so unless the knowledge of *liṅ-artha* is produced, desire cannot be produced by *liṅ-artha.*

The knowledge of *liṅ,* etc. is also not *vidhi.* Knowledge does not arise without its object *(viṣaya),* and so according to old Naiyāyikas it is the object of knowledge which is the meaning of *vidhi* or *liṅ-artha.* The *siddhāntin* advocates a different view. According to him *liṅ-artha* is the *abhiprāya* of God, which is the cause of the knowledge of that object.

इष्टहानेरनिष्टाप्तेरप्रवृत्तेर्विरोधतः।
असत्त्वात् प्रत्ययत्यागात् कर्तृधर्मो न सङ्करात्॥८॥

It (vidhi) *cannot be an attribute of the agent owing to less-pervasion and over-pervasion, or owing to absence of* pravṛtti, *or owing to inconsistency, or owing to absence (of the presence of desire), or owing to the rejection of the accepted cause, or owing to confusion.*

8.i. *Liṅ-artha* is a *dharma* or attribute, since it induces activity. It has to be decided whether it is *kartṛdharma,* or *karmadharma,* or *karaṇadharma*, or *niyoktṛdharma,* viz. *abhiprāya.*

Kartṛdharmas are *spanda, kṛti, icchā* and *jñāna. Spanda* cannot be *liṅ-artha,* because *liṅ-artha* induces *pravṛtti,* whereas *spanda* does not induce any *pravṛtti.* For example, after hearing the text 'one should know the Ātman' one should have *pravṛtti,* but if *spanda* were *liṅ-artha,* there would be no *pravṛtti* in that case, as in knowing the Ātman one does not have any such activity as *spanda.* Therefore, if *spanda* were *liṅ-artha,* there arises *iṣṭahāni,* i.e. there arises the contingency of absence of *pravṛtti* in the case where *pravṛtti* should arise *(iṣṭa).*

Again, on hearing a sentence 'he goes to the village' there arises *pravṛtti,* because 'goes' *(gacchati)* implies *spanda.* But in this case *pravṛtti* should not arise, because the verbal root *gam* does not have any such suffix *liṅ,* etc. which are *vidhi-pratyaya.* Thus, if *spanda* were *liṅ-artha,* there arises not only *iṣṭahāni,* but also *aniṣṭāpti,* i.e. presence of *pravṛtti* in a case where *pravṛtti* should not arise.

Kṛti or *prayatna* is also not *liṅ-artha. Kṛti* is implied not only by *liṅ,* but by all other *pratyayas,* such as *loṭ, liṭ,* etc. Just as on hearing other *pratyaya, pravṛtti* does not arise so also on hearing *vidhi-pratyaya,* such as *liṅ,* etc., *pravṛtti* will not arise. That is why the *siddhāntin* says, if *kṛti* were *liṅ-artha,* there will arise *apravṛtti.* Therefore, *kṛti* is not *liṅ-artha,* which gives rise to *pravṛtti.*

If *icchā* were *liṅ-artha,* the following difficulty arises. *Icchā* ensues from the knowledge of *liṅ-artha;* so if *icchā* were *liṅ-artha,* it follows that the knowledge of *icchā* gives rise to *icchā*. Again, the object of the knowledge of *icchā* is *icchā*, the object is the cause towards its knowledge, and so *icchā* will be the cause towards the knowledge of *icchā*. The outcome of this argument is that *icchā* is the cause of the knowledge of *icchā*, and the knowledge of *icchā* which is *liṅ-artha* is the cause of *icchā* which is the cause of *pravṛtti*. It is obvious that this argument suffers from the fallacy of mutual interdependence, which is signified by the word *virodataḥ* in the *kārikā*. Therefore, *icchā* cannot be *liṅ-artha*.

It may be said that the cause of the knowledge of *icchā* is not *icchā* but *liṅ*, that is, the knowledge of *icchā* is produced from hearing *liṅ*. This is untenable, for even if the knowledge of *icchā* is produced from the *liṅ*, *pravṛtti* will not arise, as *svarūpasat-icchā* which is the cause of *pravṛtti* is not present at the time when *liṅ* is heard.

There are two types of causes *svarūpasat* and *jñāta*. When it is said 'eating is the cause of satisfying hunger', the cause 'eating' is *svarūpasat*, as it being present by itself causes satisfaction of hunger. *Svarūpasat-icchā* means an *icchā* which being present by itself prior to *pravṛtti* gives rise to *pravṛtti*. Thus, as *svarūpasat-icchā* is the cause of *pravṛtti* and not *jñāta-icchā*, there will be absence of *pravṛtti* even after the knowledge of *icchā* is produced from *liṅ*. This is expressed by the *hetu-asattvāt* in the above-mentioned *kārikā*.

It is not proper to maintain that *liṅ* is the cause of *icchā*. Knowledge is universally accepted as the cause of *icchā*. The knowledge of cause gives rise to the knowledge of effect. When the knowledge of the cause gives rise to the knowledge of the effect, it is called *jñāta* cause. For example, 'smoke is a cause of fire being inferred.' The knowledge that the action enjoined is a means to the desired end is the cause of *icchā*, and if *liṅ* is admitted as the cause of *icchā*, universally accepted cause of *icchā*, viz. knowledge, has to be thrown overboard. The *hetu-pratyayatyāgāt* in the *kārikā* expresses this view.

8.ii. It may be said that the knowledge of *phalecchā* (desire for attaining the fruit) which arises after hearing the *liṅ* is *liṅ-artha*.

Liṅ gives rise to the knowledge of *phalecchā* and from this is generated *upāyecchā* (desire for attaining the means).

This view is untenable. The *hetu* of *upāyecchā* is the knowledge of *iṣṭasādhanatā* (the knowledge that a particular action is a means to a desired end), the cause of which is the *liṅ*. If the above-mentioned view is accepted, then the *liṅ* would give rise to the knowledge of both *iṣṭasādhanatā* as well as *phalecchā*. The knowledge of *iṣṭasādhanatā* is accepted as the invariable cause of *pravṛtti*, and if knowledge of *phalecchā* also gives rise to *pravṛtti*, there arises confusion of two causes, viz. knowledge of *iṣṭasādhanatā* and that of *phalecchā*. In order to avoid this position, knowledge of *phalecchā* should not be accepted as the cause of *pravṛtti*, since even in the absence of the knowledge of *phalecchā*, *pravṛtti* arises in the presence of the knowledge of *iṣṭasādhanatā*. The *hetu-saṅkarāt* in the *kārikā* points out the confusion of two causes of *pravṛtti*, viz. knowledge of *iṣṭasādhanatā* and that of *phalecchā*.

It has been already stated that according to the *siddhāntin prayatna* is implied by all the *pratyayas*, viz. *loṭ*, *liṭ*, etc. The *pūrvapakṣin* does not admit this view. According to him in *karoti*, etc. the *dhātu* signifies *prayatna*, and if apart from *the dhātu pratyaya* also signifies *prayatna*, there arises the defect of repetition *(paunarukti)*. Therefore, it must be admitted that *loṭ*, *liṭ*, etc. express only *saṅkhyā* (number) and *vidhi-pratyaya*, i.e. *liṅ*, etc. implies *prayatna*. This view stands rejected, because in *kuryāt* the *dhātu* signifies *prayatna*, and apart from the *dhātu*, if *liṅ* also signifies *prayatna*, there will be repetition.

It may be said that though both the *liṅ* and the *dhātu* signify *prayatna*, yet there is a difference in their signification. *Prayatna* signified by the *lin* refers to the agent *(kartā)*, whereas *prayatna* signified by the *dhātu* has no such reference. This view is not accepted, because both *dhātu* as well as *pratyaya* have the same meaning. Thus, both refer to the same number, viz. one, two, etc. If *pratyaya* refers to the agent, the *dhātu* too will refer to the same thing.

The opponent raises an objection that if the *pratyayas*, like *loṭ*, etc. signify *prayatna*, then such instance as 'the chariot goes' cannot be explained. *Prayatna* is an attribute of sentient being, the chariot which is insentient, cannot have *prayatna*. Therefore, it is not proper to maintain that *loṭ*, etc. signify *prayatna*.

In reply the *siddhāntin* says that the instance 'the chariot goes' *(ratho gacchati)* must be explained in the same manner as the instance 'the threads produce the cloth' *(tantavaḥ paṭam kurvanti)* is explained. It is the weaver who produces the cloth out of threads, the weaver is the substratum of *prayatna*, and not the threads. Similarly, in the instance 'the chariot goes' charioteer is the substratum of *prayatna* and not the chariot. *Prayatna* is here the secondary significative meaning.

It may be said that the fact that *liṅ* signifies *prayatna* is proved from such usage as 'bring the cow', because 'bringing' is an activity which implies *prayatna*. But no such usage is there to prove that *lot*, etc. signify *prayatna*. This contention is untenable, because just as 'bring the cow' implies *prayatna*, so also *pacati* which means *pākam karoti* implies *prayatna*.

The Vaiyākaraṇa (grammarian) hǫlds that *dhātu* signifies particular *phalānukūla-vyāpāra*, and *ākhyāta* or *la-kāra* (verbal termination) signifies *phalānukūla-vyāpāra* in general. Thus, *pac-dhātu* signifies *phalānukūla-vyāpāra*, and *pacati* which is explained as *pākaṁ karoti* too refers to the same *vyāpāra*. The *siddhāntin* does not admit this view. If *ākhyātārtha* and *dhātvartha* express *sāmānya* and *viśeṣa phalānukūla-vyāpāra*, then like the expression *jyotiṣṭomena yajeta* which conveys particular *yāga-vyāpāra*, there should be *pākena karoti* expression too. In fact, *pacati* is explained by the Nyāya as *pākaṁ karoti*, the *dhātu pac* is explained by the word *pāka* and *tip* is explained by *karoti* which signifies *prayatna* or *kṛti*. Therefore, according to the Nyāya, *kṛti* or *prayatna* is *ākhyātārtha*.

According to the Mīmāṁsaka, *phālānukūla-vyāpāra* or *bhāvanā* is *ākhyātārtha*. From the Mīmāṁsaka standpoint it may be said that *karoti* in *pākam karoti* signifies *anukūla-vyāpāra*. *Pratyaya* implies *prayatna* indirectly through *anukūla-vyāpāra* but it does not signify *prayatna* directly. Thus, *anukūla-vyāpāra* is *ākhyātārtha* and not *prayatna*, otherwise postulation of both as *ākhyātārtha* violates the law of parsimony. On this view the instance, viz. 'the chariot goes', can also be explained, as *ākhyātārtha* here is *gamanānukūla-vyāpāra* which exists in insentient thing.

कृताकृतविभागेन कर्त्तृरूपव्यवस्थया।
यत्न एव कृतिः पूर्वा परस्मिन् सैव भावना।।६।।

In accordance with the rule in applying the term 'agent' based on the distinction made between the usages 'done by me' and 'not done by me', activity primarily means volition only, secondarily it signifies bhāvanā (anukūla-vyāpāra, *i.e. action conducive to producing the desired result).*

9. The *siddhāntin* does not subscribe to this view. He refutes this view in the following manner. In everyday life we observe such usages as 'the jar is made by me', 'sprout is not produced by me', etc. Thus, in respect of the jar the usage is *ghaṭo mayā kṛtaḥ* and about sprout *nāṅkuarḥ mayā kṛtaḥ*. This division of *kṛta* and *akṛta* usage in respect of jar and sprout respectively is determined by *prayatna* on the part of the agent (*kartā*). *Kṛ-dhātu* signifies *prayatna*, in respect of jar *kṛtaḥ* is used, because a jar is produced by the volitional activity *(prayatna)* of an agent, whereas in respect of sprout *kṛtaḥ* is not used, because human volition cannot produce a sprout. From this it is concluded that *kṛ-dhātu* signifies *yatna* or *prayatna* and the *ākhyātārtha* is *phalānukūla-yatna*. But in such usage as *tantavaḥ kurvanti, kṛ-dhātu* signifies *yatna* secondarily, it primarily signifies *anukūla-vyāpāra*. Just as the word *śrotriya* primarily means a pure *brāhmaṇa* who has studied the Vedas, and yet it also means a pure *brāhmaṇa* (who has not studied the Vedas) in secondary sense, so also in *tantavaḥ kurvanti* primary *ākhyātārtha* is *anukūla-vyāpāra* and *anukūla-yatna* is secondary. The secondary meaning must be accepted. Instances are there where secondary sense is admitted unquestionably. For example, sometimes there is usage of first person and second person in respect of inanimate objects, though by first person and second

person only sentient beings are meant. Here such usage should be taken in secondary sense.

Again, when *jānāti* is used in respect of a third person, it does not signify *yatna*, because it means *jñānavān* which does not refer to *yatna*. Here *jānāti* implies *yatna* secondarily.

It is not proper to maintain that just as the word *'hari'* has various meanings, such as Viṣṇu, etc., so also the verb *jānāti*, etc. have got various meanings. For, when things can be explained in secondary sense, it is not reasonable to attribute various meanings to the same word.

Pacati, etc. signify both *anukūla-vyāpāra* as well as *yatna*. Otherwise, if *pacati* means only *anukūla-vyāpāra*, then *pacati* may as well refer to the guest for whom cooking is being done. Since *pacati* refers to the person who cooks and not the person for whom cooking is done, it may be concluded that *ākhyātārtha* is *anukūla-yatna*.

In verbal usage *kṛ-dhātu* signifies *vyāpāra* of a sentient *kartā* (agent). For example, in Devadatta *kāṣṭhaiḥ odanaṁ pacati, pacati* refers to the *pāka-vyāpāra* of Devadatta who is a sentient being. This fact also supports the contention that *dhātu* signifies *yatna* and *ākhyātārtha* is *anukūla-yatna*. It cannot be said that Devadatta is admitted as *kartā* here, because *pac-dhātu* and its *pratyaya* convey the *vyāpāra* of Devadatta. For, *pāka-kartā* is always admitted as a sentient being, even when its *vyāpāra* is not conveyed. Nor can it be maintained that only *pacati* has *yogyatā* to signify *vyāpāra* of a sentient agent, because *kāṣṭhaiḥ* too has that *yogyatā* equally. In fact, all the *kārakas* (grammatical cases) such as *karaṇa-kāraka*, etc. refer to *phalānukūla-vyāpāra*, but only a sentient agent is called *kartā* and not others. It is also not that a particular *kāraka* is called *kartṛ-kāraka* according to the will of the speaker, because even in the absence of that will *(vivakṣā)* a sentient agent is called *kartā*. Nor is it that Devadatta is called *kartā*, because he is the substratum of his own *vyāpāra*. For, all the *kārakas* are substrata of their own *vyāpāra*; so in this sense all the *kārakas* would turn out to be *kartā*.

As a matter of fact, *kārkas* are called *karaṇa*, *karma*, etc.

not in reference to their own *vyāpāra*, but in reference to *pradhāna-kriyā*. Among many *kāraka-vyāpāras* there is one main *kriyā*, and other *kāraka-vyāpāras* are determined by this *kriyā*. Thus, in reference to this *kriyā* Devadatta becomes *kartā*, *kāṣṭha karaṇa*, and so on.

In verbal usage *vyāpāra* of a sentient being is admitted as main *kriyā*, and a sentient being is called *kartā*, and so forth with reference to that *kriyā*. The question arises by what criterion such usage is determined. The fact is that a sentient being is called *kartā*, because he possesses *jñāna* (knowledge), *cikirṣā* (will) and *prayatna* (volitional effort), as possession of these qualities determines *kartṛtva* (agenthood). Thus, it is established that *kartṛtva* (agenthood) means *yatnavatva* (possession of volition) and the *dhātu-kṛ* signifies *yatna*, which is also *ākhyātārtha*.

भावनैव हि यत्नात्मा सर्वत्राख्यातगोचरः।
तया विवरणध्रौव्यादाक्षेपानुपपत्तितः।।१०।।

Verbal termination (ākhyāta) *always means volitional activity conducive to producing the desired result; as this explains the fact,* ākṣepa *cannot be established.*

10. The *pūrvapakṣin* says that *ākhyātārtha* is only *phalānukūlatva*, and *prayatna* is obtained from *ākṣepa*. By *ākṣepa* is meant explanation of some fact through *arthāpatti-pramāṇa*, which fact cannot be explained otherwise. The *siddhāntin* does not admit this view. *Yatnatva* cannot be obtained from *anukūlatva*, because only *vyāpaka* can be obtained through *ākṣepa* and not *vyāpya*, and there is no *vyāpya-vyāpaka* relation between *anukūlatva* and *yatnatva*. *Anukūlatva* may exist even in insentient thing, such as wood, etc., but *yatna* exists only in a sentient being, that is why there is no invariable *vyāpti* relation between them.

Yatnatva cannot be obtained through *saṅkhyā* (number) also. For, *saṅkhyā* belongs to *saṅkhyeya* (a thing which has *saṅkhyā*) and so from *saṅkhyā saṅkhyeya* can be obtained, but not *prayatna*, as *prayatna* has got no *saṅkhyā*.

Nor can *prayatna* be obtained through *kartā*, because by *kartā* is meant only substance which is devoid of *prayatna*. Obviously *prayatna* cannot be through something which does not possess *prayatna*.

Prayatna cannot be obtained through *dhātvartha* (root meaning) also, because the *dhātu vid*, etc. do not refer to *prayatna*. Let us examine whether *prayatna* can be obtained through *kāraka* (a grammatical case). To take an example *caitraḥ odanaṁ pacati; caitra* is not a *kāraka*, because according to Pāṇini *prātipadika-pratyaya* is devoid of *kāraka*-meaning. Therefore, *prayatna* cannot be implied by *caitra*. The word *odanaṁ* is *karma-kāraka*, and since all the *kārakas* imply *kriyā*, *prayatna* which is a *kriyā* too is implied by the *kāraka*. That *prayatna* is implied by the word *odanaṁ* is provided by the fact that after hearing that word such expectancy as *kiṁ bhuṅkte pacati vā* arises in the mind.

But the mere utterance of the verb *pacati* too gives rise to similar expectancy, even when *kārakas* are not heard. Since *kriyā* gives rise to *kāraka-ākāṅkṣā*, it is not proper to maintain that *prayatna* which is *kriyā* is understood from *kāraka*. Therefore, it must be admitted that *ākhyāta* or *la-kāra* implies *prayatna*, and *ākhyātārtha* is *anukūla-prayatna*.

The Naiyāyika explains *pacati* as *pākaṁ karoti*, and on the basis of this maintains that the *dhātu kṛ* implies *prayatna* which is also *ākhyātārtha*. From the standpoint of the Vaiyākaraṇa the *pūrvapakṣin* may say that *pacati* may be explained as *pākayatnavān*, and since *pākayatnavān* implies *kartā*, *kartā* may be admitted as *ākhyātārtha*.

आक्षेपलभ्ये सङ्खयेये नाभिधानस्य कल्पना।
सङ्खयेयमात्रलाभेऽपि साकाङ्क्षेण व्यवस्थितिः।।११।।

Verbal termination (ākhyāta) *does not signify (an agent),as this which is connected with number can be obtained by inferenc; even though that which is connected with number may be obtained (in this way), yet it is regulated by expectancy.*

11.i. This view of Vaiyākaraṇa does not find favour with the *siddhāntin*. *Kartā* is implied by the number *(saṅkhyā)* which is expressed by *ākhyāta*. Since *kartā* which is also called *saṅkhyeya* can be obtained through number, *kartā* cannot be *ākhyātārtha*. Since *kartā* is obtained by an inference from number expressed by *ākhyāta*, it is called *ākṣepalabhya*. The expression *ākṣepalabhya* means *prathamāntapadopasthāpyatve sati ākhyātārthaviśeṣyatvam*. That is, that which is *prathamānta-pada* and is the *viśeṣya* of *ākhyātārthakṛti*, is called *ākṣepalabhya*. In such verbal usage as *caitraḥ taṇḍulaṁ pacati* which is in *kartṛvācya*, *caitra* is *prathamānta-pada* and in *caitrena taṇḍulaḥ pacyate* which is in *karma-vācya*, *taṇḍulaḥ* is *prathamānta-pada*. So, number expressed by both the *ākhyāta* is related with *caitra* in the first case and with *taṇḍulaḥ* in the second case. Both these *prathamānta-padas* are also *viśeṣya* of *ākhyātārthakṛti*, i.e. both are subjects of *pākānukūlakriyā*.

The self-same *ākhyāta* that expresses *yatna*, expresses *saṅkhyā* too, and *saṅkhyā* is related with that with which *yatna* is related. In other words, the *viśeṣya* of *yatna* is the *viśeṣya* of *saṅkhyā* also. *Yatna* which is a *vyāpāra* has expectancy for its substratum *(āśraya)*, and its *āśraya* is that which without being related to any other *vyāpāra* prior to the relation with *yatna* has expectancy *(sākāṅkṣā)* for *vyāpāra*. The relation between *yatna* and its substratum depends on the expectancy on both the sides, viz. *yatna* as well as *āśraya*.

Yatna has expectancy for *śuddha-prātipadikārtha*. For example, in *caitraḥ taṇḍulaṁ agninā pacati*, the word *caitraḥ* which is *śuddha -prātipadika*, i.e. which is devoid of any *kāraka*, expects for *vyāpāra*, and the *ākhyātārthayatna*, viz. *pākānukūla-kriyā*, expects for its *āśraya*. So, on the basis of expectancy on both the sides, i.e. *vyāpāra* as well as *aśraya*, they are mutually related. *Yatna* and *āśraya* are never related on the basis of *karaṇa-vyāpāra*, etc., e.g. in *pacyate taṇḍulaścaitrena*, *caitra* has *karaṇa-vyāpāra* and *yatna*, viz. *pākānukūlakriyā*, has no relation with it. The word *taṇḍula* does not have any *kāraka-vyāpāra*, hence, *yatna* is related with it. *Saṅkhyā* expressed by the respective *ākhyāta* is also related with *caitraḥ* and *taṇḍulaḥ* following *yatna*.

In the instance *supyate caitrena*, there is no *saṅkhyā* relation either with *kartā* or with *karma*, because neither there is expectancy on the part of *kartā*, nor is there any *karma* present. In this case *saṅkhyā* is related with *dhātvartha* (root meaning).

In *kartṛvācya*, *yatna* is related with its *āśraya* through *āśrayatā*-relation, in *karmavācya* through *uddeśyatā*-relation and in *bhāvavācya* through *janyatva*-relation.

Yatna cannot be related with other *vyāpāra*, because if one *vyāpāra*, viz. *yatna*, is related with another *vyāpāra*, there arises the fallacy of *ātmāśrayatva*. Nor can it be maintained that *yatna* is related with other *vyāpāra* through *samavāya*-relation. Just as *yatna* has expectancy for *śuddha-prātipadikārtha*, so also its *samavāya* has expectancy for *śuddha-prātipadikārtha*. The fallacy of *ātmāśrayatva* cannot be evaded by saying that as *yatna* is different from other *kāraka-vyāpāra*, the former can be related with the latter existing in the substratum. For, *yatna* and other *kāraka-vyāpāra* never exist in the same substratum, so there is no ground for any relation between two different kinds of *vyāpāra*.

It has already been mentioned that *saṅkhyā* expressed by *ākhyāta* is connected with what *yatna* implied by *ākhyāta* is connected. Thus, as *yatna* is connected with *viśeṣya*, so also is *saṅkhyā* connected with it. The fact that *saṅkhyā* is connected with *yatna-viśeṣya*, can be clarified by an example. In the instance *caitro na gauro na spandate*, *caitra* is *viśeṣya* and *gaura*, etc. are *viśeṣaṇa*, and the negative word *na* is connected with *viśeṣaṇa*. Just as *na* is connected with *viśeṣaṇa* aspect and not with *viśeṣya* aspect, so also *saṅkhyā* has connection only with *yatna-viśesya* and not with *yatna-viśeṣaṇa*.

The *pūrvapakṣin* may say that the negative sign *na* is related with *viśeṣaṇa* and not with *viśeṣya*, due to the reason that if it is related with *viśeṣya*, *viśeṣaṇa* becomes useless. For, by negating *viśeṣya*, the *viśeṣya* qualified by *viśeṣaṇa* is negated, and so there is no need of negating *viśeṣana* again. If both *viśeṣya* and *viśeṣaṇa* are negated separately, then *na* will be related with both separately; consequently, there arises the contingency of repetition. Without repetition when the relation with *viśeṣaṇa* itself satisfies expectancy, there is no need for the relation with the *viśeṣya*. The *siddhāntin* says that it is the same principle of expectancy on which is based the relation of *saṅkhyā* expressed by *ākhyāta* with *yatna-viśeṣya* also. *Saṅkhyā* is related with *yatna-viśeṣya*, because that satisfies expectancy.

It may be argued that if *ākhyāta* does not imply *kartā*, *karma* etc. there being no *abhihita-anabhihita-vyavasthā*, there may be in vogue such usage as *kaṭaḥ karoti caitram*, etc. as well. The question arises whether such expression is used by trustworthy people or not, that is, whether it is *āpta-prayoga* or *anāpta-prayoga*. To take the first alternative, is *caitram* a *prathamānta-pada* or a *dvitīyānta-pada?* If the first, it is not proper, because the word *caitra*, which is a masculine gender, in first-case-ending would be *caitrah* and not *caitram*. If the second, being a *karma-kāraka*, it satisfies the expectancy for *karma*, and *yatna* or *kriyā* which is implied by *karma-kāraka* expects for the *kartā*. But *kaṭaḥ* cannot be the *kartā*, because experience shows that only *caitra* can be the *kartā* of *kaṭaḥ* and not vice versa.

If such expression, on the other hand, is *anāpta*, then it cannot be taken into consideration, as there is no criterion for such usage.

It cannot be said that third-case-ending *(tṛtīyā)* refers to *kartā* and second-case-ending *(dvitīyā)* refers to *karma*, and therefore, such expressions as *caitrena kaṭaḥ karoti, caitraḥ kaṭaṁ kriyata*, etc. are possible. In fact, such expressions are not in existence, even untrustworthy people do not use such expression. The rules that regulate the use of words and language are not formulated by Pāṇini or by others according to their will and pleasure, but they are the expression of what governs the verbal usage. According to verbal usage we have such expressions as *caitraḥ kaṭaṁ karoti, caitrena kaṭaḥ kriyata*, etc., and so any deviation from such usage is not acceptable.

11.ii. The *pūrvapakṣin* says that though *yatna* is implied by all the *ākhyāta*, there is a difference between *yatna* implied by *laṅ*, etc., and that implied by *liṅ*. *Yatna* implied by *laṅ*, etc. refers to past tense, etc., but *yatna* implied by *liṅ* has no reference to any tense. Owing to this difference between *liṅ* and *laṅ*, etc. the former is *vidhi* and not the latter. This view is untenable, because just as after hearing the word *yatna* no *pravṛtti* arises, so also no *pravṛtti* arises after hearing the *liṅ*. As *yatna*, so *yatna-pratiti* too is not the *hetu* of *yatna*. Were it so, it would contradict the fact that desire is the cause of *yatna*.

Therefore, it is decided that *yatna* which is a *kartṛ-dharma*, is not *vidhi-pratyayārtha*. Similarly *icchā* is also not *vidhi*. The knowledge of *icchā* does not give rise to *yatna*. *Icchā* being previously present produces *yatna*, but such *icchā* is not present at the time when the *liṅ* is heard. The *liṅ* does not produce that *icchā* also, the *liṅ* produces the knowledge of the sentence-meaning that coveys *icchā*. After that there arises *icchā*. The *liṅ* can be said to produce *icchā* only in this sense, not in any other sense. If the *liṅ* gives rise to *icchā* without producing the knowledge of the sentence-meaning, then *icchā* may arise even in those who do not understand the sentence-meaning. In that case understanding of the sentence-meaning becomes useless. Therefore, *icchā* cannot be the meaning of *vidhi-pratyaya*.

It may be said that since *pravṛtti* arises from one's own *saṅkalpa*, such as 'I should do,' etc., *saṅkalpa* is *liṅ-artha*. The question arises as to what this *saṅkalpa* means. If by *saṅkalpa* is meant *prayatna*, it cannot be *liṅ-artha*.As the same thing cannot be its own cause, *prayatna* cannot be the cause of *prayatna*. If by *saṅkalpa* is meant *icchā*, then also it cannot be *liṅ-artha* for the reason already stated.

It may be argued that when the *liṅ*, viz. *yajeta svargakāmaḥ*, is heard, though *yāgecchā* is not present, yet *phalecchā*, viz. the desire to attain heaven, is present, and so *phalecchā* is *liṅ-artha*. This view is refuted on the ground that *yatna* towards *yāga* (sacrifice) cannot be produced by the desire for the fruit (*phalecchā*), since the object of *phalecchā*, i.e. *svarga*, is different from the object of *yatna*, i.e. *yāga*. Moreover, *phalecchā* may arise from the knowledge of the *phala;* there is no need of the Śāstra for producing *phalecchā*.

It cannot be maintained that Śastra is needed for producing

the knowledge of the *phala,* for knowledge, pleasure, pain, etc. are apprehended directly by mental perception. The utility of the Śāstra consists in its producing such knowledge which is not produced by any other means of knowledge. Since *phalecchā* may be known by mental perception, the Śāstra is not needed for its knowledge.

Again, if *phalecchā* were *liṅ-artha,* the word *svargakāmaḥ* in *yajeta svargakāmaḥ* is a repetition, as in that case the word *yajeta* itself implies *svargakāmaḥ* and further use of the word *svargakāmaḥ* is a mere repetition, Further, if *phala-viṣayecchā* gives rise to *sādhanaviṣaya-prayatna*, then *ghaṭecchā* (desire for a jar) will give rise to *paṭa-prayatna* (volitional effort for producing cloth), since there is no criterion to restrict the *ghaṭecchā* to *ghaṭa-prayatna* and not to *paṭa-prayatna.* To maintain that *phalecchā* produces *yāga-pravṛtti*, as there is *sādhya-sādhana-bhāva* between them, is untenable. For, if the *sādhya-sādhana-bhāva* itself, and not the knowledge of it, determines the causal relation between *phalecchā* and *yāga-prayatna,* then even in the absence of the *liṅ*, *phalecchā* may produce *pravṛtti.* Therefore, it is clear that *phalecchā* cannot be *liṅ-artha.*

If, however, the knowledge of *sādhya-sādhana-bhāva* were the determinant of the causality of *phalecchā* and *prayatna*, then as the knowledge of *phalasādhanatā* is the cause of *pravṛtti*, *phalasādhanatā-icchā* or *upāyecchā* may be admitted as the cause of *prayatna.* Thus, it may argued that *upāyecchā* is *liṅ-artha.* This position is refuted on the ground that *upāyecchā* is not present at the time of hearing the *liṅ*, and so it cannot be the cause of *prayatna.* Nor can it be said that the knowledge of *upāyecchā* which is produced by the *liṅ* is the cause of *prayatna.* Were it so, the knowledge of the sentence *svargakāmo yāgacikīrṣāvān* should give rise to *pravṛtti,* but *pravṛtti* does not arise from such knowledge. It is only *icchā* which is the cause of *pravṛtti.*

It may be said that the knowledge of one's own *icchā* is the cause of *pravṛtti*, since one's own *icchā* and its knowledge arise almost simultaneously. It is not so, because in *yajeta svargakāmaḥ* one's own *icchā* is not conveyed, as the word *svargakāmaḥ* refers to third person. Therefore, if the knowledge of one's own *icchā* were the cause of *pravṛtti*, the *liṅ* which refers to third person, would be devoid of *vidhyartha*. Again, in the instance *odanakāmastvaṁ pākacikīrṣāvān* though the knowledge of one's own *icchā* is expressed, yet no *pravṛtti* arises. This instance also shows that the knowledge of one's own *icchā* cannot be the cause of *pravṛtti*.

Saṅkalpa-jñāna cannot be the cause of *pravṛtti*. *Saṅkalpa* is *icchā* which arises from the knowledge of *iṣṭasādhanatā*. *Saṅkalpa* gives rise to *pravṛtti* only after being produced from the knowledge of *iṣṭasādhanatā*, and the knowledge of *saṅkalpa* to *pravṛtti*. Therefore, *saṅkalpa-jñāna* cannot be the cause of *pravṛtti* or *prayatna*. Consequently, it is not the meaning of *vidhi*. The knowledge of *iṣṭasādhanatā* is also not *vidhi*, because knowledge without its object is not the cause of *pravṛtti*. The object of knowledge is *iṣṭasādhanatā*, and so *iṣṭasādhanatā* can be *vidhyartha*, but not the knowledge of *iṣṭasādhanatā*.

As *spanda*, *pravṛtti*, *icchā* and *jñāna*, which are *kartṛ-dharmas*, are not *vidhi*, it is decided that *vidhi* is not *kartṛ-dharma*.

अतिप्रसङ्गान्न फलं नापूर्वं तत्त्वहानितः।
तदलाभान्न कार्यञ्च न क्रियाऽप्यप्रवृत्तितः।।१२।।

The fruit cannot be the meaning of vidhi *(injunction), because it would be over-pervasive; nor can* apūrva *be the meaning, because it would lose its very nature;* kārya *also cannot be the meaning, because it is not present ; nor can* kriyā *be the meaning, because there is absence of volitional effort.*

12.i. The *siddhāntin* now examines the view that *vidhi* is *karma-dharma* and ultimately refutes it. *Karma* is derived as *kriyate iti karma*, that is, *karma* means that which is done. According to this derivation *karma* may stand for *phala* (fruit), e.g. *svarga*, etc. or *apūrva (adṛṣṭa)* which produces the *phala* or *kriyā* (action), e.g. *yāga* (sacrifice), etc. which produces *apūrva*.

The *dharma* of *phala* is *phalatva*, which is *icchā-viṣayatva*. *Apūrva* has two *dharmas*, viz. *apūrvatva* and *kāryatva*. The *dharma* of *kriyā* is *kāryatva.*

Phala-dharma cannot be *vidhyartha,* because *phalatva* which is nothing but *phalecchā* does not give rise to *yāga-prayatna,* as object *(viṣaya)* of *phalecchā,* viz. *svarga,* is different from the object of *yāga-prayatna,* viz. *yāga.* Moreover, if *phalecchā* gives rise to *yāga-prayatna* in spite of their objects being different, then as already has been stated *ghaṭa-icchā* will give rise to *paṭa-prayatna,* leading thereby to the fallacy of *atiprasaṅga* (overpervasive).

Vidhi or *liṅ-artha* is not *apūrva-dharma* too. If *liṅ-artha* were *apūrva-dharma,* then as *apūrva-dharma*s are two, viz. *apūrvatva* and *kāryatva, liṅ-artha* would be either *apūrvatva* or *kāryatva* or both. The first alternative is untenable. The knowledge of all words or sentence-meaning arises from empirical usage. *Apūrvatva* is not known by any means of valid knowledge, so if *apūrvatva* were *liṅ-artha,* knowledge of the *liṅ*-sentence-meaning will not arise at all. In the understanding of sentence-meaning, the relation between words and their meanings is comprehended. Hence, understanding of sentence-meaning requires the knowledge of words and their meanings.

In the case of the *liṅ*-sentence, *apūrvatva* which is supposed to be the meaning of *liṅ,* being incomprehensible the relation between the *liṅ*-sentence and its meaning will not be comprehended. Consequently, the knowledge of the *liṅ*-sentence will not arise at all.

If it is said that *apūrva* is known from *liṅ-pada,* then there arises the contingency of mutual interdependence, as it amounts to saying that the knowledge of the *liṅ-śakti* gives rise to the knowledge of *apūrva* and from the knowledge of *apūrva* arises the knowledge of *liṅ-śakti.*

Moreover, the very word *apūrva* as compounded *na* (not) and *pūrva* (previously existing) implies that it was not known before the verbal knowledge was produced from hearing the *liṅ.* Therefore, if it is known before, it will cease to be *apūrva.* That is why the *siddhāntin* says if *apūrva* were *liṅ-artha,* it would lose its very nature *(nāpūrvaṁ tattvahānitaḥ).*

It may be said that just as the word 'earth' signifies a substance that possesses smell, though smell is not necessarily perceived, so also *liṅ* implies *apūrvatva,* though *apūrva* is not perceived.

This position is unsound. It is possible for the substance possessed of the special nature of earth to be known to possess smell by means of memory or inference from previous perception. But such a supposition is precluded in the present case by the very name *apūrva* which implies absence of any previous perception.

The second alternative that *liṅ* signifies *kāryatva* also stands rejected. The Prābhākara Mīmāṁsā maintains that the denotation of a word or a sentence is cognizable only through its connection with a particular action. Each sentence appertains to some action, and the denotation of the sentence is comprehended by observing the usage of older persons. This usage lies in the form of an injunction addressed by one person to another for doing certain act. Thus, the denotation of all words or sentences lies, directly or indirectly, with something to be done. Similarly, it is said that the Vedic injunctions or commands, e.g. *yajeta svargakāmaḥ,* also imply *niyoga* (what is to be done). *Niyoga* is related with *niyojya* (the fruit for the attainment of which the injunction is enjoined). Hence, the performance of the sacrifice is connected with *svarga* (heaven) which is the fruit of the sacrificial act. The heaven, however, is not attained by the mere sacrificial act. The sacrificial act is destroyed the very moment it is performed. So, it cannot give rise to the fruit. Therefore, it is believed that the sacrificial act which is *kārya* gives rise to an imperceptible subtle force, viz. *apūrva,* which produces the fruit. Thus, *apūrva* is implied by *kārya,* viz. sacrificial act *(kāryatvenopalakṣyam),* which is the sense of the *liṅ*-sentence or *liṅ-artha.*

This position is not accepted by the *siddhāntin.* If *kāryatva* implies *apūrvatva,* it must be pointed out in what way does the former imply the latter. This implication cannot be either in the form of recollection or of inference, because *apūrvatva* can neither be recollected nor be inferred from *kāryatva.* When something stimulates the recollection of some other thing or leads to the inference of something else, it requires that the relation between the two things should be known beforehand. Since *apūrva* is not known before, the relation between *kāryatva* and *apūrvatva* can never be comprehended. Therefore, *apūrvatva* can never be recollected or inferred.

It may be said that sometimes recollection of something arises from *saṁskāra* or *adṛṣṭa*, even though the relation of that thing with *saṁskāra* or *adṛṣṭa* is not comprehended. Similarly, recollection of *apūrvatva* may arise from *kāryatva*, though their relation is not known.

To this the reply is that *saṁskāra*, etc. being *svarūpasat* (itself present) become the cause of recollection, but in the latter case the knowledge of *kāryatva* is supposed to give rise to the recollection of *apūrvatva*. When the knowledge of something gives rise to the recollection of something else, the relation between the two must be comprehended. For example, when it is said *dhūma eva dhūmadhvajam*, the relation between *dhūma* and *dhūmadhvaja* is comprehended beforehand. Therefore, as the relation of *kāryatva* with *apūrvatva* is not known, the former cannot imply the latter. Consequently, the view that *apūrvatva* implied by *kāryatva* is the *liṅ*-sense also stands rejected. And the Mīmāṁsaka position that one does not engage in any activity unless that action gives rise to the desired fruit also stands futile.

It is experienced that in the case of secular *liṅ* usage, such as *gāmānaya*, etc. *kāryatva*, viz. act of bringing, etc. is the cause of *liṅ-pravṛtti*. If *apūrvatva* were the *liṅ*-sense, the opponent must explain how *kāryatva* becomes the cause of *liṅ-pravṛtti* in secular *liṅ* usage and not *apūrvatva*.

To this the opponent replies that due to non-discrimination *(bhedāgraha)* between *kāryatva* and *apūrvatva* the former appears as the cause of *liṅ-pravṛtti*. In fact, it is *apūrvatva* which is *liṅ*-sense and not *kāryatva*. This position is untenable, because *bhedāgraha* is not possible in respect of a thing that is unknown. When a shell is mistaken for a piece of silver and *pravṛtti* arises in respect of that, it is the non-discrimination between the silver which is seen beforehand and recollected at that time and the shell which is perceived is the cause of *pravṛtti.* But non-discrimination between *kāryatva* and *apūrvatva* cannot be the cause of *liṅ-pravṛtti,* because *apūrva* being never known, non-discrimination between *kāryatva* and *apūrvatva* is not possible.

Moreover, *apūrva* may be the cause of *liṅ-pravṛtti* so far as Vedic injunctions *(vidhivākyas)* are concerned. But in the case of secular *liṅ* usage, viz. *gāmānaya,* etc., *apūrva* is not the cause of *pravṛtti.* So, if *apūrva* were the *liṅ*-sense, secular *liṅ* usage would become meaningless.

If it is said that in secular *liṅ* usage *kāryatva* is the cause of *pravṛtti,* then the same should be the cause of *pravṛtti* in the case of the Vedic *vidhi-vākyas* too. Otherwise, if *kāryatva* were the cause of *pravṛtti* in the former case and *apūrvatva* in the latter, there arises the contingency of postulating many *śaktis,* viz. *kārya-śakti* and *apūrva-śakti.*

It may be further said that *kāryatva* is the cause of *liṅ-pravṛtti* in secular usage; *kāryatva* means *ghaṭa-kārya,* etc., but in the Vedic *liṅ* usage *kāryatva* means *apūrvatva.* In the Vedic *liṅ* usage *kāryatva* means *apūrvatva,* because in the case of *yāga-kārya,* etc. the performance of *yāga* does not give rise to heaven directly, so *apūrva* which is produced from the *yāga-kārya* is inferred as the intermediary cause of heaven.

This view is beset with certain difficulty. According to the Mīmāṁsaka there are four kinds of *karmas* (ritual acts) — *nitya* (obligatory), *naimittika* (occasional), *kāmya* (optional) and *niṣiddha* (prohibited). *Kāmya* - and *naimittika-karmas* are performed out of desire for some fruit which is produced by *apūrva* or *adṛṣṭa* that is ensued from the *karma. Nitya-* and *nisiddha-karmas* do not give rise to *apūrva. Niṣiddha-karma* implies abstaining from the prohibited actions enjoined by the Śruti. Abstention means indifference or absence of effort, so no *apūrva* ensues from this *karma. Nitya-karma* implies morning and evening prayer, etc. If such *karmas* are not performed, there arises demerit, but the performance of them do not give rise to merit *(apūrva).*

The opponent's argument is that in the case of the Vedic *liṅ* usage *(vidhi-vākya)*, *kāryatva* means *apūrvatva,* because the fruit of the *yāga-kārya,* viz. heaven, is attained through *apūrva.* Against this the *siddhāntin* points out that as no fruit is attained by the performance of *nitya-* and *niṣiddha-karmas, kāryatva* cannot mean *apūrvatva.* In such cases *kāryatva* stands for *kriyā* and not *apūrva.* It cannot be argued that some fruit is attained by *nitya* - and *niṣiddha-karmas,* because such a thing is never heard of and that which is never heard cannot be presumed.

12.ii. The *siddhāntin* now enquires about the reason behind Mīmāṁsaka's presumption of fruit which is produced by *apūrva.* Is the fruit presumed, because *apūrva* cannot be explained otherwise, or because the knowledge of the *yāga,* etc. cannot be explained otherwise? It is not the first, because according to the Prābhākara Mīmāṁsaka it is *apūrva* which is the only fruit to be desired, and there is no other fruit excepting *apūrva.* When it is proclaimed that heaven is attained by the performance of *yāga,* etc. there also, according to the Mīmāṁsaka, what is desired is the *apūrva.* As the Mīmāṁsaka's sole concern is for *apūrva; apūrva* turns out to be the fruit and not heaven. Therefore, it is clear that the ground for the presumption of fruit by the Mīmāṁsā is not that the *apūrva* otherwise remains unexplained.

The second alternative is also untenable. It may be asked: 'Is the fruit *(iṣṭa)* presumed because the knowledge of *iṣṭasādhanatā,* viz. the fact that the performance of the sacrificial rites is the cause of attaining heaven, gives rise to the knowledge of *kāryatva,* viz. *yāga-kārya,* etc., or is it because *kāryatva,* viz. the performance of the sacrificial acts, cannot be without being accompanied by the fruit?' The first alternative is rejected on the ground that this is tenable only when the knowledge of *kārya* arises from the knowledge of *iṣṭasādhanatā.* For example, after knowing that cooking gives rise to what is desired *(iṣṭa),* viz. food that satisfies hunger, acts of cooking *(kārya)* is known from the knowledge of *iṣṭasādhanatā.* In this case the knowledge of *iṣṭasādhanatā* gives rise to the knowledge of *kārya.* But in the case of the Vedic *vidhi-vākyas, liṅ* usages themselves produce the knowledge of *kāryatva,* so there is no need of the knowledge of *iṣṭasādhanatā* for producing the knowledge of *kāryatva.* Otherwise, if the *liṅ-śabda* gives rise to the knowledge of *kāryatva* after producing the knowledge of *iṣṭasādhanatā*, *iṣṭasādhanatā* itself can be admitted as the *liṅ*-sense, and *kāryatva* the knowledge of which arises after the knowledge of *iṣṭasādhanatā* will not be considered as the *liṅ*-sense.

The second alternative is untenable, because there is no universal concomitance to support the fact that wherever there is *kāryatva*, there is *iṣṭasādhanatā*. Happiness undoubtedly implies *iṣṭasādhanatva*, but it is not a *kārya*.

Thus, *kāryatva* cannot mean *apūrvatva* on the ground that *kārya* produces fruit through *apūrva*, for according to the Mīmāṁsaka *apūrva* is the only fruit which should be desired and no such *apūrva* ensues from *nitya*- and *niṣiddha-karmas*. Therefore, *apūrvatva* cannot be the *liṅ*-sense.

The opponent may argue that the *liṅ* does not imply *apūrva* in *nitya*- and *niṣiddha-karmas*, but it implies *apūrva* in *kāmya*- and *naimittika-karmas*. The *siddhāntin* maintains that the *liṅ* does not imply *apūrva* even in *kāmya*- and *naimittika-karmas*.

Liṅ does not mean a *kārya* qualified by *phalasādhanatva* (i.e. an action accompanied by the means that brings forth the fruit), so *apūrva* cannot be established on the ground that it is the means of attaining the fruit. It cannot be said that the word *svargakāma* implies *svargasādhanatva*, just as in the instance *brāhmaṇo yajeta* the word *brāhmaṇatva* is *kartṛ-viśeṣaṇa*, so also the word *svargakāma* may be taken as *kartṛ-viśeṣaṇa*.

Nor can *sādhanatva* be inferred in the form—that desiring which one ought to do an act is one's *iṣṭasādhana*. This inference is fallacious. Happiness is desired, but it is not *kārya*. Hence, the reason is *anaikāntika* (inconclusive).

If, however, the *kārya* is done out of desire for some fruit other than the *kārya* itself, then *kāryatva* or *kartavyatva* will be *aupādhika* and not *svābhāvika*. *Svābhāvika-kartavyatva* means when something should be done for that thing's sake. *Aupādhika-kartavyatva* means when something should be done for attaining something else. According to the Mīmāṁsā *apūrva* is the only thing to be attained through sacrificial acts, so there is *svābhāvika*-

kartavyatva for *apūrva*.

With the Nyāya *apūrva* or *adṛṣṭa* is necessary for producing the fruit; so according to the Nyāya *kartavyatva* in respect of *apūrva* is *aupādhika*. *Apūrva* is not something that should be acquired for its own sake, but for the sake of fruit. If *kāryatva* turns out to be *aupādhika*, in other words, if it is conditioned by the desire of attaining some other fruit, then the knowledge of *kāryatva* will be preceded by the knowledge of *iṣṭasādhanatva*. Therefore, *iṣṭasādhanatva* cannot be inferred from the *hetu-kāryatva*.

12.iii. The *pūrvapakṣin* maintains that one need not probe into the nature of *kāryatva*, whether it is *svābhāvika* or *aupādhika*. The *śruti*, viz. *yajeta svargakāmaḥ*, implies that one who is desirous of attaining heaven should perform sacrificial rite. Since *kāryatva* is preceded by the knowledge of *iṣṭasādhanatva*, *iṣṭasādhanatva* can very well be inferred from *kāryatva*. Moreover, there is no *anvaya* (relation) between *svargakāma* and *yāga-kartavyatva*, as the object of *svargakāma*, viz. *svarga* and that of *yāga-kartavyatva*, viz. *yāga*, are different. Consequently, *yāga-kartavyatva* does not have syntactic compatibility *(yogyatā)* towards *svargakāma*. As there is no *anvaya* between the words, *śābdabodha* will not arise at all. So, in order to have relation between *svargakāma* and *yāga-kartavyatva*, the knowledge of *iṣṭasādhanatva* must be admitted. The knowledge that the performance of *yāga* leads to the attainment of heaven gives rise to the knowledge of relation between *svargakāma* and *yāga-kartavyatva*. Therefore, *iṣṭasādhanatva* can be established through inference as a linking factor between the two words, viz. *svargakāma* and *yajeta*.

The *siddhāntin* does not subscribe to this position. According to him the *liṅ* here may convey *iṣṭasādhanatva-samānādhikaraṇaka-kartavyatva*, but there is no possibility of *iṣṭasādhanatva* being established by inference. Moreover, there is no need of postulating *iṣṭasādhanatva* in order to relate the words *svargakāma* and *yajeta*. There can be relation between the two words, if *svargakāma* is taken as *kartṛ-viśeṣaṇa*, just as *brāhmaṇatva* in the instance *brāhmaṇo yajeta*. Thus, as *kāryatva* does not imply *kāmanā* (desire for fruit), there is no reason for inferring *iṣṭasādhanatva.* In that case *apūrva* cannot be established as *phalasādhana.* Therefore, *apūrva* is not the *liṅ*-sense in *kāmya*- and *naimittika-karmas.*

Further, *apūrva* cannot be established on the ground that it is *pravṛttinimitta* of *kāryatva.* For example, *apūrva* is not *kārya. Kāryatva* is either *kṛtivyāpyatva,* i.e. the substratum of the *vyāpāra* of volitional action *(kṛti),* or *kṛtiphalatva* i.e. produced by the *kṛti,* or *kṛtyuddeśyatā.* In the first case, *kāryatva* is posssible in respect of established thing *(siddhadravya),* viz. *brihi,* etc. That is, *brihi,* etc. can be the substratum of volitional activity. *Apūrva* is not an established thing, so no *kāryatva* is possible in respect of it. In the second case, *kāryatva* is possible in respect of *yāga,* because *yāga* is accomplished directly by the *kṛti.* In the third case, it is heaven in respect of which *kāryatva* is possible, for heaven is *iṣṭa* by itself. *Apūrva* is not *iṣṭa* by itself according to the *siddhāntin,* and so *kāryatva* is not possible in respect of *apūrva.*

The opponent may point out that the *yāga* does not last till the attainment of fruit, so it cannot be the *sādhana* for attaining the fruit. To this it may be replied that as in secular actions, so in the Vedic sacrificial acts also some mediate *(vyavahita)* factor is presumed as the *sādhana* for the fruit.

According to the Prābhākara Mīmāṁsā sometimes a cause directly produces the effect, sometimes it produces indirectly through some *vyāpāra.* When a cause produces the effect through *vyāpāra,* the cause must be present at the time of the production of the effect, because according to him a cause must be present immediately before the production of the effect *(avyavahitapūrvakālavarti).* Otherwise, if a cause is not present immediately before the production of the effect, a deceased father may become the cause of the murder of a *brāhmaṇa,* which is committed by his son.

The *siddhāntin* refutes this argument in the following manner. According to him a cause need not be present immediately before the production of the effect. *Kāraṇatva* means only *pūrvabhāvaniyama,* that is, the cause must exist before the effect. A factor which is generated from the action and helps in producing the effect, is called *vyāpāra (kāryānuguṇāvāntara-kāryasyaiva vyāpāratvāt).*

The *vyāpāra* can produce the effect even when the cause is destroyed. Just as *anubhava* is the cause of *smṛti* through *saṁskāra-vyāpāra*, though *anubhava* is destroyed at the time when *smṛti* arises; similarly, *kṛti* becomes the cause of *apūrva* through *yāga-vyāpāra*. It cannot be said that a mediate factor, i.e. *vyāpāra* is not the primary cause, because there is nothing to invalidate it as primary cause. When a cause cannot produce an effect without producing some intermediary factor, that factor is the *vyāpāra* of that cause in respect of that effect. A deceased father cannot be the *vyāpāra* of the murder of a *brāhmaṇa,* which is committed by his son, because there is no *anvaya-vyatireka* relation between a deceased father and his son's committing the murder of *brāhmaṇa.* A son may not commit the murder of a *brāhmaṇa* and a father without giving birth to a son may commit the same.

Further, if causality consists in the fact that the cause must be present immediately before the effect, then whatever is present at the time of the production of the effect must be admitted as the cause. Thus, if a person immediately after discharging an arrow towards a *brāhmaṇa* is killed by a tiger, he will not be considered as the cause of murder of the *brāhmaṇa*, because he does not exist at the time when the *brāhmaṇa* is killed by the arrow shot at him. And his father who may be present at that time will become the cause of the death of the *brāhmaṇa*, though he has nothing to do with the murder.

Therefore, *yāga* is *kārya* because of its being the *sādhana* of the heaven, and *apūrva* is not *kārya. Apūrva* is *vyāpāra.* The *siddhāntin* admits *apūrva* as *vyāpāra*, but what he denies is its *kāryatva* and *liṅ-arthatva,* that is, *apūrva* is neither the *kārya* nor the *liṅ*-sense. As neither *apūrvatva* nor *kāryatva* is *liṅ-artha,* the third alternative, mentioned earlier, that both *apūrvatva* and *kāryatva* together is *liṅ-artha,* stands rejected. Thus says the *kārikā* that *kārya* is not the meaning of the *vidhi*, because it is not present.

Kāryatva in the sense of *kriyā-dharma* is also not the *liṅ-*

sense *Kāryatva* may be either *kṛti-adhyavasāya* which means *saṅkalpa,* or *kṛti-arhatva.* All the *vidhi-vākyas* lay down what one should do or what is one's *kartavya. Kartavyatva* is expressed in the form *kuryām,* i.e. 'I should do', etc. In this sense *kartavyatva* may be taken as *saṅkalpa.* But this position, according to the *siddhāntin,* is untenable, because there is a difference between *saṅkalpa* expressed in *kartavyatva* and that expressed in *kuryām. Kartavyatva* is *vyavahita-kārya-saṅkalpa,* that is, in the case of *kartavyatva* action may take place at a later time. But *kuryām* is *śīghrabhāvi-kārya-saṅkalpa,* that is, in this case action takes place immediately, because the expression *kuryām* is immediately followed by action. This kind of *saṅkalpa,* however, cannot be the *liṅ*-sense, which has already been discussed in connection with *kartṛ-dharma.*

Kṛti-arhatva means capability or *yogyatā* to execute *(utpādanārhatvam)* which has *prāgabhāva* or prior non-existence. *Kṛti-arhatva* may be either *puruṣa-śakyatva* or *puruṣopakārakatva.* If it is the first, all the injunctions should be carried out, since *puruṣa-śakyatva* is there in respect of all the injunctions. The second alternative, viz. *puruṣopakārakatva* stands for *iṣṭasādhanatva* which will be discussed in the passage that follows.

असत्त्वादप्रवृत्तेश्च नाभिधाऽपि गरीयसी।
बाधकस्य समानत्वात् परिशेषोऽपि दुर्लभः।।१३।।

Abhidhā *(primary signification) cannot be the meaning of the* liṅ, *because it cannot be proved to be present in the* liṅ *(and even if it were present) the knowledge of it does not give rise to volitional effort; it* (abhidhā) *cannot be the meaning of* liṅ *through elimination too, because it is subject to the same invalidating factor which vitiates the other alternatives.*

13.i. The *siddhāntin* has shown how the *liṅ*-sense can neither be *kartṛ-dharma* nor *karma-dharma* and *kriyā-dharma*; now he examines and ultimately rejects the view that the *liṅ*-sense is *karaṇa-dharma.*

The word *karaṇa* may either stand for word(*śabda*), i.e. the instrument of verbal knowledge *(śābdabodha)* or *bhāvanā*, that is, *yāga-vyāpāra* which is the instrument of heaven, etc. The *dharma* of *śabda* is *abhidhā* (primary signification) and that of *yāga-vyāpāra* is *iṣṭasādhanatā*.

The *siddhāntin* proves that *karaṇa-dharma* in either sense cannot be the *liṅ-artha*.

The reasons for rejecting *abhidhā* as the *liṅ*-sense are as follows. It cannot be proved that *abhidhā* is present in the *liṅ* usage, and even if it is present, the knowledge of it does not give rise to *pravṛtti*, Therefore, there is no ground for admitting *abhidhā* as the *liṅ*-sense. It is futile to argue that *abhidhā* stands out as the *liṅ*-sense through elimination *(pariśeṣa)*, as all other alternatives have been rejected as conveying the *liṅ*-sense. For, this alternative too is subject to the same invalidating factor *(bādhaka)* owing to which other alternatives were rejected.

Let us examine this alternative in detail. Pārthasārathi Miśra, a Bhāṭṭa Mīmāṁsaka, advocates that *abhidhā* is the *liṅ*-sense. According to Maṇḍana Miśṛa *Iṣṭasādhanatā* is the *liṅ*-sense. Both the views do not find favour with the *siddhāntin*. *Abhidhā* is a *vyāpāra* through which word-meanings are comprehended. This *vyāpāra* is in the form of the recognition *(pratisandhāna)* of the *śabdārtha-śakti*.

Recognition resides in the soul; therefore, *abhidhā* is an attribute of the soul. Since *abhidhā* is an attribute of the soul, it cannot be a *vyāpāra* of *śabda*, because *abhidhā* cannot be proved to be present in *śabda*.

It may be argued that just as *apūrva* which inheres in the soul is *yāga-vyāpāra* even though it is not present in the *yāga*, similarly, *abhidhā* which is inherent in the soul is the *vyāpāra* of *śabda*, though it is not present in *śabda*.

To this it may be rejoined that the performance of the sacrificial rites (*yāga*) depends on the knowledge of the *yāga*, which inheres in the soul. *Yāga* may be the cause of *apūrva-vyāpāra*, as both knowledges of *yāga* and *apūrva* are the attributes of the soul. But *liṅ*-sentences consist of words *(śabdas)* which are not attributes of the soul; therefore, they cannot be the cause of *abhidhā-vyāpāra* which is an attribute of the soul.

It may be further observed that just as the perception of the probans, e.g. smoke which is not inherent in the soul, gives rise to *liṅ-parāmarśa,* viz. the knowledge of the invariable concomitance between smoke and fire, which inheres in the soul, similarly, *śabda* which is not inherent in the soul may give rise to the knowlege of *śabdārtha-śakti* or *abhidhā* which inheres in the soul. In this sense *abhidhā* may be called *śabda-vyāpāra.*

The *siddhāntin,* in reply, says that even if *abhidhā* is admitted as the *śabda-vyāpāra,* it cannot be the *liṅ*-sense. For, the knowlege of *abhidhā* does not give rise to *pravṛtti. Pravṛtti* arises in children even in absence of the knowledge of *abhidhā.* Hearing of the word *abhidhā* too does not give rise to *pravṛtti.*

It cannot be maintained that *abhidhā* becomes the *liṅ*-sense indirectly *(vilakṣaṇatayā),* because the *liṅ* produces the knowledge of *abhidhā* indirectly and not directly as *abhidhā* as such. For, in that case the question may arise, 'Does *abhidhā* become the *liṅ*-sense in that way because otherwise the meaning of the *liṅ* usages will not be comprehended, or because otherwise *pravṛtti* will not arise?' If the first, the meaning of the *liṅ* usages also must be indirect, because the meaning of the *liṅ* usages can be comprehended indirectly only when their meaning is indirect. Since the meaning of the *liṅ* usages is not indirect, this alternative that *abhidhā* becomes the *liṅ*-sense indirectly, stands rejected. The second alternative is also untenable, because the knowledge of *abhidhā* does not give rise to *pravṛtti.* As the knowledge of *abhidhā* is not the cause of *pravṛtti,* there is no reason to hold that *abhidhā* becomes the *liṅ*-sense indirectly, as otherwise *pravṛtti* will not arise.

The opponent may say that the *liṅ* gives rise to the knowledge of *abhidhā* indirectly, because *abhidhā* has contiguity towards the *liṅ.* This contention is unsound, because it is not only *abhidhā,* but root-meaning *(dhātvartha),* number *(saṅkhyā)* conveyed by the *liṅ*-words, etc. also possess contiguity towards the *liṅ.* So, there is no sound reason for maintaining that the *liṅ* gives rise to the knowledge of *abhidhā* only indirectly or in a different manner and not that of root-meaning, etc. If, however, the *liṅ* gives rise to the knowledge of *abhidhā* as well as root-meaning, etc. in a different manner, then root-meaning also can become the cause of *pravṛtti*.

It may be argued that the *liṅ* does not produce the knowledge of root-meaning, etc. in a different manner, because there is no evidence for it. So, by elimination it stands that it is the knowledge of *abhidhā* which is produced by the *liṅ* in a different manner. If so, enquiry may be made about the reason on the basis of which the knowledge of root-meaning, etc. is eliminated as not being produced by the *liṅ* in a different manner. The reason may consist either in the fact that the knowledge of root-meaning, etc. is not the cause of *pravṛtti*, or that they do not have contiguity towards the *liṅ*. Both these alternatives stand rejected. In the first case, even the knowledge of *abhidhā* does not give rise to *pravṛtti*, so the knowledge of *abhidhā* too can be eliminated as not being produced by the *liṅ* in a different manner. In the second case, the fact that the root-meaning, etc. lack contiguity towards the *liṅ* is an impossibility.

Further, the contention that *liṅ* usages give rise to the knowledge of *abhidhā* which, in turn, gives rise to *pravṛtti*, is untenable. The knowledge of *abhidhā* cannot be attained by any other means of knowledge, and so it cannot be proved that *abhidhā* possesses *liṅ-śakti*. The knowledge of what a word signifies does not arise from the utterance of the same word, it is comprehended by other means of knowledge. For example, the signification of the word 'jar'is not comprehended from the utterance of the word 'jar'. From hearing the word 'water' and knowing what it signifies, one may come to know what the word 'jar' signifies, since jar holds water. As there is no such means of comprehending the signification of *abhidhā*, the fact that it possesses *liṅ-śakti* is doubtful.

It cannot be held that the signification of *abhidhā* is comprehended from the fact of its being the cause of *pravṛtti*, as otherwise *pravṛtti* will not arise, that is, *abhidhā* is comprehended as a *vyāpāra* which gives rise to *pravṛtti*. This position stands rejected on the ground that the knowledge of *vyāpāra* which gives rise to *pravṛtti* is not the cause of *pravṛtti*. The *vyāpāra* that induces *pravrtti* cannot be the cause of *pravṛtti*.

Therefore, as the knowledge of *abhidhā* is not the cause of *pravṛtti*, the view that *abhidhā* is the *liṅ*-sense stands rejected.

13.ii. The *siddhāntin* now discusses how *iṣṭasādhanatva*, the other alternative of *karaṇadharma*, cannot be the *liṅ*-sense.

Experience shows that the knowledge of *iṣṭasādhantā* gives rise to *pravṛtti* in man. After reflecting that performance of certain act will lead one to achieve what one desires, one resolves to act, and then engages oneself in the activity. From this it is inferred that the knowledge of *iṣṭasādhanatā* that arouses desire, desire in turn leading to *pravṛtti* is the cause of *pravṛtti*. But the knowledge of *iṣṭasādhanatā* can be the cause of *pravṛtti*, provided the action which is *iṣṭasādhana* is *kṛti-sādhya*. An action may be *iṣṭasādhana*, but unless it is within the reach of human ability to be performed, the knowledge of its *iṣṭasādhanatā* cannot be the cause of *pravṛtti*. Therefore, it may be said that the knowledge of *iṣṭasādhanatā* accompanied by *kartavyatā* is the *liṅ*-sense.

It is not proper to argue that only *kartavyatā* is the *liṅ*-sense, since *pravṛtti* depends on the knowledge of *kartavyatā*, and the knowledge of *iṣṭasādhanatā* need not be admitted as the *liṅ*-sense. The knowledge of *kartavyatā* without the knowledge of *iṣṭasādhanatā* does not induce *pravṛtti*. For example, consumption of poison may be a *kartavya*, but it does not give rise to *pravṛtti*, because it is not *iṣṭasādhana*. Therefore, not only *kartavyatā*, but the knowledge of *iṣṭasādhanatā* is also the cause of *pravṛtti*. The object *(viṣaya)* of *kṛti*, viz. sacrifice, etc., is expressed by the root *(dhātu)*, *kṛti* is expressed by *ākhyāta*, and *iṣṭasādhanatā* is known as *kṛti-viṣaya* through reflection that the particular *kṛti* leads to achieve what is desired. Thus, the cause of *liṅ-pravṛtti* is the knowledge of *iṣṭasādhanatā* accompanied by the knowledge of *kartavyatā*.

The opponent may say that if *iṣṭasādhanatva* were the *liṅ*-sense, then *yāgakaraṇa* which is *iṣṭasāddhana*, being denoted by the *ākhyāta*, the word *jyotiṣṭoma* in *jyotiṣṭomena yajeta* must be in the nominative case-affix and not in the third case-affix. To this the *siddhāntin* replies that when any number is denoted by a word, nominative case-affix applies to that word. In other cases, following the *Pāṇini-sūtra, kartṛkaraṇoyaḥ tṛtīyā*, third case-affix applies to an instrument. Though *karaṇatva* is denoted by the *ākhyāta*, yet no number denoted by the word *jyotiṣṭoma* is expressed and it is the instrument for heaven also; so, the third case-affix applies to it.

Moreover, the *liṅ* usages do not denote *yāga-iṣṭasādhanatā*. In the injunction *svargakāma yajeta, iṣṭasādhanatā* is not mentioned by the *liṅ*. Through reflection one comes to know that there is *iṣṭasādhanatā* in the performance of the sacrificial rites, since it leads to the attainment of heaven, etc. Therefore, there is nothing wrong in the usage *jyotiṣṭomena*, the third case-affix being applied to *jyotiṣṭoma* in the instrumental sense.

It may be argued that the knowledge of *iṣṭasādhanatā* is not the *liṅ*-sense, because *pravṛtti* does not arise even from the knowledge of *iṣṭasādhanatā*. For example, the knowledge of *iṣṭasādhanatā* arising from such *siddhopadeśa* as *yāgaḥ svargasādhanam* does not give rise to *pravṛtti*. In reply to this argument it may be said that knowledge of *iṣṭasādhanatā* arising from such *siddhopadeśa* gives rise to *pravṛtti* only in those who have intense desire for fruit, in the absence of such desire no *pravṛtti* arises. In fact, this criticism holds good also in respect of the other two alternatives that advocate that the *liṅ*-sense is *abhidhā* and *kāryatva* respectively. Surely the knowledge of *abhidhā* or *kāryatva* does not give rise to *pravṛtti*. To evade this charge if it is said that the knowledge of the injunctive text, which arises from the *liṅ*-sense, is the cause of *pravṛtti*, then the same holds good in the case of *iṣṭasādhanatā* also.

Having set forth the view that *iṣṭasādhanatā* is the *liṅ*-sense, the *siddhāntin* now rejects it. It is true that the knowledge of *iṣṭasādhanatā* gives rise to *pravṛtti*. The knowledge of *iṣṭasādhanatā* may be either direct or indirect. It is direct, when, for example, it is inferred that mother's milk is *iṣṭasādhana* for an infant, because it drinks mother's milk. The knowledge is indirect, when *iṣṭasādhanatā* is inferred from the *āptābhiprāya* (intention or desire of a trustworthy person), that is, when an action is known as *iṣṭasādhana* from the fact that a trustworthy person intends it to be done. According to the *siddhāntin, iṣṭasādhanatā* is not directly implied by the *liṅ*, but is known indirectly from the *āptābhiprāya*. He advances five reasons to prove that *iṣṭasādhanatā* is not implied by the *liṅ*.

हेतुत्वादनुमानाच्च मध्यमादौ वियोगतः।
अन्यत्र क्लृप्तसामर्थ्यान्निषेधानुपपत्तितः।।१४।।

Iṣṭasādhanatā *(means for attaining the desired end) is not the meaning of* liṅ, *because sometimes it is put forth as the* hetu *(reason) of the* vidhi, *because the* vidhi *may be inferred from (eulogy from which* iṣṭasādhanatā *is known), because* liṅ *usages in first and second person do not convey* iṣṭasādhanatā, *because the desire of the speaker is conveyed in other cases of* liṅ *usages, because (if* iṣṭasādhanatā *were the meaning of* liṅ*) interdicted injunction cannot be established.*

14.i. The first ground for rejecting *iṣṭasādhanatā* as the *liṅ*-sense consists in the fact that sometimes *iṣṭasādhanatā* is put forth as the *hetu* of the *vidhi*. For example, the *hetu* of such secular command as *agnikāmo dārūṇī mathnīyāt* (i.e. let him who desires fire, rub two pieces of *araṇi* wood) is *manthanamagnisādhanam* (i.e. rubbing of wood is a means of producing fire), since *manthanamagnisādhanam* which is *iṣṭasādhanatā* is advanced in this case as the *hetu* of the *vidhi*, viz. *agnikāmo dārūṇī mathnīyāt*, the former cannot be the sense of the latter, because the cause of something cannot be the sense of that thing. Therefore, *iṣṭasādhanatā* cannot be the *liṅ*-sense or *vidhyartha*.

Iṣṭasādhanatā cannot be the *liṅ*-sense for the following reason also. Sometimes the knowledge of *iṣṭasādhanatā* arises from the *arthavāda* (eulogy). For example, hearing the *arthavāda*, viz. *tarati brahmahatyāṁ yo'śvamedhena yajate* (i.e. he crosses over *brāhmaṇicide* who offers *aśvamedha*), one gains the knowledge that the performance of *aśvamedha* is desirable, because it helps one overcoming the sin incurred by *brāhmaṇicide*. Since there is no *pratyakṣa-śruti* in respect of *aśvamedha-yāga*, on the basis of *arthavāda* the *vidhi* is inferred in the form *aśvamedhena yajeta mṛtyubrahmahatyātaraṇakāma* (let him who desires to cross over death and *brāhmaṇicide*, offer the *aśvamedha*).

The point at issue is that since *iṣṭasādhanatā* in respect of *aśvamedha* is known from *arthavāda*, it must not be held that the *vidhi-vākya* which is inferred from *arthavāda* implies *iṣṭasādhanatā*. The inferred *vidhi* is expected to convey some special sense other than *iṣṭasādhanatā* which is already known from the *arthavāda*. And this sense, according to the *siddhāntin*, is *āptābhiprāya*.

The third reason for rejecting *iṣṭasādhanatā* as the *liṅ*-sense is as follows. The *liṅ* usages in first and second person do not convey *iṣṭasādhanatā*. The expression *kuryāḥ* in the second person and *kuryām* in the first person convey *ājñā* (command) and *anujñā* (permission) respectively. *Ājñā* and *anujñā* are but the expression of *abhiprāya* of the speaker. Since the *liṅ* usages in first and second person convey *abhiprāya* of the speaker in the form of *anujñā* and *ājñā*, it may be said that *liṅ* usage in third person also conveys *abhiprāya* or *icchā* of the speaker. *Icchā* of the speaker is acknowledged also in such case of the *liṅ* usage that expresses *adhyeṣaṇā* (respectul solicitation), *sampraśna* (courteous inquiry), *prārthanā* (prayer), *āśīḥ* (good wishes), etc. Therefore, it is concluded that the *vidhi-liṅ* conveys the *icchā* of the speaker, which is the *liṅ*-sense. The last reason for rejecting *iṣṭasādhanatā* as the *lin*-sense is that if *iṣṭasādhanatā* were the *liṅ*-sense, then such interdicted injunction as *na brāhmaṇaṁ hanyāt* (a *brāhmaṇa* should not be murdered) cannot be established, because by negative sign *'na' iṣṭasādhanatā* is negated here.

The *siddhāntin* opines that the interdicted injunction cannot be established in any of the alternatives that maintain that *abhidhā*, etc. are the *liṅ*-sense. In the alternative of *abhidhā*, the word *hanyāt* in *na brāhmaṇaṁ hanyāt* denotes *hananabhāvanā*, but the negative sign *na* negates this *hanana-bhāvanā*. Thus, *abhidhā* (primary significative power) is negated in this injunction. It is obvious, therefore, that *abhidhā* cannot be *lin-artha* (injunction-sense).

If by *vidhi* is meant *kālatrayāparāmṛṣṭā-bhāvanā*, i.e. *bhāvanā* having no reference to any time—past, present and future, then the above-mentioned prohibitive injunction will signify *atyantābhāva* (absolute non-existence) of *hanana-bhāvanā*. But absolute non-exitence of *hanana-bhāvanā* is not possible, since sometimes *hanana-bhāvanā* does exist.

14.ii. To take the other alternative, according to which *vidhi* implies *kartavyatva*, *kartavyatva* may refer either to *kriyā* or to *phala*. If *kartavyatva* refers to *kriyā*, the interdicted injunction will mean that *hanana* is not *kartavya* or *kṛtisādhya*. This is not tenable, because *hanana* in fact is *kṛtisādhya*. If *kartavyatva* refers to *phala*, the injunction will mean that the act of killing is devoid of fruit *(phala)*. This is also untenable, as an act of killing has its fruit in that one does killing with a view to getting rid of suffering or to deriving pleasure. If *vidhi* is admitted to imply *apūrva*, it may be enquired whether the negative sign *na* in *na brāhmaṇaṁ hanyāt* is related with *apūrva* or with root-meaning *(dhātvartha)*.

If the first, the injunction means that *no apūrva* arises from the act of killing. Consequently, nobody will hesitate to indulge in *brāhmaṇicide*, since no *apūrva* ensues from that act. But this thwarts the very import of the scriptural injunction.

If it is the second alternative, and *na* is related with the root-meaning, the injunction may mean either of the following alternatives: *apūrva* ensues from not-killing or *apūrva* ensues from the *saṅkalpa* of not-killing.

The first alternative stands rejected, because not-killing being the *prāgabhāva* of killing is beginningless; and if *apūrva* arises from not-killing, that too being an effect of *prāgabhāva* of killing is beginningless. And beginningless *apūrva* is not *kṛtisādhya*.

The second alternative is also untenable. If *saṅkalpa* of not-killing has to be taken throughout the life, there will be no time for discharging other duties.

If it is said that there is no need of taking *saṅkalpa* throughout the life, it is enough if it is taken once, then also there arises a difficulty. After taking *saṅkalpa* once, one may indulge in killing, since nothing binds one to refrain from killing.

The opponent may say that prohibited injunction means that whenever there arises the *saṅkalpa* of killing, *saṅkalpa* of not-killing must be taken then and there. In this case, there will never arise the *pravṛtti* of killing and also one can find the time of discharging other duties as well.

This view is rejected on the ground that there is no verbal injunction to this effect, that is, there is no injunctive text that states that whenever there arises the *saṅkalpa* of killing, it should be followed by the *saṅkalpa* of not-killing.

The opponent may argue that prohibition aims at condemning *pravṛtti*. So in this sense, i.e. in the sense of condemning the *pravṛtti* of killing, it is said that whenever the *saṅkalpa* of killing arises, the *saṅkalpa* of not-killing should be taken immediately. This explanation, however, holds good when something is condemned. Condemnation *(pratiṣedha)* establishes the absence of some counter-correlate. But the *saṅkalpa* of not-killing is taken with a view to attaining *apūrva*, according to the Prābhākara Mīmāṁsā. Therefore, its condemnation is not possible.

Moreover, it is not the fact that people refrain from killing for attaining *apūrva*. People do kill just to achieve some immediate fruit. The Prābhākara says that the *apūrva* is the highest fruit that can be desired, but one may not desire *apūrva* at all, just to achieve some immediate end one may take to killing.

Admission of *iṣṭasādhanatā* as the *liṅ*-sense also cannot account for the interdicted injunction. The knowledge that act of killing is *aniṣṭasādhana* (i.e. means to undesirable end) is what is *iṣṭasādhana*. But this is not the sense of the injunction *na brāhmaṇaṁ hanyāt*. The injunction seems to suggest that the act of killing has absence of *iṣṭasādhanatva*, but it does not in any way mean that killing is *aniṣṭasādhana*.

It is not proper to say that *aniṣṭasādhanatva* is inferred from the *hetu*, absence of *iṣṭasādhanatva*, because there are things which have absence of *iṣṭasādhanatva*, but they are not also *aniṣṭasādhana*.

Nor can it be said that prohibition *(pratiṣedha)* itself can be the *hetu* for inferring *aniṣṭasādhanatva*, meaning thereby that *pravṛtti* which is prohibited, must be admitted *aniṣṭasādhana*. According to the Mīmāṁsā that which is prohibited is the absence of *iṣṭasādhanatva*. But prohibition of the absence of *iṣṭasādhanatva* does not give rise to the inferential knowledge of *aniṣṭasādhanatva*. Moreover, when prohibition *(pratiṣedha)* itself is unestablished, it cannot be the *hetu* of inferring *aniṣṭasādhanatva*.

Absence of *iṣṭasādhanatva* cannot be admitted as *iṣṭa*, because the act of killing, as has already been mentioned, may be *iṣṭasādhana* in that it may lead to achieve some immediate result. Therefore, it cannot be definitely said that the act of killing has absence of *iṣṭasādhanatva* which is *iṣṭa*.

It may be argued that though killing is *iṣṭasādhana*, its *aniṣṭasādhanatva* is implied by its being prohibited. Were it so, it may be counter-argued that the prohibition of the sacrificial acts by the atheist may also imply absence of *iṣṭasādhanatva*. Surely, the Mīmāṁsā will not subscribe to this position. To say that there is no proof for the latter implication does not solve the problem, because the same can be said about the former implication too.

It may be maintained that though the interdicted injunction implies absence of *iṣṭasādhanatva*, yet its import *(tātparya)* is *aniṣṭasādhanatva*. This view is also not convincing. Both commendatory injunctive texts as well as interdicted injunctions are *vidhi-vākya*. They do not convey any other meaning except the primary meaning. The prohibited injunction *na brāhmaṇaṁ hanyāt* may have *aniṣṭasādhanatva* as its import, only when *na hanyāt* is inseparably connected with *aniṣṭasādhanatva*, or *aniṣṭasādhanatva* is the intended meaning *(abhiprāya)* of the speaker. In fact *na hanyāt* has no inseparable relation or universal concomitance with *aniṣṭasādhanatva*, since absence of killing is not concomitant with *aniṣṭasādhanatva*. Nor is *aniṣṭasādhanatva* the intended meaning of the speaker, because the opponent, viz. the Mīmāṁsaka, does not admit it.

Thus, as interdicted injunctions do not imply *aniṣṭasādhanatva* which is *iṣṭasādhanatva* in the case of interdicted injunctions, *iṣṭasādhanatva* cannot be the sense of the interdicted injunction. If, however, *iṣṭasādhanatva* is admitted as the *liṅ*-sense, interdicted injunctions remain unestablished.

विधिर्वक्तुरभिप्रायः प्रवृत्त्यादौ लिङादिभिः।
अभिधेयोऽनुमेया तु कर्त्तुरिष्टाभ्युपायता।।१५।।

Vidhi *signifies the will of the speaker conveyed by the* liṅ *in the form of an injunction enjoining activity or cessation therefrom; the fact that (the performance of sacrificial rites) is the means to a desired end for the performer is inferred (from the will of the speaker).*

15. After the discussion of various current opinions on the nature of *vidhi* (injunction), Udayana puts forth his own view. *Vidhi-pratyaya,* according to him, conveys the *abhiprāya* of the speaker about *pravṛtti* (enjoining activity) or *nivṛtti* (cessation from activity). And *iṣṭasādhanatā* about *yāga*, etc., that is, the fact that the performance of sacrificial rites is the means to what is desired by the performer, is inferred from the *abhiprāya* of the speaker.

Iṣṭasādhanatā, says the *siddhāntin*, is not implied by the *liṅ* directly. When an injunction is uttered by a trustworthy person, the intention of the speaker is that the hearer should either do something or refrain from doing something. And the hearer infers from this that doing the thing will enable him to achieve the object he desires or refraining from doing the thing which is prohibited, will save him from reaping certain undesirable consequence. This is the case with injunctions that have human author. Similar is the case with the Vedic injunctions also. The use of *vidhi-pratyaya* in the Vedas presupposes a trustworthy person as their author whose *abhiprāya* is conveyed through the *vidhi-pratyaya.* And it must be God whose *abhiprāya* is conveyed by the injunctive verbal suffix in the Vedas. The *siddhāntin* intends to say that the existence of command in the Vedas proves the existence of a commander.

It has already been said that *iṣṭasādhanatā* is inferred from the *abhiprāya* of the speaker In the case of *vidhi-vākya* the inference takes the following form: 'Performance of the sacrificial rites is my *iṣṭasādhana*, because the speaker intends me to do it'. In the case of *niṣedhavākya* the inference is as follows: 'Killing is not my *iṣṭasādhana*, because in spite of its being *iṣṭasādhana* the speaker intends me not to do it'.

The Mīmāṁsaka does not admit any authorship of the Vedas. He argues that the meaning of the *vidhi-pratyaya* advanced by the *siddhāntin* may apply to secular injunctions. But it does not apply to the Vedic *vidhi-vākya*. Udayana observes that the authority of the Vedas consists in the Vedas being composed by an omniscient author who is God. The use of the *vidhi-pratyaya* in the Vedas can be explained only on the supposition that the Vedas are composed by an omniscient author. To maintain that the use of *vidhi-pratyaya* in secular injunction conveys the intention of the speaker and in the Vedic injunctions does not convey so, will go against the Mīmāṁsaka dictum that words have the same connotation in all sentences, whether secular or the Vedic.

Vidhi-pratyaya certainly does not convey the intention of the teacher of the Vedas, because they merely repeat the Vedic injunctions. A parrot too may repeat an injunctive text, but the *vidhi-pratyaya* therein does never convey the intention of the parrot. Again, a royal proclamation made by an announcer does not convey the intention of the announcer, but it conveys the intention of the king.

Thus, it is decided that the Vedic *vidhi-pratyaya* or the *liṅ* usage conveys the *abhiprāya* or *icchā* of an independent speaker who is God. The inferential argument in support of this view is that 'the *vidhi-pratyayas* used in the Vedas convey the *abhiprāya* of an independent speaker (viz. God), because they are *vidhi-pratyayas*, like the *vidhi-pratyayas* used in the secular injunction'.

कृत्स्न एव हि वेदोऽयं परमेश्वरगोचरः।
स्वार्थद्वारैव तात्पर्यं तस्य स्वर्गादिवद्विधौ।।१६।।

All the Vedas refer to the existence of God; it (passage that refers to God) can have its import in vidhi *only by conveying its primary signification, like the word 'heaven' etc.*

16. Udayana now resumes the interpretation of the first *kārikā* interrupted by long discussion on *vidhi.*

Śruteḥ is the next ground for God's existence. The second interpretation of *śruteḥ* is as follows. All the Vedas mention God and sing His glory. The *Puruṣasūkta* praises God as the creator of the universe, the *Rudrasūkta* relates His splendour *(vibhūti).* He is *Śabda-Brahman* in the *Maṇḍalabrāhmaṇa.* The Upaniṣads after describing the nature of the external world say that He transcends the external world, and so on. As *śruti* mentions God, sings His glory, and lays down His characteristics, God must exist. Hence, *śruti* is adduced by the *siddhāntin* as one of the proofs for God's existence.

The Pūrva-Mīmāṁsā school advocates that the Vedas do not establish any accomplished fact *(siddhārtha).* The Vedas always teach *bhāvyārtha* or *kārya,* i.e. an act to be accomplished. Since the *Puruṣasūkta,* etc. do not deal with *bhāvyārtha,* the Mīmāṁsaka may object that they are not authentic. In reply to this objection, it may be observed that the Bauddha scripture advocates *bhāvyārtha,* but the Mīmāṁsaka does not admit it as authentic. Therefore, there is no reason whatsoever to maintain that only that scripture which teaches *bhāvyārtha* is authentic and not others. That which make a scripture non-authentic are error, etc. and the latter may equally exist in any scripture irrespective of its dealing with *bhāvyārtha* or *siddhārtha.*

The opponent may argue that the Vedic passages that mention God do not convey their primary meaning; they have their *tātparya* (import) in *vidhi-vākya* which enjoins *pravṛtti* or cessation therefrom, i.e. *nivṛtti.* Against this Udayana points out that the passages that mention God can have their import in *vidhi-vākya* only by conveying their primary signification. The Vedic passages about God possess an authority to establish God which they mean primarily in the same way as the passages declarative of heaven, hell, etc. establish the objects which they primarily mean. Thus, it is not porper to say that these passages being eulogies of a previous *vidhi* have their import in the *vidhi* elsewhere expressed and do not convey their primary signification.

If it is argued that the passages that mention heaven, etc. establish the object denoting heaven, etc. because there is nothing to invalidate it, the same argument holds good about the passages that mention God also.

To say that there is no perceptual evidence for God's existence is no good. For, no perceptual evidence is there for heaven, hell, etc. also.

It may be said that if there is no heaven, such *vidhi-vākya* as *svargakāmaḥ yajeta,* etc. will be meaningless, and people will abstain from performing sacrificial rites. To this Udayana rejoins that if there is no God, the *vidhi-vākya* that enjoins the worship of God too will be meaningless. If God does not exist, then such fruits as living in the abode of God *(sālokya),* enjoying godly bliss *(sāyujya),* etc. will be meaningless. As a result, all will shun worship of God. Thus, as there is no valid ground for disproving God's existence, God must exist.

16.ii. The other ground for God's existence is *vākyāt.* A sentence *(vākya),* according to Udayana, is an expression of a particular relation between words It is a fact that the author or the speaker of a sentence possesses the knowledge of the relation between words beforehand independently of the composition of the sentence. The sentences that form the Vedas also express particular relation between words. Therefore, there must be an author who had the knowledge of this particular relation independently of the composition of the Vedas. And this author, according to Udayana, is God. This conclusion is in conformity with the Nyāya presupposition that the validity of knowledge depends on the excellence of the speaker *(vaktṛguṇa),* and as the Vedas are acknowledged as the source of valid knowledge, their author, viz. God, must have had the valid knowledge of the Vedas before composing them.

स्यामभूवं भविष्यामीत्यादिसङ्ख्या च वक्तृगा।
समाख्याऽपि न शाखानामाद्यप्रवचनादृते।।१७।।

Number expressed by the verbal forms 'let me be', 'I was', 'I will be', etc. refers to the speaker; the names of the different recensions of the Vedas also could not be without (one who) taught them for the first time.

17. *Saṅkhyāviśeṣāt* is given as the last ground for God's existence. The word *saṅkhyā* has two meanings. It may mean number one, etc., or it may stand for *samākhyā* (name). In secular usage number (*saṅkhyā*) expressed by the verbal form in the first person refers to the speaker. Similarly, in the Vedas the first person singular number, expressed by the verbal forms *syām, abhūvam, bhaviṣyāmi,* etc., refers to the original speaker. And it must be God who is the original speaker of the Vedas.

The other meaning, viz. *samākhyā,* stands for the names like Kāṭhaka, Kālāpaka, etc. by which different recensions of the Vedas are known. Different recensions of the Vedas are known by such names as Kāthaka, Kālāpaka, etc., because God after embodying Himself as Kaṭha, Kalāpa, etc. taught those recensions for the first time to the Ṛṣis. The names of these recensions are according to the different names God assumed when He embodied Himself as Kaṭha, and others.

The different names cannot be after the names of those teachers who taught them, because they are innumerable in number, Moreover, were it so, as Kāṭhaka, etc. are being taught even now, they must as well be called by the names of the present-day teachers also.

It may be said that these recensions are named after those teachers who are proficient *(prakarṣa)* in teaching. If that be the case, it may be enquired what is meant by proficiency. If by proficiency is meant adding *svaravarna* to those *svaravarṇa* already spelt by other teachers, then there is possibility of the recensions being distorted.

If, however, teachers merely repeat what has already been taught by other teachers, then there arises no question of proficiency.

It may be further argued that though all the teachers teach the same thing, yet teaching of some is marked by such proficiency as adequate pronunciation, neither fast nor slow movement, sweetness, etc. But this kind of proficiency is not the special feature of only Kaṭha, Kalāpa, etc.; innumerable teachers are there who possess the same proficiency. That being the case, it cannot be determined after whom the names Kāṭhaka, etc. are given. Nor can it be maintained that Kāṭhaka, etc. carry the names of those who taught them for the first time, because the Mīmāṁsaka does not believe in creation and dissolution and in the absence of creation and dissolution the order of priority cannot be determined.

If creation and dissolution are admitted, then one who taught these recensions for the first time, must be God in the form of Kaṭha, etc. The opponent may say that it is not that the recension Kāṭhaka, etc. were named after Kaṭha who taught them. By Kāṭhaka is meant those who study Kaṭha and observe the same. All the *brāhmaṇas* study Kaṭha, and by *kaṭhatva* is meant *brāhmaṇatva.* This contention is untenable, because not only *brāhmaṇas* but *kṣatriyas* and *vaiśyas* also study Kāṭhaka, therefore, there is no proper reason to maintain that by *kaṭhatva* is meant *brāhmaṇatva.* Nor can it be maintained that Kāṭhaka, etc. are studied by *brāhmaṇa*s only, and *kṣatriya* and others study some other recension. It is also not that by Kāṭhaka is meant exclusive study of Kāṭhaka by Kaṭhas. *Brāhmaṇas* do study other branches of the Vedas and observe them. It may be argued that previously *brāhmaṇas* never used to study any other branch except Kāṭhaka, it is only now that they study other branches. But if present-day *brāhmaṇas* study other branches, then it can no longer be maintained that *brāhmaṇas* study only Kāṭhaka and not other branches. Further, there is no sound reason to corroborate the fact that past *brāhmaṇas* were studying only Kāṭhaka, and not other branches.

After considering all the possible explanations of the problem under discussion, which ultimately stand rejected, Udayana is drawn to the belief that it is God who at the beginning of each creation, assuming a particular body, viz. Kaṭha, Kalāpa, etc., taught these recensions, viz. Kāṭhaka, Kālāpaka, etc.

The five *stabakas* of the *Nyāyakusumāñjali,* specially the fifth one offers inferential proofs for the existence of God who has been referred to by the revealed scripture (Śruti), non-revealed scripture (Smṛti), and epics (Itihāsa) etc. The same God is also directly perceived by a few, because God is an object of knowledge who can be described about and who is an object like a pot. The counter-argument that may be advanced against this argument is that 'God cannot be directly perceived, because He is not subject to the causal aggregate of direct perception'. When analysed, it is argued that causal aggregate may pertain to either external sense-organs or to the mind. The counter-argument may intend proving that God is not subject to the causal aggregate of either the external sense-organs or that of the mind; hence, the direct perception of God is not possible. Udayana does not lend support to this position. According to him it is a fact that God is directly perceived, and hence, the direct perception of God definitely comprises the causal aggregate of perception. Udayana is of the view that the direct perception of God comprises the causal aggregate of mind. To argue that as the mind does not operate outside the body it occupies, the perception of God who is outside the body occupied by the mind is not possible, is not acceptable to the *siddhāntin*. For, in dream the mind does operate on the external objects even though the external sense-organs which are legitimately to operate on the external objects, do not function at that state. As in dream so in the case of God also the mind gives rise to the perception of God, an object other than the body it occupies, in assistance with other auxiliary factors.

In dream the mind perceives many distant past and future things which the external sense-organs cannot come in contact with. It is not proper to argue that dream lacks perceptual character and is only a memory on the ground that it is caused by impression with the help of strong residues of previous experience. For, dream-experiences are not recollected as 'I remember' or as 'it was remembered'. At the state of dream the awareness that 'I see the chariot' or at the awaking state the awareness that 'I saw the chariot' definitely establish the perceptual character of dream. It is also not that perceptual character of dream-experiences at the state of awakening is an imposition, for there is no proof to establish this fact. Dream-perceptions are not memory, because in dream one cognizes things differently and not in the way they appear in the dream, e.g. in dream one cognizes the cutting off of one's own head as one would have cognized the cutting off of another man's head. Granted that the cognition of the cutting off of one's own head is erroneous, even then it has to be admitted that in dream cognition arises in the form of perception which is not an imposition and is not memory.

Moreover, it is absurd to consider dream as memory. In memory one recollects "I cognize that jar". Memory recollects the thing in that particular form in which it was perceived before. Memory does not give rise to cognition of a thing in 'this' form, such as "I cognize 'this' jar". On the contrary, in dream one always recollects a thing in 'this' form, such as "I cognize 'this' jar". Owing to this very particular difference between 'this' character of dream and 'that' character of memory, dream cannot be equated with memory.

It may be argued that dream-perception being illusory defective causal aggregate of mind gives rise to perception of exernal object which is devoid of external sense-contact. So, there arises in the shell the perception of silver with which the external sense is not in contact. Similarly, it may be said that the perception of God is also not valid owing to lack of the causal aggregate of mind which is the cause of perceptual knowledge. This position is untenable, for dream in general cannot be levelled as illusory; sometimes dream generates valid perception of past and future objects.

It may be objected that future object of perception appears as present object in dream and hence, dream-perception is illusory. In reply to this objection it may be said that it is true that present character of future object of dream-perception is not real, but in practice perception in general gives rise to knowledge of present object only and following the same logic presentness of dream — object may be presumed. But the presumption of the present character of the object of dream does not render the perception of the object illusory. Dream-perception does not take place owing to any defect; it is generated by the causal aggregate of mind in assistance with the merit produced by the previous actions. In the case of dream the merit produced by the previous actions gives rise to perceptual knowledge, whereas in the perception of God it is the merit produced by *yoga* etc. that assists the causal aggregate of mind to generate the perception of God. The *yogīs'* direct cognition of God is valid knowledge, because it is produced by the causal aggregate of mind assisted by the merit generated by *yoga*. Since such cognition is direct perceptual awareness, it is perception. Knowledge which is produced by mere impression unassisted by merit is not valid. Thus, the objection that the perception of God is not possible owing to the absence of causal aggregate of perception stands refuted.

To argue that the direct perceptual knowledge of God is produced by erroneous cause is beset with contradiction. For, in that case the Vedic injunction, viz. 'two Brahmans are to be known' referring to direct knowledge of *jīvātmā* and *paramātmā*, will turn out meaningless, as it would enjoin something not possible to accomplish. It is also not that the particular injunction is meant only for mere knowledge of the sentence-meaning and not for any direct experience, owing to the reason that the Vedic text is enjoined upon for study only. The Śruti text that 'Atmā is to be seen, heard about, reflected on, meditated upon' refers to the fact that God can be meditated upon and directly seen apart from scriptural study about and reflection on Him. Similarly, atoms, etc. which are supra-sensuous can also be directly perceived by a person possessing merit accrued to from *yoga*, as *yogaja-dharma* (merit acquired by the accomplishment of *yoga*) can cause direct perception of everything, sensory or supra-sensory.

Thus, the Śruti speaks about Paramātmā, the One, possessing eternal knowledge, all-pervading, designated as Viṣṇu, possessor of infinite knowledge, unrivalled, having possessed of eternal direct perceptual knowledge of everything supra-sensuous. Smṛti proclaims that all should take refuge in the Lord Vāsudeva only; all actions should be undertaken for the sake of Vāsudeva only without desiring for fruits of actions; all actions other than those undertaken for the sake of Viṣṇu cause bondage; actions that are done for propitiating Viṣṇu destroy bondage accrued to by accumulated *karmas*.

Patañjali in his *Yoga-darśana* enjoins meditation on God who is worshipped by *karma-yoga* with a mind purified by the performance of religious acts. According to Patañjali meditation of God also includes intuitive worship of God accompanied by recitation of *praṇava-mantra* and surrendering of all acts at the service of the Lord. The Seers also proclaimed God by performing *jyotiṣṭoma* sacrifice, etc., by building temple, etc., by practising austerity such as enduring heat and cold, etc., by the practice of *yoga* such as observing restraint through abstention from all kinds of injury to any life, etc., by cultivating the culture of purification of the body and the mind, of contentment, etc., and by practising posture, breath-control, etc. The Śruti also says that realizing that when God is known, all else becomes known, one should devote oneself wholeheartedly to that Lord. The *Gītā* commands: 'Fill thy mind with Me, be My devotee, sacrifice unto Me, bow down to Me; thus having made thy heart steadfast in Me, taking Me as the Supreme Goal, thou shalt come to Me'. Also: 'Knowing Me as the dispenser of *yajñas* and asceticisms, as the great Lord of all worlds, as the friend of all beings, he attains peace'. All these sacred texts proclaim the existence of God.

इत्येवं श्रुतिनीतिसंप्लवजलैर्भूयोभिराक्षालिते
येषां नास्पदमादधासि हृदये ते शैलसाराशयाः।
किन्तु प्रस्तुतविप्रतीपविधयोऽप्युच्चैर्भवच्चिन्तकाः
काले कारुणिक, त्वयैव कृपया ते भावनीया नराः।।१८।।

In spite of having been repeatedly washed by water of the assemblage of various śruti and logical arguments, those who do not find faith in the existence of God are indeed stone-hearted; but even those who oppose our proposition, when start keenly reflecting on Thou at the time of misfortune, 'O Merciful One, Thou in Thy mercy remove their doubts about Thy existence'.

अस्माकन्तु निसर्गसुन्दर, चिराच्चेतो निमग्नं त्वयी-
त्यद्धाऽऽनन्दनिधे, तथापि तरलं नाद्यापि सन्तृप्यते।
तन्नाथ, त्वरितं विधेहि करुणां येन त्वदेकाग्रतां
याते चेतसि नाप्नुवाम शतशो याम्याः पुनर्यातनाः।।१९।।

As for us, O naturally beautiful One, our minds, though long absorbed in Thee, the ocean of joy, being restless are yet to be satisfied; therefore, O Lord, haste to bestow Thy mercy so that our minds remain fixed on Thee only and we will not any more be subjected to the continual afflictions of Yama.

इत्येष नीतिकुसुमाञ्जलिरुज्ज्वलश्रीर्यद्वासयेदपि च दक्षिणवामकौ द्वौ।
नो वा ततः किममरेशगुरोर्गुरुस्तु प्रीतोऽस्त्वनेन पदपीठसमर्पितेन।।२०।।

May this radiantly beautiful handful of flowers of reasoning delight both theists as well as atheists; it, however, matters not, even if this does not delight anybody; let alone the Guru of Indra's Guru be pleased by this offering of mine at His feet.